TIME
the Familiar Stranger

TIME
the Familiar Stranger

TEMPUS™

J. T. Fraser

Tempus Books of Microsoft Press
A Division of Microsoft Corporation
One Microsoft Way, Redmond, Washington 98052-6399

Published by arrangement with The University of Massachusetts Press

Library of Congress Cataloging-in-Publication Data
Fraser, J.T. (Julius Thomas), 1923-
Time, the familiar stranger.
    Reprint. Originally published: Amherst : University of Massachusetts Press, 1987.
    Bibliography: p.
    Includes index.
    1. Time. I. Title.
BD638.F69   1988   529   88-28381
ISBN 1-55615-171-3

Printed and bound in the United States of America.

2 3 4 5 6 7 8 9   AGAG   3 2 1 0

Distributed to the book trade in Canada by General Publishing Company, Ltd.

Tempus Books and the Tempus logo are trademarks of Microsoft Press.
Tempus Books is an imprint of Microsoft Press.

The excerpt from C. Day Lewis, *"Is it far to go,"* is from C. Day Lewis, *Collected Poems 1954*, and
is reprinted by permission of the Executors of the Estate of C. Day Lewis, Jonathan Cape Ltd.,
Hogarth Press, and A.D. Peters & Col. Ltd.

For the Disciplined Dreamer
a Book about Time

CONTENTS

ILLUSTRATIONS

CONTEMPLATING NANOSECONDS:
PREFACE TO THE TEMPUS EDITION

The Latin word *tempus* comes from the Greek *tempo*. The early meaning of *tempo* was that of anything that has been cut out or marked off, such as the parts of the heavens marked off by the motion of the sun along its path. Hence the meaning of *tempo* and *tempus* as a period, season, or time. The association of the act of cutting with the word *tempus* survives in the English word "template," meaning a thin plate that has been cut out to serve as a guide in mechanical work.

As does *tempus,* the Latin word *templum* also comes from the Greek *tempo*. *Templum* meant an area marked out by the staff of an augur as a place where one could enter to observe omens from which the future might be told. It also meant an area of the sky marked out for the same purpose. That notion survived in astrology as the concept of a "domicile" or "house" of a planet.

What did good Greeks or Romans do when they were inside a *templum* or observed one of the "houses" in the sky? They viewed, they beheld, they surveyed and reflected. The Romans called that kind of activity *contemplatio*. (The prefix *con*- means "together," "with others.") We now call it contemplation.

Tempus Books is an imprint of Microsoft Press, a division of Microsoft Corporation, a developer of microcomputer software. It is appropriate, therefore, to use the preface to this Tempus edition for contemplating the ways in which the broad use of computers has helped shift the focus of our concern with time from long-term continuity to short-term, functional periods.

As the burdens of public administration, data transfer, record keeping, and communication are increasingly assumed by computers, people—in developed countries, at least—have become increasingly computer-friendly. As young parents adapt to their children, people of the computer age have begun to adapt to their fast-working brainchildren.

The process of adaptation includes a shift in public assessment of what is and what is not significant in the nature of time. Instead of contemplating eternity as people of other ages have done, computer cultures focus on the rapid and the immediate. To direct attention to this shift, Jeremy Rifkin described the computer age as a "nanosecond culture."[1] It is a very apt phrase.

[1] Jeremy Rifkin, *Time Wars* (New York: Henry Holt, 1987). See Chapter One.

A nanosecond is one billionth of a second. In mathematical shorthand (explained in the footnote on page 115) it is 10^{-9} second. The switching functions of an end-of-the-century computer, its transistors, and its other components can compare, store, retrieve, and perform arithmetic operations at rates that may be conveniently measured in nanoseconds.

Although impressive as a hallmark of computer technology, in the world of science, such as in biology, a nanosecond is not an extraordinarily short period of time. Our eyes and nervous systems, for example, can distinguish between shades of color whose waves, at the end of a single swing, are out of phase by 10^{-17} second, or one-hundred-millionth of a nanosecond.

If you designed and made a computer system that worked in the nanosecond realm you could be justly proud of it. But do not brag about it to the human nervous system, for it is going to laugh at you. And, because it is a well-educated nervous system, it might even quote Puck from *A Midsummer Night's Dream:*

> Shall we their fond pageant see?
> Lord, what fools these mortals be!

There is a lot here to reflect upon. We may ask, for instance, whether the time of switching speeds or the time by which two light waves are out of phase is the same *kind* of time as that in which Puck could laugh and in which we are mortals.

No, it is not.

It is not, because time is structured. Time is not a homogeneous flow in which all the furnishings of the world—sticks and stones, living organisms, the functions of the human mind and those of society—equally partake. Systems that differ substantially in complexity carry on their functions in qualitatively different times, called temporalities. The meaning and measurement of "complexity" and the nature of the qualitative differences among temporalities are discussed in Chapter 3 ("The Times of Life, Mind, and Society") and Chapter 4 ("Time in the World of Matter").

Consider, for instance, (1) the time represented by a timeline along which we imagine nanoseconds and their ilk lined up. Such a timeline is a geometrical abstraction derived from (2) the much richer kind of time in which we, as thinking and feeling humans, experience our inevitable passage. Let me illustrate the qualitative difference between them by comparing geometrical and verbal representations of a specific human experience.

⧲ is the geometrical representation.

"O! then, dear saint, let lips do what hands do!" is the poetic version of the same experience, to wit, Romeo holding Juliet's hand.

The two statements—the geometrical and the verbal—demonstrate a qualitative difference that parallels that between the time of nanoseconds and the time of human experience.

The time of the mind and the time represented by a line do have certain properties in common. For instance, both kinds of time, if thought of as event following event, may be compared with point following point along a straight line. But the wealth of mental time is infinitely richer than that of geometrical time, as illustrated by the *Romeo and Juliet* example above. They are so different in quality that they ought to have different names. In the interdisciplinary study of time, they do. They are called nootemporality and eotemporality, respectively.

If we consider the temporalities of nature identified by physics, biology, psychology, and social science and arrange them along a scale of increasing wealth and degrees of freedom, we find another kind of time that falls between the nootemporal and the eotemporal. It is the time of living organisms, including humans insofar as we consider only their biological functions. Its name is biotemporality; I like to call it "banana-now" time.

The colorful name is suggested because the temporal reality of nonhuman life is limited almost entirely to the present. In the worlds of the most advanced species there are slight extensions of that present to the immediate future and past. Chimps, for instance, can play hide-and-seek. But even for apes, time is always "banana now." It is never "banana for my grandchildren yet to be born" or "banana last month."

As do apes, so do human infants. They live in the present. But unlike apes, growing children soon learn how to use the categories of future, past, and present to describe the world and themselves. Yet not until they are young adults do they learn to appreciate the need for, and the advantages of, living in more than the present. And only as mature adults do they acquire the most powerful weapon in the possession of our species: the ability to perceive the world in terms of long-range futures and pasts, connected by discoverable laws.

Our direct hominid ancestors left "banana-now" time when they became humans. Since then, the mental presents (the "nows") of men and women have included images of long-range futures and pasts, mixed in infinitely many ways with the thoughts, feelings, and sense impressions of the present. We describe the dynamics of this bubbling cauldron of mental life as the flow of time.

Members of our species were able to conceive and create the great cultural continuities of humankind—the arts, letters, and sciences—because the mind's reality embraced the immensity of time from distant pasts to distant futures. See Chapter 1 ("The Experience and Idea of Time") and passim.

If in our individual and collective behavior we were to let our mental present deteriorate to a "banana-now" present—which is as close to the nanosecond world as we can get in practical terms—then we should lose the very capacity that built human civilizations. We should become the time-zombies of a banana-now republic.

Could this happen?

Yes, it could.

We can observe a family of complex, interrelated changes that suggest an evolutionary reevaluation of our collective attitudes toward time. Chapter 5 ("Man the Measure and Measurer of Time") discusses the details of this reevaluation and its implications for the global community.

From among those changes, let me mention only one, because it is being powerfully promoted—although not caused—by the wide use of computers. It is a trend, observable around the globe, towards favoring stopgap measures in preference to seeking long-term solutions to socioeconomic problems. Yet, research as simple as careful reading of the daily papers demonstrates that nations, corporations, and even cultures that opt for immediate return find themselves greatly disadvantaged in competing with those that are prepared to sacrifice now and collect later.

Because of the character of the "nanosecond culture," it is difficult to oppose the preference for short-term investment of cultural, intellectual, and financial capital. But one ought to start somewhere.

For instance, instead of favoring one-liners—a practice which, if extended to all domains of human concern, is bad for mental health—how about taking the time to read and contemplate books of substance and literary quality?

Whether or not *Time, the Familiar Stranger* belongs under that heading is for the reader to decide.

J.T.F.
September 10, 1988
Hickory Glen, Connecticut

ACKNOWLEDGMENTS

I wish to thank the Library of Manhattanville College for the stack privileges I have enjoyed for fifteen years; the Burndy Library in the History of Science for its ever-ready welcome; the Westport Public Library for its fine services; Dr. Richard Martin for being an editor who knows and cares; and my fellow timesmiths of the International Society for the Study of Time.

But above all my appreciation, as always, goes to my wife, Jane. Now that the patter of little feet is gone, she has been willing to live with the clatter of a typewriter and bear with love the elsewhere-directed habits of an idea processor.

I am also grateful—to whom it may concern—to the gentle beauty of southern New England: against the greedy noise of a time-compact globe, the silent power of open passage.

J.T.F.

A CHART AND AN ITINERARY: AN INTRODUCTION

The tides rise and fall, children grow, time passes. The passage of time is intimately familiar; the idea of time is strangely elusive. This book intends to add a dimension of understanding to the intuitive knowledge of time the reader already possesses.

All species behave according to shared, genetically conditioned programs, but only humans can freely modulate their native programming by individual concerns that reach far beyond their deaths. All species have an evolutionary past, but only humans have history from which they may individually learn. Other species have been swifter, stronger, and hardier than ours, but only humans know how to make and follow long-term plans, based on long-term memories. Having these skills has made our species the unquestioned but not untroubled lords of the globe.

The ordinary view of time may be represented by an image: that of the underground cable of the San Francisco cable car. Driven by some distant and obscure machinery, the cable remains out of sight; we know that it moves because, at different points, the cable cars get attached to it and are carried along for a ride. Likewise, time in daily life is usually taken to be a universal, cosmic motion of the present—the now—driven by natural or divine powers to which matter, life, man, and society get attached and are carried along for a while.

"Time," says the old hymn, "like an ever-rolling stream / Bears all its sons away." And its daughters, as well. And everything else. It would be rather difficult to claim otherwise.

Yet, interdisciplinary studies have revealed that what we metaphorically describe as the flow of time—the motion of the now—is not a feature of the physical universe. Instead, it has its origins in the life process, in the creativity of the mind, and in social conventions and modes of communication. None of this changes the problems of aging or the finality of death, but it does open new vistas that were unknown, unsuspected and even unimaginable before.

The substance of this book derives from over three decades of research in the interdisciplinary study of time. This is a new field of scientific and humanistic undertaking that I like to describe as the work of "time-smiths." For it consists of the mental hammering out, from theoretical and experimental material, the novel shapes of our understanding of time.

I wanted *Time, the Familiar Stranger* to be an informative and interesting report, without referenced argumentation. To that end, I had to employ the technique of the mapmaker. When a mapmaker prepares a chart he imagines himself as having a bird's-eye view of the terrain, then he draws schematically what he sees. Two paths running parallel and close to each other are merged and shown as one; only the outlines of towns are shown, but not individual buildings within their boundaries. And only the highest mountain peaks are named; the lower ones are only drawn, implied, or not even implied. A traveler who knows a part of the land—a reader familiar with one or another area of the issues—will find the map (the book) incomplete but, I hope, not inaccurate. But let the necessary limitations not inhibit the enjoyment of discovery. For those who wish to continue their explorations, I have prepared a bibliographic "sampler," to be found at the end of the book.

Beyond being a survey, this book refocuses the interdisciplinary study of time on an observation with which it began in 1966:

Watching the clash of cultures [during the Second World War] and the attendant release of primordial emotions stripped of their usual niceties, I could not help observing that man is only superficially a reasoning animal. Basically, he is a desiring, suffering, death-conscious and hence, a time-conscious creature.

Temporal experience, it seemed, more than any other aspect of existence is all-pervasive, intimate, and immediate. . . . In short, it appeared to me that time must and should occupy the center of man's intellectual and emotive interest. These thoughts turned out to be far from novel. But they were sufficiently stirring to lead to the present volume.[1]

They were indeed and remained so through two decades of intense time-related research, all the way to *Time, the Familiar Stranger*.

In the arrangement of the material I wanted to be free of the stale categories of "this is what physics says," "this is what biology says," "this is what Uncle Remus says" and began searching for a suitable framework. Fate came to my side during a stay in England.

The body of the great English novelist and poet Thomas Hardy rests in Westminster Abbey, but his heart shares the grave of his first wife in the yard of the small parish church of Stinsford, Dorset. It was on a summer's day, with the tranquility of rural England in the air, that my wife and I visited that church to pay our respect to the memory of Hardy. Next to a series of Hardy family graves I came upon the tomb of another poet, Cecil Day Lewis. His stone was inscribed with five lines of his poem, "Is it far to go?"

> Shall I be gone long?
> For ever and a day.
> To whom there belong?
> Ask the stone to say,
> Ask my song.

The poet's voice called me back to his tomb again and again until I became aware that the lines form a lyric summary of man's concerns with time.

The first line is a question by the startled mind, having realized the finiteness of human life on earth. It records the discovery of time.

The second line appeals to the skill of time reckoning, made possible by our ability to count.

"To whom there belong?" is a bid to search for eternity through belonging. Because of the hierarchical organization of nature, each of us

---

1 J. T. Fraser, ed., *The Voices of Time*, 1st ed., 1966; 2d ed., with a new introduction, "Toward an Integrated Understanding of Time" (Amherst: University of Massachusetts Press, 1981), pp. xvii–xviii.

belongs with matter around the universe, with all forms of life around the globe, with all other humans of our species, and also with different social groups. Each of these organizational levels—matter, life, the human mind, and society—has its own temporality.

"Ask the stone to say" is an invitation to explore time in the physical world.

The last line of the verse, being also its bottom line in a figurative sense, tells us that the ultimate reference of all inquiries, including that on the nature of time, is man, the measurer and measure of time.

The five lines of the poem are the titles and themes of the chapters of this book. Together, they serve as a chart and an itinerary for our voyage of discovery.

Shall I be gone long?

T H E   E X P E R I E N C E   A N D   I D E A

O F   T I M E

The earth had been spinning for four and a half billion years when, on
a lazy afternoon, one of my hairy ancestors shooed away his hairy kids and
began to fashion a tool for collecting honey. He remembered earlier
honey-gathering trips, could see those honey-makers swarm and even
hear their buzz, although they were not present. But suddenly he stopped
his work because his mind conjured up an image, one he had no reason to
take less seriously than that of the bees. He put down his honey-gathering
tool, picked up a stone, and began to make a weapon to fight the
frumious Bandersnatch which, he was certain, would appear that night,
gyring and gimbling in the wabe.[1]

This hairy person was able to act in the service of a possible or impossible future condition, constructed by his mind from things and events
remembered, with the salt of imagination added. He already knew of
future and past in a concrete, pragmatic sense as well as we do. For this
reason, he was just as distant from his nonhuman ancestors as we are.
Humans are different from all other species because they are able to comprehend the world in terms of distant futures and pasts, and not only in
terms of the sense impressions of the present.

In the broad field of time we imagine ourselves as passing. It is obvious
that our passage in time, and time's own passage, are metaphors of our

---

1   My imaginative ancestor might very well have been a female. In this book the
masculine pronoun stands for humans of either sex, with only an occasional "he or she" to
remind the reader of this footnote.

language that stand for a family of rich, many-faceted experiences. This chapter sketches some of those experiences and traces the paths along which they came to take the form of abstract ideas.

The Discovery of Time: Once over Lightly

Many animal species prepare for the future but, unlike humans, they cannot *not* prepare. When their internal clock says "migrate," they begin flying, whether or not it makes sense. The most advanced species can modify their behavior by responding to unpredictable environmental conditions, but the temporal horizons of their behavior, as compared with those of man, remain severely limited. Many species can communicate their fears and plans, but, except for a few examples of limited scope, they cannot generate or receive communication about the past. I can convey to a dog the idea, "I will feed you," and the dog will respond appropriately. But there is no way I could tell Fido, "I've already fed you."

How does each of us reach the privilege of extended time horizons?

The universe of the human infant is one of ceaseless change: everything is always new. Then, out of the chaos emerge some permanent patterns: the cycles of hunger and satiety, sleep and wakefulness, day and night (see fig. 20). Stable, noncyclic continuities also emerge: mother, father, the corner of the bed, and the nose of the bear. Next come pieces of furnishings, features of the outdoors, and things one is not supposed to touch. To the early impression of ceaseless change, a universe of permanent patterns is added.

It is a guess based on the logic of the situation that the evolutionary development of the sense of time in our species followed some such a path as it does in today's infant. My ancestor and his fellows began to discern the rhythm of seasons, of animals and plants, of predators and prey, and of their own bodies. With the modicum of security granted by the recognition of rhythms, they learned to postpone their responses to present stimuli. They mastered the art of predicting future events based on past experiences. From being able to identify permanent patterns in their environment, they came to recognize stable patterns among their feelings and actions.

They learned to associate gestures and sounds with the permanent features of their external and internal worlds, without distinguishing sharply between external and internal realities. The capacity of perceiving the world in terms of its predictable and unpredictable features eventually enabled our species to create signs and symbols, use them for communication, and, with the help of communal coherence thus established, build civilizations.

The process of becoming human ought not be imagined, however, as a neat ascent. Instead, it probably comprised a number of mutually reinforcing patterns of behavior and modes of perception, improved communication, and a growing capacity to recognize a highly differentiated environment. All these developmental steps helped to widen our species' temporal horizons, thus providing an increasingly more powerful weapon in the struggle for survival.

Sooner or later, each member of our species must have become aware of an object that moved when he or she wanted to move, felt hungry, desired a mate, experienced fear and anger. Somewhere along the line it surely dawned upon these humans-to-be that the peculiar object with all those attributes must eventually die. The knowledge of inevitable passing was then added to the other elements of the developmental feedback circuit. Just as the last few atoms added to an atomic pile trigger a chain reaction, the discovery of death jettisoned the brain into functioning in ways we call mental.

I imagine one of our ancestors grabbing a sharp stone, hobbling to the cave wall, and, while gesticulating wildly, scratching a number of lines on it, all crisscrossed. He then grinned because he finally caught the flea that had been bothering him and because he succeeded in expressing what, in his world, meant the same as

> Golden lads and girls all must
> As chimney sweepers, come to dust,

in the words of Guiderius, singing in Shakespeare's *Cymbeline*.

When was all this supposed to have happened?

The oldest fossil remains of primates—the order that includes man—are over 70 million years old. The oldest known fossils of apes—the family that includes man—are around 30 million years old. The separation of our ancestral apes from Old World monkeys goes back about 20 million years. *Ramapithecus,* an extinct genus of apes that probably walked erect, is believed to be the oldest known genus of which humans, but not any modern apes, are a descendant. Its earliest remains date to about 7 million years B.P. (before the present). Next follows a gap of a few million years, from which very little fossil remains have thus far been discovered. Then at about 4 millions years B.P. human-like primate fossils resume their motionless march.

From between 4.6 and 4 million years ago to about 1.1 million years ago the trail is traced through subsequent species of the genus *Australopithecus*, the "southern ape" or "apeman." The two latest known species of this extinct genus, *A. robustus* and *A. boisei* ("of the forest"), were contemporary with the two earliest known species of the genus *Homo: H. habilis* ("skillful man") and *H. erectus*. This brings us to somewhere between 1.6 and 0.5 million years B.P.

The numbers I gave are scientific guesses subject to change, as are the classifications of the known fossils. The continuity among these species is tantalizingly clear, but it is uncertain which of the human-like/apelike creatures, walking on their hind legs, were our direct ancestors.

The earliest remains of archaic *H. sapiens* date from between 600,000 and 250,000 years ago, but the use of tools for tasks at hand goes back to a much earlier period, perhaps to 2.4 million years ago, which was the age of the apeman.

It is not possible to pinpoint precisely the beginnings and endings of the different archeological ages; those boundaries are uncertain and also vary greatly with geographical regions. A useful scale places the beginning of the Old Stone Age—the Paleolithic—at the 2.4-million-year mark mentioned above. The Neanderthal man lived as recently as 150,000 to 30,000 years ago; modern *H. sapiens* appeared 40,000 years before our age. The beginning of the Mesolithic may be taken as 18,800 years B.P., that of the New Stone Age (the Neolithic) as 10,600 years. The Neolithic lasted until about 6,500 B.C. around the eastern Mediterranean and until 3,000 B.C. in Spain and northern Europe.

During the three stone ages, crude flint and stone scrapers and axes were replaced by refined versions of their kind and were joined by knives, awls, and very small tools called microliths. Needles and thread appeared, as did clothing made of animal skin; pit houses were built and, as if from nowhere, Paleolithic art materialized. It bore witness to keen powers of observation and individualistic expression together with a trend toward the geometrical, as early humans became concerned with invisible forces that they believed controlled their destiny.

In the study of time, the hallmark of being human is the capacity for using long-term memory in the service of long-term plans, to meet possible or impossible future challenges. Such time-related behavior cannot be told from the sizes and shapes of skulls, but may be conjectured from cultural furnishings. Let me give an example, following the work of the English scholar of comparative religion, S. G. F. Brandon.

In the innermost recesses of the Trois Frères cavern near Ariège, France, there is a remarkable wall painting made sometime during the late Paleolithic. It shows a figure that has a stag's head with antlers, an owl's face, a bear's paws, and a horse's tail. Superman? No. A sorcerer, priest, or shaman in animal disguise, performing a dance of hunting magic, which is a form of sympathetic magic based on the belief that like produces like. By imitating the behavior of the animal, including its death, it was believed that the future may be influenced and the success of the hunt assured. The painting thus suggests a highly developed sense of time in the people responsible for the ritual.

But there is something even more intriguing here. This and similar wall paintings are usually found in those regions of caves that are most difficult to approach. Why would anyone go to the trouble of painting in those depths of the cavern, carrying his tools and such smoky lights as he had, chancing the loss of his way in the darkness? Even animals lay low in the dark unless they are nocturnal and tolerate discomfort only if they must. What motivation did he or they have to create that painting?

Professor Brandon's conjecture is both fascinating and convincing. The painting, he reasoned, was intended to secure for the members of the tribe the benefits of the dance in perpetuity, even when the dance was not being performed. It was not painted in response to an urgent present need but rather in response to a desire to control all future hunts. According to this interpretation, the Dancing Sorcerer—the name by which the painting is known—demonstrates that the Paleolithic tribes of that region knew how to create symbolic representations of the past in the service of their desire to influence the future. This kind of behavior is a hallmark of the species that perceives the world in terms of human time. The wedding photograph on the piano has the same purpose as the Dancing Sorcerer. It is an image of the past made for the purpose of securing in perpetuity the benefits of what it shows: the happiness of those early days.

Can the desire for creating a painted image, the wish to perform a magician's act, be powerful enough to take people into the depths of the cave, with all its discomforts and terror? Yes, if the challenge is appropriate to the beliefs of the epoch.

We know nothing about the Paleolithic artist but we do know about his latter-day colleague Michelangelo. Between 1508 and 1512 he painted the ceiling of the Sistine Chapel, lying flat on his back much of the time on a high, movable bridge that had to be reached by a series of ladders. There he also ate his soup and bread, carried up to him by a servant, the only person beside the pope who was permitted to enter the chapel. When, after four years, he left his "cave," he could not walk but only wobble with his head back; he could look only up. Twenty-three years later he returned to paint the monumental *Last Judgment* over the main altar of the chapel. We know that he was driven by the fear that he might die at any moment and, not having finished his work, would not be worthy for resurrection, that is, for an extension of his lifetime. In the design and execution of the *Last Judgment* he was inspired by the medieval hymn "Dies Irae"—"Day of Wrath"—that speaks of the devastation and terror of the Apocalypse at the end of time. So was Ingmar Bergman four centuries later, in making his movie *The Seventh Seal,* as we shall learn later.

Around 1550 we see Michelangelo, an old man, struggling furiously to give shape to his feelings in the stone of his beautiful Florentine Pietà. The statue is infused with the sorrowing love of an old man, rather than the sensual love of a young woman, as is his better known early Pietà. He would rise in the middle of the night and work, wearing a thick paper cap in which he placed lighted candles made of goat's tallow.

The Paleolithic cave painter, Michelangelo, Ingmar Bergman: each was driven by the insecurity occasioned by the knowledge of human time.

And, why was the painting of the Last Judgment commissioned in the first place? So that by seeing it, or simply knowing of it, people may do (or not do) something now and thus benefit from their behavior at a future date. The blueprint of the home one would like to build, kept on the kitchen counter, or the mental image of a small daughter graduating from college twenty-two years from now, are secular siblings of Michelangelo's *Last Judgment:* they influence present action in the service of a distant, future goal.

From among the many images of the future that influence present actions, the awareness of death is the most universal and powerful one. It is an essential ingredient of the mature human sense of time, whose horizons extend without limits into the future and the past. This kind of time, peculiar to the human mind, is called *noetic time* or *nootemporality*. Noetic comes from the Greek *noetikos* ("mental"), related to *noein* ("to think"), and to *nous* ("the mind"). Nootemporality is the thinking man's and woman's time.

All healthy, growing children make two important discoveries: the facts of birth and the fact of death. As the child grows, issues of birth mold into his growing awareness of sexuality. His interest in sex rapidly increases, then slowly decreases. His concern with the certainty of death does the opposite. At first it hardly exists, then slowly but steadily increases in importance, never to leave him.

Birth and death are symmetrical events, in that they are the limits of one's time on earth. But our relations to birth and death are not at all symmetrical: our attitudes toward the beginning of our time on earth and those regarding its ending are very different. As living beings we grow from eggs and sperms. But a fertilized egg is not a person; the egg I came from is not "me." Where "I" was before I was conceived is an interesting question, but it is easy to live with it even if unanswered. The same is not true for the question, Where will I be after I die? Answered or unanswered, this is not a question that has ever been lightly dismissed.

On the tomb of an Egyptian king buried during the Fourth Dynasty (c. 2600 to 2500 B.C.) there is a poem, known as the "Song of the Harpist."

> None cometh thence
> That may tell us how they fare,
> That he may tell us of their fortune . . .
> Until we too, depart,
> To the place wither they have gone.

Across forty-one centuries, the Harpist must have spoken to William Shakespeare, who speaks to us, through Hamlet's soliloquy, of

> The undiscover'd country from whose bourn
> No traveler returns, puzzles the will.

It does. "I will be nowhere" has been an unacceptable answer to all civilizations. Ever since the first burial of the dead some 50,000 years ago, burials have been ritual occasions, celebrating a transition in continued time and not a final ending. The Neanderthal man placed flowers with its dead; others placed food, ornaments, and weapons. Powerful people were often buried with their servants, murdered for the occasion, so that they may be ready to serve their master or mistress in another world that continued in time.

Beliefs in postmortem existence have accompanied the evolution of cultures. As soon as the capacity for identifying permanent and unchanging ("timeless") features in the world was acquired, death came to appear arbitrary. If the stars can go on living in time, why can't people? If the earth does not die, why should she or he whom I love?

The knowledge of human time thus made for a double-edged weapon that cut both ways. The ability to use long-term memory in preparing for future actions has conferred upon our species immense advantages in its struggle for survival. On the other hand, these advantages were paid for by a profound sense of restlessness, rooted in the certainty of passing and death.

An animal's world, its reality, is that of the present with only occasional openings into an immediate future. The human world is that of the future and the past, as well as the present. Living in that universe has not been an easy task, because the past includes pleasures that are gone and regrets that have remained, the future includes hopes and fears. By discovering human time, we have acquired a chronic case of inner conflicts. To help lessen the tension of those conflicts, our species has created civilizations with their opiates. The closing sentence of the following quote from the introduction to Marx's *Critique of the Hegelian Philosophy of Right* is widely known but not so its context:

> Religious suffering is at one and the same time an expression of real suffering and a protest against real suffering. Religion is the sigh of the oppressed creature, the heart of the heartless world and the soul of the soulless condition. It is the opium of the people.

Karl Marx made his list of opiates all too brief. People with both heart and soul, feeling oppressed if for no other reason than the finity of life on earth, need a limitless supply of opiates. Therefore they manufacture them diligently: mathematics and astronomy, technology and architecture, the sciences, the arts and letters, music and dance, political ideologies such as Marxism, the waging of wars, and the waging of peace. We are a restless bunch. But one should not snicker at all this. Those opiates are the stuff that civilizations are made of.

Language: A Weapon against Passing

The extension of temporal horizons, from the immediate future and past to beyond death and before birth, would not have been possible without human language.

The evolutionary beginnings of social communication—the ultimate origins of human language—may be found in the reaction of organisms to each other: all plants and animals communicate among themselves and with members of other species. Messages may be carried by pollen, scent, sound, and light waves. But the scope and content of all nonhuman communication are narrow, especially in reference to time. Only people can communicate about real and imaginary events and things here and now, in the future, and in the past. And only they know how to pass messages to their contemporaries outside the reach of sound, olfactory, or visual signals or to their successors in the distant future.

The origins of human language have been a subject of great fascination and have been surrounded by a touch of mystery. In primitive cultures, the ability to name was identified with the possession of whatever was named or at least with having control over it. Hence, the naming of children has been guided by instructions and protected by taboos as far back as naming can be traced. There is a sense of intimacy to peoples' names, an emotion still experienced when, under the right conditions, one utters the name of a person loved or shouts the name of someone hated. The changing of names involves emotions and, again, easily invokes a touch of the magic. Such a change may signify the happy occasion of a wedding or the unhappy one of escaping from authority, but is hardly ever a neutral act.

The very ability to speak a human language demands a command of human time. Words, the smallest units of language, stand for objects, feelings, memory images, fantasies that remain unchanged. Only by being able to separate out of the hubris of sense impression those elements that do not appear to change with time, such as the treeness of trees, can the word "tree" acquire meaning. The child's vocabulary develops in conjunction with his or her ability to separate the permanent and the changing in his or her experience of time.

Words describe features of the world judged stable. Something that appears to be a slice of cheese for a split part of a second, the tone of a violin for the next, then a prairie dog, a painting, a toothache, then the smell of garlic could not be given a name.

How permanent must an object or an idea be to be given a name? The more abstract the notion represented by a word, the more it may stand for something "really" permanent. Examples are the ideas of God, Nature, the Universe, or a numeral. Anything less abstract is going to be permanent only for a while. And even the supposedly stable realities these words signify change their meaning: God, Nature, and the Universe do not mean today what they did a century or even twenty years ago. No single sentence may be said twice and retain its meaning in all details, for the original context in which the sentence was uttered will have changed.

Because of its marvelous pliability to stand for ideas that are always new, language serves as our most important tool in the creation of new realities. It is also the most powerful, and the uniquely human, means that makes dissent possible in the face of overwhelming physical power.

The readiness to acquire and handle this tool is innate. The intensity with which an alert toddler points to objects and asks for their names by gestures and unformed tones never ceases to impress me: he wants to internalize those objects through their names; he wants to own, command, and manipulate them through words. His interest in trying to communicate can be much greater than his interest in eating. In the great passion of small children for learning to talk I sense the vested interest of the community; food only perpetuates the individual, whereas language helps perpetuate the group.

I joined these brief remarks to the development of the present chapter because language is a powerful and necessary weapon in the struggle against the finity of individual life. I will return to the theme of language and time throughout the book and in detail in Chapter 3 in the subsection, entitled "Language, the Architect of Time."

How to Live with the Knowledge of Time: Religions

Religions represent one of the many ways that humans have to acquiring a sense of security in the face of the insecurity conferred upon them by their knowledge of time. Religions, therefore, are necessarily coeval with the discovery of time, and there is no major world religion that does not address our awareness of the inevitability of individual death. But what they teach about the nature and importance, or unimportance, of time differs from religion to religion, as I will discuss in the examples that follow.

Hinduism comprises a large variety of beliefs and practices, without demand for exclusive allegiance. But there are some beliefs that each of the 2.6 billion Hindus tend to share. One is the belief in inborn duty, virtue, or destiny, called Dharma. Another is Kharma, a belief in a cosmic law of credits and debits for good and evil acts, resulting in a net worth at the end of life. A third one is Moksha, a belief in the transmigration of souls and in their eventual release from time. Finally, Brahma is held to be the ultimate ground of reality; whatever has no Brahma is unreal.

The cosmic background of these human values is the traditional cosmology of India, described in a collection of sacred hymns and offerings called the Vedas. The four Vedas, composed in Sanskrit during the period 1500 to 1200 B.C., see our world as one in which creation and destruction pursue their relentless labor simultaneously, hand in hand. As a defense against such a world, they advocate a belief in the insignificance of time's passage. Time, while real enough for daily chores, is judged unimportant in the economy of the universe. Figure 1 shows a traditional Indian image of timeless eternity.

Since our own age loves numbers, let me use numbers to suggest how time, implicit in figure 1, negates time.

The life of the Brahma is 100 Brahma years, each of which has 360 Brahma days. During each Brahma day Vishnu, the preserver and protector of the world, winks 1,000 times. Each time his eyelids open, a universe appears and lasts for 12,000 divine years. As he closes his eyes, the universe vanishes. Each divine year consists of 360 human years. Multiplying it all, and assuming that no time passes between the end of one wink and the beginning of the next, we get 155,520 trillion human years for a Brahma lifetime. When Vishnu's dream ends, the lotus closes and one Brahma life-

ब्रह्मा

विष्णुः   ब्रह्मा   लक्ष्मी

FIGURE I   A HINDU IMAGE OF ETERNITY. Vishnu is reposing on Sesha, the vast, thousand-headed serpent, emblem of eternity and attribute of Shiva, the lord of the beasts. The seated figure is Vishnu's wife, Sri Laksmi, who gently strokes his legs. It is related in the Scanda Purana, an ancient compilation of verses with legendary themes, that when the whole earth was covered with water, Vishnu lay in the bosom of eternity and dreamt of the creation of the world. His dream made the Brahma, sitting on a lotus, spring from his navel. In this illustration the Brahma has his usual four faces and is holding the Vedas. Plate 7 from Edward Moor, *The Hindu Pantheon* (London: Johnson, 1810).

time ends. As Vishnu's dream resumes, the lotus once again opens and the new Brahma begins its mission.

The numbers I gave are not standard figures; they depend on where one looks. As is the case with the $3.12 trillion the world has spent on arms during the last twenty years (or was it only $3.09 trillion?), the Hindu purpose was to send a message to the enemy. The enemy is time, and the message says that the world is timeless, and if it were not, the staggering cast of cycles upon cycles would make the experience of passing insignificant just the same. By coming down heavily on the side of timelessness as the ultimate reality of the world, Hinduism lessens the existential tension created by the knowledge of time.

Buddhism arose during the sixth century B.C. as a reform movement of Hinduism. It derives its teachings from the life and thought of Siddharta Gautama, the Buddha (the Enlightened)—sensitive and compassionate son of an Indian prince. He tried extreme self-mortification and also self-indulgence as ways of life, then decided in favor of a middle road between them. He began his teaching career in Benares (today's Varanasi) with a sermon on his basic doctrine that bears the same relation to Buddhism as the Sermon on the Mount does to Christianity. It speaks of the turning of the Wheel of Righteousness, a metaphor of time that remained a symbol for Buddhism somewhat as the Cross did for salvation in Christianity. The Buddha maintained that suffering is inherent in existence and can only be lessened by foresaking desires. This may be achieved by following the Noble Eightfold Path: the right views, thought, speech, action, livelihood, effort, mindfulness, and concentration. If the path is followed, after many transmigrations, the soul may reach the timeless state of Nirvana.

The traditional philosophical–religious systems of China have been Confucianism and Taoism. Their bargain with time differs from that of Hinduism and Buddhism. They do not seek timelessness but rather the temporal harmony within the person, among individuals, and between society and nature. Confucianism has been described as the doctrine of worldly social-mindedness, Taoism as the spiritual search for order in nature. Both of these philosophies of life date from the sixth and fifth centuries B.C. and both have influenced all aspects of Chinese life so profoundly that they became built into the Chinese national character. The philosophy of modern China since Mao's death has been a Marxism grown on the soil of Confucianism and Taoism.

Beneath it all, as Joseph Needham remarked in his magistral essay, "Time and Knowledge in China and the West,"

> time itself remained inescapably real for the Chinese mind. This contrasts strongly with the ethos of Indian civilization, and aligns China rather with the inhabitants of that other area of temperate climate at the Western end of the Old World,[2]

that is, Europe. But a fundamental difference between Chinese and European views of time remains. The Chinese have a preference for organic naturalism (nature and time perceived as aspects of dynamic, living systems to be explored qualitatively), whereas Westerners have preferred inorganic naturalism (nature and time perceived as aspects of nonliving systems, to be explored quantitatively).

The mythology of the Vedas is believed to share a common ancestry with Mithraism, the religion of ancient Persia, the worship of Mithra, the God of Light. At the turn of the seventh century B.C., the prophet Zoroaster preached against the Mythraic polytheism of *daivas*[3] and proclaimed the existence of a single deity. Zoroastrianism spread through Iran and influenced the later development of Judaism, Greek philosophy, Christianity, and Islam.

Of special interest to us is Zurvanism, a form of Zoroastrianism, in which Time is deified as Zurvan, progenitor of the opposing cosmic principles of good and evil, light and darkness, and creation and destruction. Within the nature of Zurvan, Time had two personifications: one was Zurvan "who for a long time follows his own laws" or "Time of the long dominion." This aspect of time brings old age, decay, and death. In the unembellished language of our own age we would call it "change." The other was Zurvan the "infinite Time," the grantor of eternal security. We would call it "permanence." The rich and varied teachings of Mithraism, Zoroastrianism, and Zurvanism were carried to Asia Minor, thence to Rome, and to many parts of the Roman empire. It is believed that during the first three centuries of Christianity, the various forms of

---

2   Needham, "Time and Knowledge in China and the West," in *The Voices of Time*, ed. J. T. Fraser, 2d ed. (Amherst: University of Massachusetts Press, 1981), p. 92.

3   Gods, heavenly ones, hence our "diva" via the Latin *divus*.

Mithraism must have appeared to many Christians as possible alternatives to their new faith because they shared with Christianity the perspective of the cosmic struggle between light and darkness, good and evil, timelessness and time. But Mithraism was subdued by Christianity and is seen today as the last vigorous expression of paganism. With that victory, the dualistic view of time in Western religious beliefs—that time is both permanence and change, both good and evil—was replaced by a very different assessment of the cosmic process.

The new view originated with Yahwism, the religion of the Hebrews at the time of Moses. The Hebrew tradition was preoccupied with a divine purpose, a love–hate relationship between Israel and its god Yahweh, a mutual agreement grudgingly but meticulously kept by both sides. As Freud perceived it, it was a state of creative tension between father and son.

Not for them the cycles upon cycles of time, eagerness for harmony with nature, worry about postmortem existence, or even the struggle between good and evil. Instead, the story of the world became a process of political liberation. During the ninth century B.C. a Yahwist writer of genius traced the providential actions of Yahweh from the creation of the world to the birth of Abraham, progenitor of Israel, then from there to the bondage in Egypt, and finally to the delivery of the Hebrews and the conquest of Canaan, the Promised Land. History ceased to be just one thing after another and became, instead, an intricately interwoven series of events that progressed from a well-defined beginning to an appointed goal. This view of history is known as salvation history, and with it, the idea of *linear time* was born.

The destruction of Jerusalem in 70 A.D. and the burden of the resulting diaspora did not change the central significance of salvation history for the Jewish ethos. On the contrary, it forced the extension of its promise to the largest stage to which it could possibly apply: to the whole of mankind and even to the universe.

The people of Israel remained tradition-bound in their religious rituals, tied as they were to the cycles of the astronomical year, but in history they preferred to see a linear, noncyclic progress toward a better future. This preference was later combined with the Platonic distinction between the temporal and the timeless and with Christian salvation history. In due course, the linear progression of history from the Fall to the birth, death, and resurrection of the Redeemer, and from there to the Last Judgment, became the foundation of Christian faith.

It has been convincingly argued by historians of science that the mental set of Western thought that emerged from this synthesis was necessary for the invention of the modern concept of scientific law and, therefore, for the emergence of modern science itself. In its turn, modern science has been supported by society because it held the promise of progressive, cumulative improvement in the life of man on earth. This attitude would have been unthinkable before the invention of the idea of salvation history.

With the meteoric advance of science and technology—made possible by the linear view of time and history—the relationship between God-the-timer and man-the-timed has changed. In our epoch, carrying out the promise of salvation history became a responsibility of the created and not of the Creator. In his novel *The Fratricides,* Nikos Kazantzakis gave an account of the new symbiosis. The following is an exchange between God and a Greek Orthodox priest.

> "Lead me!"
> "Lead You, Lord? But You are all-powerful!"
> "Yes, I am all-powerful, but only with the help of man; without you on the earth I created, I find it difficult to walk—I stumble. I stumble on stones, the churches, the people. Do not stare like that! Why did I create sharks in the ocean depths that cannot navigate without a little pilot fish to guide them? You are the pilot fish of God; get in front and lead me.[4]

There is a vibrant nativity scene, "The Adoration of the Shepherds," painted early in the seventeenth century by the minor Italian master Giovanni Lanfranco. The area around the Child is brilliantly lit in sharp contrast to the darkness around it. Through the magic of the brush, the light does not seem to come from any identifiable source: it just surrounds the Child naturally. In a way that Lanfranco and other artists of the age of Caravaggio could not have imagined, the painting represents the Renaissance passing of responsibility for history and time from God to the sons and daughters of men and women.

4   Kazantzakis, *The Fratricides* (New York: Simon and Schuster, 1964), p. 148.

Ideas about Time: Themes and Variations

This section is a sketch of representative philosophical and literary thought on the nature of time.

Philosophy (the love of wisdom, in its original meaning) is an invention of Greek antiquity. It is a search, by systematic speculation, for order in the world and in human affairs. The rules of those speculations are not set, however, but change as peoples' preferences for modes of reasoning change. Literature shares with philosophy the striving for excellence in form and content, and also the desire to seek out ideas of universal and lasting interest. The rules of literature also vary with place, age, and prevailing values.

I am going to use our interest in time as a searchlight to select a few representative views from an infinite store of philosophical opinions and recorded feelings. The searchlight is directed backward along the countryside that Western thought has traveled to get where it is now with respect to the idea and experience of time. Ideas and experiences are not independent, because everything we feel is modulated by what we know, and everything we know is influenced by how we feel about it.

Around 800 B.C. Homer narrated in the *Iliad* and the *Odyssey* a series of events that took place three or four centuries before his time. These epics are reliable records of the world of Homeric Greece, so called in honor of the poet.

The two epic poems are keen on the temporal organization of their plots, on life and death, youth and aging, and rosy fingered Dawn following rosy fingered Dawn (Homer reckoned days by dawns). But time is always something concrete and not a universal aspect of the world at large. The flask shown in figure 2 illustrates the notion of time in the daily life of Greece.

Three centuries after Homer, Pythagoras of Samos and his followers began to advocate a view of the world that was anything but concrete. They used the most abstract of all ideas, that of number, as their vehicle in the search for truth. They believed that ultimate reality was harmony according to number and that the task of the philosopher was therefore to search for and identify the rules of number and measure. They were the intellectual ancestors of what, millennia later, came to be known as the mathematized laws of nature. Figure 3 is a traditional portrait of Pythagoras, carved by a fifteenth-century artist.

FIGURE 2   A GREEK IMAGE OF DAY AND NIGHT. Black-figured lekhytos (oil flask) ca. 500 to 490 B.C., attributed to an artist known as the Sappho Painter. Helios (the sun) is seen rising from the sea while Nyx ("Night," daughter of Chaos) and Eos (the goddess of Dawn) are seen disappearing. In ancient Greece, time was known as duration, as events taking place one after another, but not as an abstract property of the world at large. Courtesy, The Metropolitan Museum of Art, New York, Rogers Fund, 1941.

FIGURE 3   PYTHAGORAS: REALITY IS THE TIMELESS RULE OF NUMBER.
The Pythagoreans believed that number and measure are the keys to absolute,
timeless truth. On this fifteenth-century woodcarving, Pythagoras is shown playing the lute, for he and his followers also discovered that musical harmony is
connected with ratios of numbers. The carving is the work of Jörg Syrlin the Elder,
one of the many woodcarvings decorating the choir stalls of the Cathedral of Ulm.
Courtesy, Deutscher Kunstverlag, Munich.

The desire of Pythagoras to find the rules of number stemmed from the same need as the cave artist's wish to record a magic dance on the cave wall: They both sought a shelter against the unpredictability of change. For the Pythagoreans it was not the Sorcerer but numbers and geometrical figures that performed the magic dance that could control the future. Pythagoreanism became an essential part of Western thought, and is very much alive in the preferences of our own age. In modern physics, for instance, relativity theory sees time as geometry; in human relations, the use of calculations, graphs, statistics, and computers is preferred to qualitative reflection. More will be said about the Pythagorean aspects of our age in the context of time in contemporary society.

Parmenides, another ancient maker of modern thought, lived around 540 B.C. and came from a rich and distinguished family of Elea, a village in what is now southern Italy. He wrote on many subjects, and exclusively in hexameter verse, as was appropriate then for learned men. Central to his teachings was the denial of the reality of passing. Only a thing or event of which we can say that it *is,* is real; if we can only say that it *was* or *will be,* it is not. It followed, according to Parmenides, that time could not be real, because time demanded that we speak of things and events that were and will be. Change itself must then be an illusion. The true world is one of permanence. Final reality, he wrote in his poem "The Way of Truth," is something "motionless within the limits of mighty bonds, it is without beginning or end. . . ." His philosophical rejection of the importance of change is consistent with a conservative view of the world.

For Heraclitus of Ephesus, a contemporary of Parmenides, ultimate reality was ceaseless change, while permanence was an abstraction created by the mind. For him the world was a place of continuous struggle between opposites: day and night, winter and summer, peace and war, hunger and satiety. But these opposite forces or conditions worked in unity and were governed by Logos, which meant reason, word, or speech. His views are known from 125 fragments that survive of a book he wrote around 500 B.C. "Upon those who step into the same rivers, different and different waters flow," states one of the frequently quoted segments.

Heraclitus and Parmenides are often paired and associated with opposing beliefs as to the nature of time: either only change is real or only permanence is. While neither of the two men expressed himself exactly in these words, it is a valid way of simplifying and contrasting their views.

Their opposing doctrines will come up again and again, even as we consider the role of time in Newtonian and Einsteinian physics.

Zeno of Elea was a student of Parmenides. I fancy him as a bright-eyed, dark-skinned young man, full of life, fast on the uptake, spending his days in whatever in his time corresponded to a sidewalk café. He would eye the young women, then blurt out some of his thoughts, to the amusement of his fellow gogglers.[5] Here is one of his thoughts, known as the paradox of the flying arrow. It is stated in my words, not in those of Zeno.

Imagine a flying arrow. At every instant the shaft of the arrow occupies a region of space equivalent to its length, but never longer than that. Obviously, it has no room to move. Therefore, it does not move. What I perceive as the flight of the arrow is a play of the senses.

The paradox resides in the fact that the logic of the argument appears unassailable, as does the reliability of our sense impressions that tell us of the arrow's flight, and yet they contradict each other. Zeno's purpose in identifying this paradox was to support his teacher's view that only permanence is real.

During the twenty-five centuries since Zeno's time, the paradox of the flying arrow has been solved, praised, refuted, and ridiculed innumerable times, according to the changing notions of different epochs. The fact that it has had to be continuously reconsidered suggests that there is something fundamentally wrong with the question itself. Perhaps the paradox, as usually interpreted, has been hiding some false assumptions about nature and for that reason it cannot have a solution. I will return to the paradox of the flying arrow in the context of quantum theory.

Here is a more amusing and equally powerful story of recent vintage. It is called the prisoner's paradox. Its author is unknown.

On a Saturday, a man is sentenced to death. The sentence is to be carried out at noon, on one of the seven days that follow. But, the

---

5   According to Antiphon, a professional Greek speech writer (480 to 411 B.C.), Parmenides and Zeno once came to Athens. "Zeno was then nearly forty, and tall and handsome; he was said to have been Parmenides' favorite. They were staying at Pythodorus' house. . . . Socrates was still very young at that time" (Plato, *Parmenides,* 127A).

judge declares, "you will not know which day until you are told so the morning of that day. It will be an unexpected hanging."

Later that day, the prisoner is visited by his lawyer.

"This is a very honest judge," beams the lawyer, "he always says the truth. And that's good news for you. If you are not hanged by next Saturday, then you would have to be executed on Sunday; therefore it would not be an unexpected event. For this reason, it cannot be on Sunday. If you are still alive on Friday, and since you cannot be executed on Sunday, it must then take place on Saturday. But in fact it cannot, because you would again be able to foretell it. Saturday is out, as are all other days, including tomorrow. And right now you are very much alive. Have a nice day."

The prisoner, who was trained in mathematical logic, believed that his lawyer's argument was logically correct and for that reason, had to correspond to reality. He did, indeed, have two nice days. But Wednesday morning, an official arrived and declared that this was the morning of his last day on earth. This was an unexpected piece of bad news that ruined his day.[6]

I will offer a solution to this paradox before page 360 of this book, but the reader will not know on which page, until he is told; it will be an expected–unexpected event. But, consider this: if my solution were not found by the time the reader finished reading page 359 then it would have to be on page 360 and hence would not be unexpected. And so forth, back to this very page. And since it is not on this page, by the lawyer's argument it cannot be in the book. Or can it?

Some universities are old: Harvard was founded in 1636. Some are older. The Academy of Athens was founded by Plato in 387 B.C. and flourished for over nine centuries. Its exact location is unknown except that it was near a hill where many owls nested. Since those at the Academy sought wisdom, philosophy came to be—by association—a profession for wise owls.

Plato's thought—precise, profound, and broad—has informed Western

---

6   The reader may reset the story in any number of ways. For instance: "The enemy plans a surprise war before the end of the century." If the logic of the lawyer's argument is correct, can there be such a war? If it is invalid, why so?

civilization down to our own days. Mathematized science, the hallmark of our age, is unequivocally Platonic. Its formal power derives from the sharp bisection of the world into the categories of time and the timeless, a bifurcation of reality previously mentioned.

The universe, wrote Plato, is divided into two classes of things. Members of the first class follow patterns that are "intelligible and always the same." These eternal patterns are called ideas: they are permanent and timeless. Members of the second class are only imitations of those of the first class: they are material, tangible, and sensible; they can change; and, most importantly, they can be generated, that is, made. Plato asserted that all things that can change are only imperfect copies of corresponding, unchanging—eternal—ideas. Even time itself is but an imperfect image of timeless eternity.

The most noble of all timeless ideas and forms, so Plato reasoned, are those of geometry; out of geometrical forms the world is made. The meaning of this claim has been built into modern science in the following form: geometrical relations, expressed in equations, can describe properly and precisely the behavior of the physical world. But in Plato's time the formal machinery of modern science had not yet been invented. How therefore did he propose to construct the world of tangible objects from ideas of geometry?

Figure 4 can help to explain. It shows five geometrical bodies, the only ones that can be constructed from congruent regular plane figures. They are known as the five regular or Platonic solids. Plato associated the dodecahedron with the world at large, the tetrahedron with Fire, the cube with Earth, the octahedron with Air, the icosahedron with Water.

Fire, Earth, Air, and Water were the four elements that Greek science viewed as making up the world. In contemporary science, elements mean chemical substances out of which all other substances are made. All elements may be broken down to components common to them, such as electrons and protons, but once taken apart, they lose their distinct characteristics. The Platonic notion of the four geometrized elements is the ancestor of this concept. Thus, his tetrahedron, octahedron, and icosahedron may be broken down to the congruent equilateral triangles common to them, but if so disassembled, the distinct characteristics of the bodies vanish.

For Plato, the presence of triangles in the regular solids that stood for water, air, and fire was consistent with the fact that these elements could

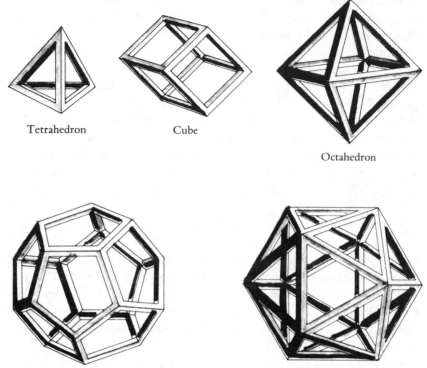

Tetrahedron

Cube

Octahedron

Dodecahedron

Icosahedron

FIGURE 4   PLATO'S GEOMETRIZATION OF NATURE: TIME IS A MIRAGE.
The five regular or Platonic solids bear certain mathematical relationships to each
other. Plato held that the relationships among the four elements recognized by the
Greeks (Fire, Earth, Air, and Water) and the relation between them (as a group)
and the world, are implicit in the mathematical relations among the solids. The
way our senses recognize those four elements of the physical world are temporal,
he would say, and for that reason, inferior copies of the corresponding permanent
geometrical ideas. Modern physics expresses in mathematical symbols what our
senses recognize as physical relationships: the idea derives from Plato's. In the
Platonic structure of today's physics, time is often judged a mental impression to
which nothing in the real world corresponds.

be transformed into each other. For instance, water and heat may be turned into air (vapor), and air can become fire, as in flaming gases that escape when wood is burned.

Neither Plato nor anyone else claimed that icosahedrons could be drunk like water. What Plato created were ways through which unchanging ("timeless") ideas could be employed to describe relations among sense impressions. Our understanding of nature is quite different from that of Plato's age, but the symbols of chemistry still reflect his logic. For instance, two vertical lines connected by a third one represent a chemical element (H), as we understand elements. When that picture is followed by a small picture of a swan swimming to the left (2), then by an oval standing on its tip (O), we have the symbol of a molecule. It tells us how a water molecule (H_2O) is broken down into its constituent components. Nobody claims that the lines of H, 2, and O, when combined, become wet. They stand, instead, for a stable ("timeless") relationship among objects (atoms) that, when together in large quantities, feel wet.

It is in this spirit of searching for whatever is eternal and mathematically expressible that Plato's theory of time must be understood. Plato tells us in *Timaeus* (37d) that "the father and Creator" made the world to be an image of the eternal gods.

> Now, the nature of the ideal being was everlasting, but to bestow this attribute in its fullness upon a creature was impossible. Wherefore he resolved to have a moving image of eternity, and when he set in order the heaven, he made this image eternal but moving according to number, while eternity itself rests in unity, and this image we call time.

To restate this paragraph, God, the ideal being, was everlasting, that is, timeless. He attempted to create the world in his own image, but had to settle for something inferior to himself. Instead of making the world (of our senses) eternal, he gave us a moving (rotating) image of eternity: the sky. The elements constituting that image—the stars, the planets, the moon, and the sun—move according to number, that is, at speeds in fixed ratios. We are temporal and fragmented; the big everything is timeless and unfragmented. What we call time is only an inferior copy of the original, the "one"—eternity.

Modern science is founded on the Platonic division of the world into the timeless and time (the lawful and the unpredictable). But the same modern science, in its familiarity with all organizational levels of nature—matter, life, man, and society—also recognized that the sharp Platonic division of the world into time and the timeless is too coarse. There is a spectrum of temporalities between the atemporal (a curious state of energy to which none of our time-related notions apply) and the nootemporal, but let me not run ahead of the story.

Whenever we read an analog (a dial) watch, we use the Platonic idea of time, as may be easily seen. First, remove the large hand of the clock, then put in two extra gears so that the small hand will complete one revolution in 24 hours instead of 12 hours. Next, let the small hand be fixed so that it always points north and let the dial rotate. The dial is now the moving image of eternity, and time is being measured by that motion, "according to number."

Here are some parting words from the founder of the Academy:

> The sight of day and night, and the months and the revolutions of the years, have created number and have given us a conception of time, and the power of inquiring about the nature of the universe.
>
> (*Timaeus,* 47a)

Unlike the analog watch, the digital watch stands for a definition of time first given by Aristotle, a student of Plato. Aristotle was not concerned with the concept of time but only how time is used in the study of nature. "For time is just this," he wrote in his *Physics* (Book 4–219B), the "number of motion in respect of 'before' and 'after.' " The beforeness and afterness does not come from the outside world of motion. It must be supplied by the person using the digital watch, which only counts the "number of motion," such as the number of oscillations by a quartz crystal. It is easy to construct a digital watch that counts hours and minutes backward: 1:32 P.M., 1:31 P.M., 1:30 P.M., and so on. But the owner could not be fooled, for we all have our built-in sense of what is meant by before or after an event.

While the Greeks debated the reality of time, to the south of them the Hebrew idea of salvation history was born, as previously discussed; to the

west of them the Roman civilization had its own way of encountering
time.

Lucretius, a Roman poet of the first century B.C., asked in "The Nature
of Things,"

> perceivest not
> How stones are also conquerèd by Time?
> Not how lofty towers ruin down,
> And boulders crumble? Not how shrines of gods
> And idols crack outworn?
> (Book 5.1, trans. W. E. Leonard)

As did the Egyptian harpist, Lucretius must also have talked to
Shakespeare:

> When I have seen by Time's fell hand defac'd
> The rich-proud cost of outworn buried age;
> When sometimes lofty towers I see down raz'd
> And brass eternal slave to mortal rage. . . .
> (Sonnet 64)

Lucretius had been dead for some eighty years when Christ was
crucified. Another ten or twenty years passed until the day when a Roman
patrician named Saul of Tarsus, on his way to Damascus to seek out
Christians and bring them to trial for heresy, underwent a complete
reversal of belief. He became a disciple of Christ, was baptized Paul, and
during the remaining thirty years of his life became the founding father of
Christianity. He changed the Hebrew salvation history from a single- to
a two-phased plan of divine purpose. The first phase was a preparation for
the coming of Christ, through the history of Israel; the second phase was
the carrying out of the mission of the Church, which was to prepare
mankind for the Last Judgment. The two phases were separated by the
ministry of Christ.

The reader may recall that the goal of Hebrew salvation history was the
betterment of the fate of Israel or, perhaps, of mankind, here on earth.
With St. Paul, the promise of a better life on earth was transformed to the
promise of a postmortem, timeless existence: "For the things that are seen

are temporal, but things that are unseen are eternal." Sixteen centuries later, during and after the Renaissance, the theology of salvation became secularized. The journey along the road to the salvation of the soul was replaced by the journey along the road of progress. The bridge between heaven and earth that used to be the person of Christ was replaced by the responsibility of man for his own fate.

St. Paul laid the foundations of the theology and politics of Christianity; his labor signified the transition from pagan to Christian antiquity. The thought of St. Augustine, Bishop of Hippo in Roman Africa, signaled the transition from Christian antiquity to the Middle Ages. Born in 354 in what is now Algeria, he witnessed almost eighty years of the social transformation that is often described as the decline of the Roman empire. In his mind the religion of the New Testament fused with Platonic tradition, creating a synthesis of new ideas that included many about the nature of time. Augustine's concepts were embraced by medieval Christianity, were transmitted to Renaissance Protestantism, and have filtered through it to our own days. His reflections focused on time in human experience.

Consider, he wrote in his *Confessions,* the syllables of the hymn, "Deus creator omnium." These syllables are not attached to a moving body one can observe; they are only fleeting voices.[7] I measure the length of one syllable against another:

> It is in you, my mind, that I measure time. . . . As things pass by, they leave an impression on you. . . . It is this impression which I measure. Therefore this itself is time or else I do not measure time at all.
>
> (Book 11. sec. 27)

Modern science can fill in the details and add that St. Augustine was talking only about noetic time.

---

7   If not at ease with Latin, consider *McGuffey's Eclectic Reader* of nineteenth-century America, a bestseller of its time. Here are five lines from a poem that children might have learned; each line can illustrate St. Augustine's point: "Oh, a wonderful stream is the river of Time, / As it runs through the realm of tears, / With a faultless rhythm and a musical rhyme, / And a boundless sweep and a surge sublime, / As it blends with the ocean of Years."

What then, is time? If no one asks me, I know. If I wish to explain it
to someone who asks, I know it not.

(Ibid., sec. 14)

Augustine's dilemma is due to the qualitative difference between *time
felt* and *time understood*. Time felt is the temporal reality or ambience of the
world as interpreted by the older regions of the brain and, hence, by the
hidden levels of the mind. Time understood is the temporal reality or
ambience of the world as interpreted by the newer regions of the brain
and, hence, by the easily accessible levels of the mind. The contents of the
deeper regions of the mind is not generally available for cognitive
examination and, hence, our feelings cannot easily be put into words. And
vice versa, the contents of the newer regions of the mind—such as time
understood—is alien to the world of the older, in a sense more primitive,
mind. In sum, we are all at two minds about time: a feeling one and a
thinking one. If no one asks us to explain our feelings about time, we feel
ourselves on safe grounds. If someone asks us to explain those feelings, we
do not.

St. Augustine was also an early Christian witness to the Western love
affair with number. The laws of numbers, so he held, are higher than the
laws of human reason. But since man's soul does contain the secrets of
number—thus sharing a bit of the Divine—we can use numbers, such as
those in the numerical organization of music and metric poetry, to become
aware of ourselves as beings that exist in time.

Seven centuries after Augustine's death, Islamic science and philosophy
flowered, advocating its own assessment of the nature of time. The most
articulate speaker of that assessment was a twelfth-century astronomer,
medical man, and legal scholar, Abu-al-Walid Mohammed ibn Rushd,
known in the West by his latinized name, Averroës. He thought of the
universe as an organic structure, of humanity as having a single intellectual
soul, and of time as demonstrating man's partnership with the sensible
cosmos. The human soul may, however, reach outside the cosmos and
into timelessness through such means as the ecstasy of the dance, a belief
sympathetic to Islam but also present in other religions (see fig. 43).

Unlike the name of Averroës, the name of his contemporary, Omar
Khayyám, is probably familiar to the reader because of his ever-popular

Rubayyat. He was a Persian mathematician, astronomer, and poet who, as poets are wont to do, put his thoughts and feelings into words. About a hundred of his thousand quatrains are available in English in the magistral rewriting of Edward FitzGerald. Many attest to a passionate, fatalistic, effusive sense of the world and time and record the puzzlement and hedonistic melancholy of the time-knowing creature. Here is a sample:

> One moment in annihilation's waste,
> One moment of the well of life to taste—
> The stars are setting and the caravan
> Starts from the dawn of Nothing—
>   Oh, make haste!

A little more than a century after Omar Khayyám, Chaucer caught the passing of time in a more cheerful stanza in the "Introduction to the Man of Law's Tale":

> Lost money is not lost beyond recall
> But loss of time brings on the loss of all;
> It can return to us again, once sped,
> No more than can poor Molley's maidenhead.

In yet another century that superb synthesis of ideas about man, time, and the world that was provided by medieval Christianity began to come apart. It ended a violent millennium of semicivilization: pockets of beautifully refined cultural life, surrounded by a great deal of squalor but not necessarily darkness, because it was an age of spiritual order in which time had meaning, though life had little value. Through the alchemy of history the veneration of custom gave way to a new intellectual temper that, four and a half centuries after it was born, was named the Renaissance.

The Schwartze Katze wine sold in the local liquor store comes from the village of Kues (today's Bernkastel-Kues) on the Moselle River. So did the fifteenth-century churchman, Nicholas of Cusa (of Kues), the original Renaissance man who helped lessen the hold of earth-centered cosmology on the medieval mind. Until his time, it was held that the universe was finite in its extent. If the universe were infinite, wrote Aristotle nineteen centuries earlier, then those stars that circle the earth at infinite distances would have to traverse infinite distances in finite times. Since this is impossible, he concluded that the universe must be finite. In a very non-Aristotelian fashion, Nicholas of Cusa produced a synthesis of finity and infinity. In his *Of Learned Ignorance* he maintained that "though the world is not infinite, yet it cannot be conceived as finite, since it has no limits within which it is enclosed." This curious unity is known as the coincidence of contraries.

If living with this idea of a finite–infinite universe as mystically envisaged by Nicholas of Cusa is difficult, the reader may turn to the cosmology of general relativity theory—to be discussed later—which teaches that the world is finite but unbounded. This does not mean that Nicholas of Cusa ought to be credited with inventing relativistic cosmology. Rather, his thought was one of the many components of the Western intellect from which modern science derived its system of understanding, including that of time.

At the turn of the fifteenth century, Nicolaus Copernicus showed that it made more sense to think of the earth as spinning about its axis, revolving around the sun, and also precessing like a spinning top than to imagine the earth at rest with the heavens revolving about it. In one fell swoop, he removed us from the center of the universe.

In ancient thought the heavens, unlike the earth, were perfect. So were circles, because unlike straight lines, they were beautifully symmetrical— and endless. While no one could draw an infinite straight line, everyone could draw an endless circle. Motion around a circle could go on forever, motion along a straight line had to come to an end sooner or later. Eternity was judged good, finity bad; circular motion was perfect, linear motion imperfect. Putting it all together, the orbits of planets and stars had to be circular. If in doubt, look up at the sky at night: those shiny objects revolve endlessly round and round, making for a moving image of eternity.

But there were problems. The planets did not move at uniform speeds and some of them sometimes backtracked, in what is called retrograde motion. These problems, astronomers believed, would be resolved in due course, for those orbits *had* to be circular.

Copernicus maintained that although the center of the universe is the sun and not the earth, "the movement of the celestial bodies is regular, circular and everlasting. . . ." In 1543, the year his *On the Revolutions of the*

Celestial Spheres was published, the starry heavens were still perfect, eternal, and timeless. That situation changed with the work of Johannes Kepler.

Kepler was the son of an innkeeper, an infant born two months prematurely, a delicate child who contracted smallpox, which led to permanently impaired eyesight. His life was one of trials and tribulations made worse by the Thirty Years' War, which was a fifty years' war for the European balance of power. But Kepler could still feel enthusiasm for great scientific insights and express them, as he did in the dedication of Book 5 of his *On the Most Perfect Harmony of Celestial Motion*, published in 1619:

> What I have prophesized twenty-two years ago as soon as I had discovered that the relationship among the five planetary orbits corresponded to the five regular solids [see fig. 4] . . . what I have promised my friends in the title of this book . . . what I have set for myself as a goal sixteen years ago . . . I have brought to light at last and have seen it to be true beyond my fondest hopes. . . . Let nothing confine me, I will indulge in my sacred ecstasy. I scornfully defy all mortals with the open acknowledgment: I have stolen the golden vase of the Egyptians, to raise a tabernacle to my God far from the land of Egypt. If you forgive me, I rejoice. If you are angry, I cannot help it. I cast the die and write a book for the present or for posterity. It does not matter. Perhaps it will have to wait a century for a reader. God awaited six thousand years [from the Old Testament date of Creation] for a thinking observer.

Kepler was a genius of astronomical calculations; he could find order where others saw only disorder. His ideas on the correspondence of planetary orbits to the Platonic solids is intelligible but wrong: there is no such correspondence. However, his three planetary laws remain fundamental to all calculations of orbital motion in the gravitational field of a central body, such as those of planets, spacecrafts, and missiles.

However, Kepler's most revolutionary achievement was none of the above. Rather, it was his genius to suspect, his know-how to demonstrate, and his courage to insist that contrary to tradition and belief, the planetary orbits were elliptical and not circular and the planets did not move at constant speeds (fig. 5). His insistence on elliptical orbits and changing speeds was as revolutionary a challenge for his age as Darwinian evolution has

PROTHEOREMATA.

I.

SI intra circulum deſcribatur ellipſis, tangens verticibus circulum, in punctis oppoſitis; & per centrum & puncta contactuum ducatur diameter; deinde a punctis aliis circumferentiæ circuli ducantur per pendiculares in hanc diametrum: eæ omnes a circumferentia ellipſeos ſecabuntur in eandem proportionem.

Ex l. 1. Apollonii Conicorum pag. XXI. demonſtrat COMMANDINVS *in commentario ſuper* V. *Sphæroideon* ARCHIMEDIS.

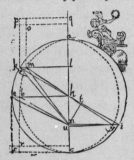

Sit enim circulus A E C. *in eo ellipſis* A B C *tangens circulum in* A C. *& ducatur diameter per* A. C. *puncta contactuum, & per* H *centrum. Deinde ex punctis circumferentiæ* K. E. *deſcendant perpendiculares* K L, E H, *ſecta in* M. B. *a circumferentia ellipſeos. Erit ut* B H *ad* H E, *ſic* M L *ad* L K. *& ſic omnes aliæ perpendiculares.*

II.

Area ellipſis ſic inſcriptæ circulo, ad aream circuli, habet proportionem eandem, quam dictæ lineæ.

Vt enim B H *ad* H E, *ſic area ellipſeos* A B C *ad aream circuli* A E C. *Eſt quinta Sphæroideon* ARCHIMEDIS.

III.

Si a certo puncto diametri educantur lineæ in ſectiones ejusdem perpendicularis, cum circuli & ellipſeos circumferentia; ſpacia ab iis reſciſſa rurſum erunt in proportione ſectæ perpendicularis.

FIGURE 5   KEPLER'S BREAK WITH THE ARISTOTELIAN BELIEF THAT THE HEAVENS WERE PERFECT. This is a page from *Astronomia Nova,* published in Prague in 1609. It is the beginning of Kepler's demonstration that the orbit of Mars is elliptical. The planet is represented by the god Mars on a chariot. Though a god of war, Mars was originally a god of agriculture; a farmer first, then a soldier. By Kepler's time, astronomers had to use a minimum of 34 circles, intricately moving with respect to one another, to account for the observed motions of the planets. But circles they had to be, for circles and the heavens were perfect. Kepler could account for the observed planetary motions by using one single ellipse for each planet. Along these ellipses the planets moved at changing speeds, their velocities being functions of their changing distances from the sun. Henceforth, heavenly motions would have no privileges. Courtesy, The Burndy Library, Norwalk, Connecticut.

been for ours. For if his reasoning was correct, then the physics of the sky was not different from that of the earth—as it was believed—and the world could not be separated into a timeless heaven and a temporal earth.

Copernicus removed us from the center of the world, Kepler democratized heaven and made it temporal. The medieval distinction between the heavens and the earth began to vanish, and the age of reason and doubt began to emerge.

The mathematician and philosopher René Descartes, a younger contemporary of Kepler, maintained that we were made of a spatial part that he called the extended or spatial substance, and a temporal part, the thinking substance. This stark division between body and mind, as if they were totally independent, is not defensible. But the sharp distinction between space and time appeared to have been obviously correct until modern physics showed that it was not universally so. There are organizational levels of the natural world where the distinctness between distances and times has no reality.

Descartes was also the inventor of the scientific method of doubt. Do not accept anything as a fact, he wrote, until after it has withstood a careful examination that demanded clarity and distinctness. The strength of the scientific method, so clearly useful in the exact sciences, derives partly from this Cartesian instruction. But another son of the sixteenth century, William Shakespeare, insider to all human feelings, made in *Hamlet* a very different assessment of the role of doubt in understanding the world that includes ourselves.

> Doubt thou that stars are fire,
>   Doubt that the sun doth move,
> Doubt truth to be a liar
>   But never doubt I love.

For Shakespeare, as for most poets and writers, and for people dealing with life rather than with the inorganic, time was feeling and the creation of the unpredictably new. His plays and sonnets show his fascination with the kind of dynamic process that has no place in exact science: conflict arising from harmony, and harmony emerging from conflict; life issuing from matter, then returning to matter. In *The Winter's Tale*, Paulina, high priestess of creation, makes life and change come about from unchanging, dead stone.

> Music, awake her: strike! [*Music*]
> 'Tis time; descend; be stone no more; approach;
> Strike all that look upon you with marvel. Come.
> I'll fill your grave up; nay, come away;
> Bequeath to death your numbness. . . .

The stone became a living, thinking, feeling woman. Should we examine this poetic fact with Descartes' scientific method, or should we examine the supposed factuality of science with the methods of literary criticism because that factuality is only poetic play? Beneath these alternatives lurks a duality of temperaments. One favors permanence and insists that anything worth knowing must be predictable; this is the temperament of exact science. The other favors change and insists that anything worth experiencing must be an example of creativity; this is the temperament of humanistic knowledge.

Seventy years after Shakespeare's death, Isaac Newton laid the foundations of modern physics in his *Mathematical Principles of Natural Philosophy* or briefly, *Principia*. If research on the ultimate nature of time were to be conducted today by means of a Gallup poll, using 10,000 randomly selected telephone customers, it would surely conclude that time flowed and thus confirm Sir Isaac's views as correct.

He held that "absolute, true and mathematical time of itself, and from its own nature, flows equably without relation to anything external, [anything other than time itself] and by another name is called duration . . ." (Scholium in the Definitions of the *Principia*). Gone were the attempts to relate time to the motion of the stars (as proposed by Plato), to the "number of motion" (Aristotle), to the mind (Augustine), to the world and mankind (Averroës), or to life and feeling. Time became a type of universal order that existed by and in itself, regardless of what happened in time.

The postulate of absolute time, as this idea is known, was a stroke of Newton's genius. It made possible the formulation of scientific laws expressed in equations, wherein the symbol t stood for time, defined as "you know what I mean." The physicist's *t*, as it is called, bypasses all questions about the nature of time. Neither Newton's absolute time nor Einstein's relativistic time, as we shall see, says anything about what we are to mean by the present, the future, or the past. They both assume that we already know what is to be meant by time. For that reason they are not the same *kinds* of ideas about time as the ones I have been describing. They belong, instead, to a new breed of notions that, for pragmatic reasons, judge life, mind, and society as irrelevant to those aspects of time that are really important.

Newton, using his powerful idea of absolute time (and space), succeeded in formulating the first truly universal principle of science expressed in strict mathematical form: the law of gravitation. With it, he planted the seeds of doubt concerning the identity of the ordainer of cosmic order: was it God or was it Nature? In our Declaration of Independence, a document composed five decades after Newton's death, the two appear on equal footing as "the Laws of Nature and of Nature's God." The respect paid to the laws of God was to be shared, henceforth, with an equal respect paid to those of nature.

Time, said Newton, was independent of man; it flowed by itself.

Not at all, it was nothing like that, said the eighteenth-century German philosopher Immanuel Kant, a great admirer of Newton's work. If time were independent of man, it should be possible to get noncontradictory answers to some basic questions about time, such as whether or not it had a beginning. But it is not possible to get such answers.

For let us say that time did not have a beginning. I can say the words but what do they mean? It is impossible to imagine a no beginning to time. Let us say, therefore, that it did have a beginning. But what am I to mean by, "there was a time when there was no time?" Clearly, to assume that time somehow flows by itself leads to contradictions. But nature cannot be contradictory. Kant's conclusion: time is a mode of human understanding.

> The idea of time does not originate in the senses but is presupposed by them. . . . Time is not something objective. It is neither substance nor accident nor relation, but a subjective condition, necessary owing to the nature of the human mind, of the coordination of all sensibles [experienced stimuli] according to a fixed law. . . .
>
> (*Inaugural Dissertation*, sec. 3.14)

In other words, time is a part of our mental apparatus. Yet we must assume it to be real because only by doing so can we make sense of the world. In the jargon of our own, post-Darwinian days, the Kantian stance says that time is a way we perceive the world because that kind of perception has been programmed into our genes. It is a development of thought and behavior that evolved because it is useful in the struggle for survival.

In Kantian thought, it is meaningless to ask whether time is really *real*, as it were, whether there is anything in the world to which the idea corresponds. Human time (nootemporality) is intersubjective.[8] That is, time is subjective for each person involved, but since it is an idea necessary for survival, time also becomes something objective.

Before concluding this time-lapse description of the expansion of Western ideas about time, two more people should be mentioned: Hegel and Marx. Both were intellectual heirs of the Judeo-Christian notion of salvation history, and both equated time almost entirely with the history of mankind. But they differed substantially about the road humanity should follow to achieve salvation: Hegel's was the road of the spirit, that of Marx, the road of materialism.

Georg Wilhelm Friedrich Hegel, a younger contemporary of Kant, perceived of time as our awareness of the way the human spirit works on its destiny. History consisted of a series of spiritual conflicts that, when resolved, gave rise to a synthesis that produced its antithesis to fight with—and so on. He called the process the dialectic of the spirit.

Karl Heinrich Marx, born in 1818, became a Hegelian who saw in history the dialectic of matter. Marx equated human life with making a living and thus held that the most important science was economics. The basis of economic production and of currency (the article of bartering) is the time of labor. It follows, on this view, that time is entirely a social convention.

All this is a rather far cry from my hairy ancestor making a weapon to fight the frumious Bandersnatch, but that is where it all began. And with the dawning awareness of the parting known as death, so sensitively sketched by Emily Dickinson, eons after the Bandersnatch episode.

---

8   A definition of the concept of nootemporality or noetic time can be found in the first section, "The Discovery of Time," or in the Glossary.

> Parting is all we know of heaven,
> And all we need of hell.

The enterprise of trying to stop time has been a most profitable wild goose chase. For as people strived to understand time, they improved their understanding of themselves and of the world. The new insights were then put to use in the struggle for survival. Each step along the path of inquiry changed the precepts upon which those inquiries were based, which made the task of understanding time open ended. Each victory revealed a yet larger world to conquer. In the study of time there can be no final point of arrival, but only discoveries of new points of departure.

Since about the mid-nineteenth century, the sciences have taken on an increasingly larger role in determining human values, and through them, they came to determine the preferred ways for learning about the world. Along that path, and true to the poet's instructions that serve as our chart and itinerary, I want to turn now to the essentials of time measurement.

For ever and a day

T H E   R E C K O N I N G   O F   T I M E

Homo sapiens—man the thinker—responded to his discovery of passing and time by attempting to overcome the finality of death. He, she, and they created languages to help their societies survive the individual finity of their members. Using language they created religions, philosophies, and the magic of the recited and written word, so as to lessen the anxiety conferred upon them by the discovery of time. All this happened in Chapter 1.

This chapter is about *Homo faber*—man the tinker—alter ego by birth to his wisdom-seeking self. He put his intuitive, gut knowledge of time in the service of daily existence. He learned how to count and invented many means of time reckoning: clocks, calendars, chronologies, and cosmologies. By means of such metaphysical bookkeeping, he helped put order into and find order in human life and the world at large.

The Stuff that Clocks Are Made Of

To measure anything is to make a comparison and express the results in number: my driveway is 100 times as long as my meterstick. The meterstick does not make the comparison; I am the one who does. The meterstick simply *is*. In contrast, a clock always *does* something: it ticks, hums, strikes, blinks, points; it is said to measure time. It, and presumably not me, must be continuously comparing itself with something and expressing the results of its findings in number: 2:35 A.M. To what does the clock compare itself when it measures time?

To say that it compares itself to another clock only passes the buck, because we then must ask, to what does that other clock compare itself? To get an answer one must survey some of the many methods that have been used for time reckoning and ask what is common to all of them. This is what the present section does.

An Eddic poem recorded in the thirteenth century describes sternly and beautifully the predicament of knowing time:

> woe's in the world, much wantonness;
> axe-age, sword-age—sundered are shields—
> wind-age, wolf-age, ere the world crumbles;
> will the spear of no man spare the other.

The time of axe, sword, wind, and wolf come and go with deadly regularity but exactly when the axe will fall or wind will blow, there is no way of telling. This combination of certainty and uncertainty has kept our species on its toes since the Jabberwocky affair. Or was it the Bandersnatch affair? The past, as we shall see in this chapter and later, is a changing map that is continuously being redrawn both in its individual and collective dimensions. In any case, since *illo tempore,* timekeeping has consisted of the use of reliable rhythms for the prediction of future events.

Notations on bones 28,000 to 30,000 years old, showing the waxing and waning of the moon for three lunations, have been identified by Alexander Marshack, an anthropologist and pioneer researcher in archaic calendars. A bone plaque dating from around 10,000 B.C. shows lunations extending to three and a half years, with notations indicating the four milestones of the solar year: the two equinoxes and two solstices. Although the interpretation of evidence is still being debated, research in the disciplines of archeology and ethnoarcheology suggests that as long as 8,000 to 9,000 years ago our ancestors recorded cycles of fertility, birth, and the menses. They were able to identify stable patterns in their experience of time and represent them by engraved symbols.

Originating at a much later date, some time between 1800 and 1400 B.C., are the monumental stone circle calendars of England. Stonehenge, the most famous of them, has a legendary history. Different tales speak of the stones having been brought from the farthest reaches of Africa, from Kilimanjaro, or from Ireland. One source insists that Merlin, King Arthur's magician, supervised the erection of the stones.

Since the circle is 100 meters across and some of the boulders weigh
50 tons, attributing the construction of Stonehenge to superhuman beings
is understandable. In a more realistic vein, the techniques used in carving
the stones suggest a Greek and Cretan influence. Present thought holds
that Stonehenge was an observatory built to determine the crucial days of
a lunisolar calendar for use in Druid worship.

Five centuries after Merlin built his rather heavy calendar, someone in
Egypt made a very light and small shadow clock from green slate; it may
be inspected at the Ägyptisches Museum in West Berlin. It has a raised
crosspiece and a straight base inscribed with six time division marks; it
is the surviving ancestor of the most ancient method of timekeeping, that
by the sun's moving shadow. At first, the shadows of trees and people
were most likely used to indicate the passage of the day. Later, as in the
Egyptian shadow clock, it was a horizontal crossbar. Even later, it was the
sundial.

How does one construct a sundial?

Take a fencepost and hammer it vertically into flat, open ground.
Observe the changing direction and length of the shadow of the sun as it
moves from morning to night and changes its orbit throughout the year.
You have the simplest of sunclocks: a stick in the sun. If instead of hammering the fencepost vertically into the ground, you hammer it at such an
angle as to make it parallel with the earth's axis, then the direction of the
shadow at any given hour will be independent of the season. You will have
made a clock and a calendar of very great sophistication.

The fencepost is now a gnomon, getting its name from the Greek where
it means adviser, inspector, or one who knows. The gnomon of a sundial
is a knowledgeable translator or interpreter: it translates the rules of
heavenly motion into the language of earthly behavior. The motion of the
shadow is a part of the heavenly world, the divisions on its calibrated dial
are useful to the human world on earth (fig. 6). The two are connected by
our trust in the regularity of solar motion.

Now that we have what we judge to be a uniformly moving shadow, how
may it be used to organize our time?

For the Babylonians, Greeks, and most primitive tribes the day began at
sunrise. The ancient Egyptians counted it from midnight to midnight, as
we do. In traditional Judaism, it started at sunset. The Teutonic tribes
counted nights instead of their days, which is why two weeks make a fort-

FIGURE 6   WHITE MARBLE SUNDIAL, THIRD CENTURY B.C. The face of the
sundial is a cone, seen here from the front. The Greek inscription is on the plane of
the sundial seen here at an angle, slanting down and back. At the latitude where the
device was used, the plane would be placed parallel with that of the equator. The
gnomon, now missing, protruded along the axis of the cone and was parallel with
the axis of the earth. The gridwork of lines meeting at the center corresponds to the
changing direction of the shadow, demonstrating that the dial was a clock: it
showed the time of day. The concentric circles, used to measure the length of the
shadow, tell us that it was also a yearly clock, familiarly known as a calendar. The
circles mark the length of the shadow at noon as the sun entered the consecutive
signs of the zodiac. The sundial was found in 1873 at the foot of Mount Latmus,
Turkey. Courtesy, Musée du Louvre, Département des Antiquités Grecques et
Romaines, photo by M. Chuzeville.

night; in eighteenth-century England people still spoke of a se'nnight. Until 1925, astronomers counted their days from noon to noon; since that year they have counted them from midnight to midnight. Hotel days often begin "after 2 P.M." and end "before 10 A.M." The reader could probably name a dozen other ways of counting days.

There have also been many different methods for dividing the day into hours. The earliest ones provided 3, 6, or 12 equal parts to the daylit day and the same number of parts for the length of the night. Such divisions are called unequal hours, because as the lengths of days and nights change, so do the lengths of the hours. Medieval monastic time was divided into 12 equal—and therefore seasonally changing—periods of day and 12 of the night. Only after the introduction of the mechanical clock could the whole day be conveniently divided into equal, unchanging hours.

The traditional Chinese horary system counted the day from midnight to midnight and divided it into 12 double hours, a practice already well established by the second century B.C. Each double hour was given the name of an animal: 5 to 7 A.M., hare; 7 to 9 P.M., bear. Midnight was contained in the hour of the rat, cutting the double hour 11 P.M. to 1 A.M. into two halves. The Chinese also divided the day into 100 *k'o,* called quarters by Western writers, because they turned out to be 14 minutes 24 seconds long. Simultaneously, the Chinese maintained a division of the night into five unequal night watches. In the context of medical problems, they used units of four double hours called by the names of the seasons: 3 to 11 A.M. was called spring; 11 A.M. to 5 P.M., summer; 5 P.M. to 11 P.M., autumn; and 11 P.M. to 3 A.M., winter. The names illustrate the Chinese preference for organic naturalism. Western names of the hours are inorganic, mathematical: 7 to 8 A.M. is called "seven"; 10 to 11 P.M., "ten" or "twenty-two."

The earliest devices that helped keep the time of day without using the sun's motion were water clocks; they were trusted because of the belief that the rate at which water flows through an orifice was uniform. Bucket-shaped, outflow-type water clocks dating from the fifteenth century B.C. survive from the Temple of Karnak in Egypt, and there is evidence of their use in thirteenth-century B.C. Assyria. Water clocks are known today by their Greek name, clepsydra, which means water thief.

Three decades before the birth of Christ, the Roman architect Vitruvius described a clock that had a rotating dial, moved by a float, as the float sank in an outflow clepsydra. The pointer was fixed, its dial showed the signs of the zodiac: the clock was a moving, mechanical image of the rotating sky. Its display (the circle of the zodiac) was the ancestor of our clock dial. Today's dials, instead of showing a Bull, a Virgin, an Archer, and the other nine signs of the zodiac show pictures, such as 1, 2, 7, and 11, that stand for one, two, seven, and eleven, or they just show straight lines. How unimaginative!

Chinese clepsydra techniques date from the sixth century B.C. Instead of using a single water clock, they often used a series of them, one emptying into the next so as to make the head of the penultimate one steady. This made its outflow rate even and guaranteed that in the very last one, from which time was read, the water accumulated at a uniform rate. In some arrangements a steel yard held the last compensating tank at a set weight, with the weight varied as the length of the days changed, so as to assure the division of day or night into six equal (but changing through the season) double hours. Since water froze in the winter, the Chinese sometimes used mercury in jade containers and pipes.

In seventeenth-century Europe, there was a resurgence of interest in water clocks, as a part of the renaissance of interest in harnessing natural forces. Some were simple and interesting, others only cute, some ornate and gaudy. The sandglass of our kitchen is a combination of an outflow clepsydra and an inflow clepsydra that uses specially prepared sand instead of water. It first appeared in the fourteenth-century; its inventor is unknown.

The dripping faucet, which reminds the would-be sleeper of the curious, expected–unexpected nature of future events, is a water clock of long and distinguished ancestry.

If water could be used, why not fire? In Europe we find a family of clocks that used controlled rates of burning for the measurement of time. There were graduated candles and oil lamps, like the one invented by Geronimo Cardano, a contemporary of Copernicus and inventor also of the universal joint (used today in every car). Alfonso X, King of Castille (in the mid-thirteenth century), owned or was aware of a candle clock, described at the King's request by "Samuel L. Levi, the Jew of Toledo," in the King's Book of *Astronomical Knowledge*. The clock is said to have consisted of a candle carried on a movable platform counterbalanced by a pulley system. As the candle was consumed and hence lost weight, the counterweights raised the platform—so we are told—and moved a tablet upon which the hours were marked.

The Greek biographer Plutarch remarked once that when the candles are out, all women are fair. But the use of sweet-smelling incense sticks as timers in geisha houses, for the services of the flower girls, was not a Greek but a Japanese custom. The burning of incense for time measurement was a widely practiced Oriental art; figure 7 shows a fine example of that tradition.

From moving shadows, flowing water, and the scent of incense, let me backtrack to a technical development in the measurement of time.

The Athenian mathematician Archimedes—the same who jumped out of his bath, streaked down the street hollering "Eureka! I found it!" (meaning his Principle)—also built a machine of bronze. This is what the Roman author Cicero said about it in his *Tusculan Disputations:*

> When Archimedes fastened on a globe the movements of moon, sun and five wandering stars [planets], he, just like Plato's God who built the world in the *Timaeus,* made one revolution of the sphere control several movements utterly unlike in slowness and speed. (1:63)[1]

Thus, we learn that Archimedes made a model of the world which showed the planets circling the earth at dissimilar speeds; today we call such a machine an orrery or a planetarium. No other details of Archimedes' device are known. But early in our century, near the island of Antikhytera, fishermen found four lumps of corroded pieces of bronze together with art objects and pottery. As reconstructed by the historian of science Derek J. de Solla Price, the fragments revealed themselves to be parts of the gearwork of a planetarium made around 87 B.C.

A hand-driven device, it displayed the positions of the sun and the moon, showed the length of the synodic month (new moon to new moon) and of the lunar year (twelve lunations). It was a mechanical calendar with the relative rates of planetary, solar, and lunar motions determined by its

---

1   Cicero, *Tusculan Disputations,* trans. J. E. King (Cambridge: Harvard University Press, 1966), pp. 74–75.

FIGURE 7   CHINESE INCENSE CLOCK, EARLY FOURTEENTH CENTURY. The
illustration shows a groove carved into a hardwood disk. The length of the groove
is about 240 inches. Incense, made from a variety of aromatic powders, was placed
into the groove and lit, probably in the center. It burned for perhaps six double
hours. The hours were recognized by their different scents. We can still tell time by
whiff: bacon and eggs is morning; perfume is night. The morning, midday,
evening, and midnight of the outdoors have different scents. "The Greatly
Elaborated Incense Seal," is reproduced here from *Hsin Tsuan Hsiang P'u*, or
"Newly Compiled Handbook of Incense," popular in medieval China, courtesy
Silvio A. Bedini. The interpretation of how the clock was used follows his "The
Scent of Time," *Transactions of the American Philosophical Society*, new ser. 53, pt. 5.

gear ratios. If the planetarium was cranked so as to make the revolution of the sun take a day, then all other objects moved in what computer engineers would describe as "real time." In its conception, the Antikhytera machine was the ancestor of all geared clocks.

An impressive planetarium and clockwork erected a millennium later, in a different part of the world, survives in the drawings and descriptions of its maker, a Chinese engineer named Su Sung. Figure 8 shows a pictorial reconstruction of Su Sung's "Water-Powered Armillary Sphere and Celestial Globe Tower," built in 1090. An armillary sphere—from the Latin *armilla* ("bracelet" or "iron ring")—is a skeleton sphere whose bones are the important circles of the celestial sphere: the equator, the tropics, the ecliptic, and the reference longitudes. A celestial globe shows the positions of the constellations. "The Heavenly Clockwork," a name Joseph Needham gave to this remarkable device, modeled with precision the motions of the sun, the moon, stars, and the planets. Since its solar indications were kept in phase with the actual position of the sun, it was also a clock.

The Heavenly Clockwork was driven by a waterwheel. Its control mechanism stopped it for a quarter of an hour while one of its scoops was filled from a clepsydra. When full, the scoop tripped the mechanism and let the wheel rotate so as to permit the next scoop to report for duty. Devices that alternately restrict and release the rotation of a wheel are called escapements. The story of mechanical timekeeping until the appearance of tuning forks and quartz crystals was the development of escapement mechanisms.

In the West the first stop-and-go device was the verge and foliot escapement: a horizontal rod or wheel made to swing back and forth about a vertical axis. The inertia of the rod, its resistance to being rotated now this now that way and back again, controlled the rate at which the clock hand moved. From its first appearance in the early 1300s until the invention two centuries later of the crossbeat escapement (an improvement in the way the foliot was made to swing), the verge escapement was the only known means of regulating a mechanical clock.

Early clocks were made entirely for monastic use. They struck each hour by a single strike; later, the number of chimes was made to correspond to the hour. Subsequent refinements made the display of

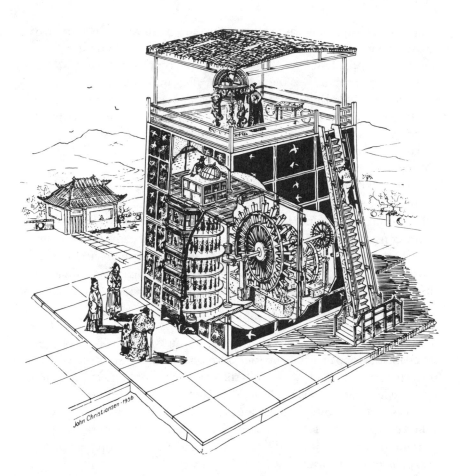

FIGURE 8   SU SUNG'S CLOCK TOWER, BUILT IN 1090. A pictorial reconstruction by John Christiansen. The water wheel with its ingenious escapement
rotated the armillary sphere on the upper platform and the celestial globe on the top
story. The double hours and the quarters were signaled by sound and the motion of
puppets. An observing tube attached to the armillary sphere would follow a star, as
do the equatorially mounted telescopes of our age. The clockwork was 10 meters
high, the celestial globe weighed some 15 tons. Courtesy, Joseph Needham,
F.R.S.

astronomical information possible, leading to astronomical clocks of marvelous ingenuity (see fig. 14).

But foliots were heavy and awkward, their rates of oscillation depended on the force used to drive them, and they could only be controlled by changing the weights hung on them. A breakthrough in assuring greatly increased uniformity in the oscillation of the escapement and in making that rate independent of the magnitude of the driving force was Galileo's discovery of the isochronism of the pendulum. Isochronism means having equal durations. Specifically, the period of pendular swing is independent of the amplitude of that swing; it depends only on the length of the pendulum. This is the reason why grandfather clocks are fine-tuned by slightly raising or lowering their bobbins. Once the length of the pendulum is fixed, changes in the driving force do not influence the frequency of its oscillations.

Seventy years after Galileo's discovery of the laws of pendular motion, the Dutch physicist Christian Huygens concluded from theory that a free pendulum, after all, is not precisely isochronous. To make it so, its bobbin must swing not along the arc of a circle, but along the arc of a curve called cycloid. In 1673 he constructed a pendulum that did just that (fig. 9).

Huygens' work is a good example of what is meant by improving the accuracy of a clock. It signifies the use in timekeeping of a more refined understanding of natural processes (exemplified by the swings of a pendulum), and the making of devices which use that improved understanding. In other words, *a clock ticks more uniformly than does another if the way it works can convince us that it ought to.* The most convincing clock, therefore, remains the most accurate one until such time that someone can think of a reason, acceptable to others, why the clock ought not be accurate.

In the social context, the coming of mechanical clocks marked a change in the way communities organized their labor. In the words of the historian of science Lynn White,

> the mechanical clock seized the imagination of our ancestors. Something of a civic pride which earlier has extended itself to cathedral building was now directed to the construction of astronomical clocks of outstanding intricacy and elaboration. No European community felt able to hold up its head unless in its midst the planets wheeled in

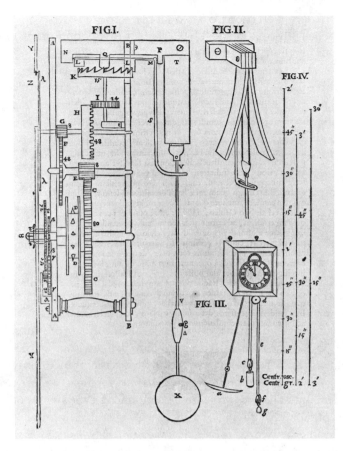

FIGURE 9 **CHRISTIAN HUYGENS' IMPROVEMENT ON THE ISOCHRONY OF THE PENDULUM**. The illustration shows the pendulum and the escapement of Huygens' clock. The crown wheel, K (in FIG. I), rotated by a weight, *b* (in FIG. III), is alternately restrained and released by two pallets, L (FIG. I), working in tandem. They are attached to a shaft, M, whose swings, driven by the crown wheel, are controlled by the pendulum. Huygens had found from theory that in order to be truly isochronous, the bobbin of the pendulum (X) should not swing along a circular arc, but along the arc of a cycloid. He provided for this motion by using the two cheeks, T (FIGS. I and II). By wrapping itself around these cheeks, the length of the pendulum changes during its swing, making the bobbin's path a cycloid. This example of improvement helps us in discovering to what a clock compares itself when it measures time. From Christian Huygens, *Horologium Oscillatorium* (Paris, 1673). Courtesy, The Burndy Library, Norwalk, Connecticut.

cycles and epicycles, while angels trumpeted, cocks crew, and apostles, kings and prophets marched and countermarched at the booming of the hour.[2]

The large clock in the church tower generated a public enthusiasm comparable to that for the computer in our own days. In the fifteenth century, ordinary soldiers took their roosters along to serve as alarm clocks.

> I have a gentil cock
>> Croweth me day;
> He doth me risen early
>> My matins for to say,

reports an anonymous early fifteenth-century poet. At home people were wakened by the bell of the tower clock, the pride and joy of the town. To get one built, planning had to begin early, for it took decades to get it finished. Extra taxes were levied for years. Once delivered, the town usually had to buy the lifetime services of the clockmaker as the clock's "governor," for the works were inaccurate and always in need of repair and of frequent winding. Everybody was involved with mechanical clocks. Astronomers, mathematicians, physicists, churchmen, soldiers, kings, merchants, and an army of artisans came and went, each having some notions of how to improve the clock or else how to use it.

Clock mechanisms came to be combined with astrolabes (instruments we shall presently meet). There were clock-driven armillary spheres, equatoriums (which computed planetary orbits), astraria (small mechanical universes), orreries (planetary machines), and a host of other instruments, all of them variations on the double theme of measuring time and the motions of the planets. The development of geared clockworks demanded skill; their functions spoke of orderliness and precision. The skills used in creating these devices made the development of scientific instruments possible; the order in nature represented by the mechanical clock came to serve as the model of modern scientific reasoning.

★

2   White, *Medieval Technology and Social Change* (Oxford: Oxford University Press, 1962), p. 124.

What is common to the many different modes of time measurement?
What is the stuff that clocks are made of, and to what does a clock compare
itself when it measures time? Let me answer these questions by a visit to
a shepherd, his dog, and his sundial in the happy land of Arcadia.

There are lines drawn on the sundial to help him identify the direction of
the shadow by number. One sunny day when the shadow points to *14,* the
shepherd's lover appears, carrying a wheel of cheese. When the shadow
points to *8,* his dog barks. These are two sets of coincidences or simultaneities: *lover & 14,* and *bark & 8.* How much time passed between the two
events?

If the dog barked on the morning of the day when the lover arrived, then
6 hours passed between the two events. If the dog barked while the lover
was leaving the morning after she and the shepherd celebrated the night
together, then 18 hours have passed. Only the shepherd, his lover, or
another human being can ultimately tell which of the two coincidences
(*lover & 14,* and *bark & 8*) happened earlier, which later.

All the examples of time measurement I have given may be represented
by this story. They all involve comparisons between two processes running side by side. In this case, one is the life of the shepherd, the other the
motion of the sun. We connect the two by the motion of the shadow. Then
we observe two sets of coincidences and mark off the time between them
in terms of the numerals on the clock, using our own awareness of passing
time. (Only we recognize what came first and what later; the clock could
and did not.) We are the ones who measured time; the clock did not. And
we measured the shepherd's time by the sun's time rather than the other
way around, because by unspoken agreement the motion of the sun is
judged a more uniform process than the life of the shepherd.

In our daily lives one of the processes used in the comparison is usually
so deeply hidden that it is not noticed. For instance, what are the two
processes in the statement "The first man landed on the moon in 1969"?
One may be selected quite freely; let us say it is the history of transportation: feet, donkeys, horses, trains, airplanes, and spacecrafts. The other is
the chronology of Western life: a mixture of religious and secular notions
we use for counting years. What are the two sets of coincidences? One is
year 1 & walking barefoot. The other is *year 1969 & moonwalk.*

Measuring time is much more than looking at a watch or calendar, but it
is done so routinely that its complex nature goes unnoticed. That my

watch may show 9 A.M. is not yet a time measurement. Walking out to the
porch at 9 A.M. is. In this case, one of the processes is the oscillation of
a quartz crystal counted by a microchip. The other is my motion. One set
of coincidences is *0 hour midnight & sleeping,* the other *9 A.M. & walking.*
Instead of thinking of a day and a watch, we may think of a year and any
appropriate other process. Thus, I could time my activities by the opening
of the pine cones in the fall or time those openings by my activities. The
two processes time each other.

What may serve as a clock? Any process judged sufficiently regular to
permit the numbering of its events: the instants when the baby wants to be
fed, days of attacks by a terrorist group, solar eclipses, atomic oscillations,
rapes in Central Park. Each of these may serve as a clock, but no clock in
itself can measure time.

*To perform a time measurement it is necessary—and sufficient—to have two
processes usable as clocks, as well as the conviction that their indications are
connectable in a meaningful way. Such a conviction, when carefully stated and
expressed in mathematical terms, is known as a scientific law of nature.*

Let me now illustrate how time measurement and scientific law relate by
turning to Geoffrey Chaucer. Chaucer is best known for his *Canterbury
Tales,* in which twenty-four pilgrims tell their spicy stories. The *Tales*
have many references to astronomy, because Chaucer was interested in the
subject; he even wrote an advanced treatise on it, preceded in 1391 by
a popular one: *A Treatise on the Astrolabe: Bred and Mylk for Childeren.*

The astrolabe is a portable instrument that can relate the motion of stars,
the planets, the sun, and the moon to the geographical location of the
observer. It can also perform a number of astronomical measurements,
such as the telling of local time from the positions of heavenly objects,
without the need for numerical calculations. It was invented during the
second century B.C., which makes it the oldest known scientific instrument, although it did not reach its mature form until after the first
millennium A.D. The Arab and Latin astronomers of the Middle Ages
called it the mathematical jewel.

The device is a mechanical data bank and an analog computer combined
(fig. 10). Frozen into the lines of its plate and the points of its rete are the
equations that connect stellar and solar motion to the time of an observer at
a specified geographical location. Chaucer's *Treatise* was an instruction

FIGURE 10 "TO KNOWE THE SPRING OF THE DAWYING": A PLENI-
SPHERIC ASTROLABE. A refined astronomical instrument popular with naviga-
tors all through the Middle Ages to the eighteenth century, when it was replaced
by a combination of sextants and printed tables. Its plate has many curves engraved
in it: the tropic of Cancer, the equator, the tropic of Capricorn, lines of equal azi-
muths, circles of equal star altitudes (called almucantars, from the Arabic), and the
hour angle lines. The plate also shows the zenith, the north celestial pole, the horizon line, and the twilight line (the official beginning and ending of daylight).

The gridwork above the plate, and rotatable with respect to it, is the rete (network, as in "retina"), also called the spider. The pointers show the relative positions of the major stars. The circle with the names of the constellations is the ecliptic circle; it shows the annual motion of the sun.

The geometry frozen into the lines and pointers of the astrolabe corresponds to
those equations of astronomy that connect heavenly with earthly processes. The
"mathematical jewel" is, therefore, a clock with a permanent data bank and
software.

This astrolabe of gilt brass was made in 1532 by Georg Hartmann, an instrument
maker of Nürnberg and friend of the painter Albrecht Dürer. Photo by J. Devinoy.
Courtesy Musée de Cluny, Paris, and The Burndy Library, Norwalk, Connecticut.

book accompanying a present, the kind Dad always wanted to have for himself. So he got it for his son and wrote an instruction book to accompany it.

> Littel Lewys my sone, I haue perceiued well by certeayne euidences thine abilitie to lerne sciencez touchinge noumbers & proporciuns; & as wel considere I thy bisi preyere in special to lerne the tretis of the astrelabie. . . .
> Thyne astrelabie hath a ring to putten on the thowmbe of thy ryht hand. . . .

The *Treatise* hath forty-six sections telling littel Lewys all he needed to know to use his astrelabie. Here are some of the measurements he could make:

> 6. To knowe the spring of the dawying [the time of the dawn] & the ende of the euenyng, the which ben called the crepuscules [twilights]. . . .
> 7. To knowe the arch of the day [the length of daylight hours] that some folk kallen the day artificial from sonne arisyng til hit go to rest. . . .
> 8. To turn howres in-equales in howres equales. . . .
> 10. To knowe the quantity of howres in-equales by day. . . .
> 15. To knowe which day is lik to whych day as of lengthe &c.
> 46. For to knowe at what houre of the day, or of night, shal be Flode or ebbe. . . .[3]

To put order into his activities and to recognize order in nature, littel Lewys used the principles fixed in the geometrical relationships of his astrolabe. The scientific measurement of time makes the implicit need for those relationships explicit and expresses them in permanent, mathematical forms. This understanding gets us ready to learn about . . .

---

3   *A Treatise on the Astrolabe Addressed to his Son Lewys by Geoffrey Chaucer AD 1391*, ed. W. W. Skeat (London: Oxford University Press, [1872]).

Trillions of Cambridge Surprise Majors: The Scientific Measurement of Time

To be able to read time from a clock's dial, it is necessary to have a reference line and a time reckoner; time is then measured by the changing angle between the two directions. On a clock the reference line connects the center of the dial to the numeral 12. The time reckoner is the short hand. (As before, let us forget about the long hand.) Imagine now that the dial is divided into 24 and not 12 equal parts and that the small hand is slowed so as to complete one rotation in 24 hours. Time may now be read from the clock throughout the day. Using this clock as our model, it is easy to construct a heavenly clock dial.

Draw a circle in the sky so that it passes through the two celestial poles and your zenith (the point in the sky vertically overhead). That half of the circle which includes your zenith is your celestial reference, called the meridian. Next, draw a similar half circle but this time let it go through your lucky star and not your zenith. This is the hour circle of the star; it is your celestial time reckoner. Figure 11 shows a top view of this general arrangement. The angle between the plane of the meridian and that of the hour circle is the hour angle of the star. Since it takes about 24 hours for the hour circle to make a complete revolution with respect to the meridian, each 15° rotation signifies an hour's time.

In 1986 there are over six billion people on earth, each of them entitled to have his or her own meridian. Fortunately, to eliminate confusion, it was agreed in 1884 to establish a common or prime meridian for the whole earth: the one that passes through the site of the Royal Greenwich Observatory in England. Also, instead of just picking any star, it was agreed to use the sun as the common time reckoner. If you live anywhere along the prime meridian, the hour angle of the sun will give Greenwich Mean Time four times a year.

Checking this celestial clock against a sandglass or clepsydra will reveal, though, that the sun moves unevenly. Perhaps the problem is with the sandglass. But if many sandglasses and clepsydrae agree among each other, then let us blame the sun: it is a nonuniformly moving clock hand. To get around this problem, the hour circle may be attached to a fictitious object called the *mean sun*. The other, real one, is called the *apparent sun*.

The mean sun is an imaginary point that rises in the east, sets in the west,

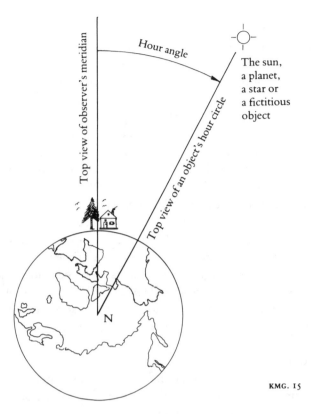

Top view of observer's meridian

Hour angle

Top view of an object's hour circle

The sun,
a planet,
a star or
a fictitious
object

N

KMG. 15

FIGURE 11 THE HANDS OF THE CELESTIAL CLOCK. Imagine being high above the north pole, then traveling far enough until the earth is but a dot near the letter *N*, where the two lines meet. The earth is shown larger than it would appear, to help explain the celestial dial. The plane of our family's meridian (perpendicular to the plane of the paper) is between our home and our yggdrasil tree. It is fixed to us, the observers. The top view of that plane is the reference line of the heavenly clock. The plane of the hour circle goes through my favorite star. The top view of that plane is the small hand of the heavenly clock. The angle between the plane of the meridian and that of the hour circle is the hour angle of the star. The time reckoner makes one complete revolution with respect to the reference line in about 24 hours. The international reference direction, however, is not our family's meridian but the one that goes through the base of a particular telescope at the Royal Greenwich Observatory in England. Neither does the international time reckoner go through my favorite star; it is attached, instead to other real or fictitious objects.

moves from west to east among the constellations, as does the apparent sun, and completes a year in exactly the same time as its nonghostly twin. When checked against a sandglass, quartz watch, or atomic clock the mean sun moves uniformly. The apparent sun sometimes runs ahead, sometimes falls behind, and four times a year coincides with the mean sun. The continuously changing difference between the apparent and the mean sun, a correction factor to be added to or subtracted from the apparent solar time to obtain local mean time, is called the equation of time.

There is yet a third point in the sky whose hour angle has been serving as a time reckoner. It is the point where the apparent and the mean sun cross the equator together in the spring. This point is known as the first point of Aries, the vernal equinox, or, simply, the equinox.

We have not gone very far but already have encountered three different celestial clocks. *Apparent solar time* is a measure of time based on the motion of the actual sun; *mean solar time* is its averaged value. The hour angle of the equinox measures *sidereal time,* based on the revolution of the fixed stars around the earth. The name is from the Latin *sidereus* ("constellation"). In practice, its time reckoner is the equinox; equinoxial time would have been a better name.

Which of these clocks shows the really true time?

Our biological clocks care little about mean solar or sidereal time; they follow the rising, the motion, and the setting of that shiny object in the sky. They go by apparent solar time.

Ordinary clocks indicate mean solar time. If they showed apparent solar time, the lengths of the hours through the year would vary. If they showed sidereal time, our midnights would gradually shift forward, because a sidereal day is about four minutes shorter than a solar day.

For astronomy, space science, geodesy, and related applications, the preferred time scale is the sidereal one. The stars are so far from us that they may be considered fixed with respect to each other. They do of course move, and had littel Lewys lived long enough, the rete of his astrolabe would have had to be redesigned. Still, the fixed stars form a stable enough framework against which planetary and solar motions may be measured.

In the course of scientific development, time measurements have and will become increasingly refined, as will be illustrated below. However, it should be clear from what has been said that there does not exist one "really true" time scale. There are only different natural processes that may be compared by time measurements.

The circular motion of the time reckoner (the hour circle) of the celestial clocks is due to the rotation of the earth. The nonuniformity of the apparent sun's travel is due to the elliptical orbit of the earth, to its varying speed, to the angle of the earth's axis to the ecliptic (the reason why all globes are mounted at an angle), and to the unevenness of the earth's spin. As astronomers, aided by other scientists, learn more about the earth's spin, they can apply increasingly finer corrections to the heavenly clocks. For instance, the celestial north pole (the spot where the earth's axis pierces the sky) describes a small circle about the ecliptic north pole (where the axis of the earth's orbit pierces the sky). The period of this precessional motion is 25,725 or 25,784 years, depending whether one takes into account the rotation of earth's orbit. And there are many other quirks to the heavenly dial.

The earth's axis, for instance, nutates: it waves back and forth about the small circle of precession, at an oscillatory rate of 18.6 years. The ecliptic wabbles, the elliptic orbit of the earth slowly turns. The north pole wanders because of changes in the earth's crust, leading to slow changes in the geographical position of all points on earth. (You might not have noticed it but, while you slept, someone moved your bed, home, and country.) There is a regular slowing down of the earth's rotation rate; a year in the Devonian period (350 to 410 million years ago) had 400 Devonian days. There are also unpredictable changes in the earth's rate of spin due to the motion of the magma. All of these changes show up in the rotation of the celestial time reckoners: the hour circles of the apparent sun, the mean sun, and the equinox.

To provide a clock free of these variations, astronomers have constructed yet another time scale, a democratic public opinion of the stars, the planets, their moons, our moon, and the sun. This is the time scale used for scientific work where very great precision and a common standard are essential. It is called *ephemeris time*. In Greek, ephemeris means diary; in the present context, it means tabulation. The following definition is taken from the *Explanatory Supplement to the Astronomical Ephemeris and the American Ephemeris and Nautical Almanac,* prepared jointly by the Nautical Almanac Offices of the United States of America and the United Kingdom.

Ephemeris time is a uniform measure of time, depending for its determination on the laws of dynamics. It is the independent variable in the gravitational theories of the Sun, Moon and the planets, and is the argument for the fundamental ephemerides of the Ephemeris.

Here is a small dictionary to use in translating the definition.

"Ephemerides" are the tabulated, assigned places of the heavenly bodies, published ahead of time in the ephemeris, which is the name of the book. The ephemeris gives the positions of the sun, the moon, the planets, a number of stars, and that imaginary point, the equinox, in terms of ephemeris time. Those positions have been calculated using "the laws of [celestial] dynamics," which are those laws of physics that govern the motion of celestial objects.

Ephemeris time is the "independent variable." If you buy a turkey by weight, weight is the independent variable, price is the dependent one. The relationship may be turned around. If you decide to have a $20 turkey, then price is the independent variable, weight the dependent one. So it is with ephemeris time: it is the variable in terms of which locations are calculated. The "gravitational theories" are those of Kepler, Newton, and Einstein. "Argument" means the same as independent variable.

Ephemeris time is not shown by any clock. It could not be shown. It is a calculated time that is being continuously computed. Let me give a sketch of how it is done.

There are some 150 time services around the globe connected by radio links and continuously transmitting time signals: WWV and WWVB are in Ft. Collins, Colorado; WWVH is in Hawaii; CHU in Ottawa; GBR in Rugby, England. These signals are controlled by atomic clocks (to be discussed later in this chapter) that were found reliable in earlier tests. At many cooperating laboratories around the world astronomical time determinations are made every night. An astronomical time observation consists of the determination of simultaneities between two instants: the crossing of the center of a telescope by a star, the moon, or planet is one. The indication of the local clock is the other. (Remember *lover & 14,* and *bark & 8?*) Each night several crossings are photographed together with the digital indications of the local atomic clock.

Later on, after many cups of coffee and doughnuts, using the equations of celestial mechanics and the published ephemerides, it is determined whether the local clock was slow or fast and by how much. This step assumes that the accumulated knowledge represented by the equations of astronomy and by the figures of the ephemeris is more reliable in determining time than any actual clock would be. The corrections obtained at each laboratory are averaged for every ten days. They are then reported to and published by the Bureau International de l'Heure in Paris, about six weeks after the measurements were made. From Circular D of that office, dated November 3, 1986, one may learn, for instance, that for September 6, 1986, the master clock of the laboratory in Torino, Italy, was ahead of computed time by 0.000,000,29 ephemeris seconds, the one in Ottawa was behind by 0.000,008,27 ephemeris seconds. Earlier calculations that used provisional time indications may then be corrected ex post facto.

The same Circular D also carries corrections to international atomic time (see below), coordinates of the earth's pole in terms of ephemeris time, and the lengths of past days. For instance, September 10, 1986, was 86,400.00125 seconds long, despite whatever else the reader's watch might have indicated.

Ephemeris time could be calculated from the consecutive positions of the sun or from the orbital motion of any planet. But in practice it is determined from the motion of our moon, of the planets Mercury and Venus, and of the moons of Jupiter.

Since 1984 ephemeris time has been replaced by *dynamic time,* which is its finely tuned version; it takes into account certain relativistic corrections. For most uses, however, ordinary ephemeris time remains satisfactory, and therefore the general term, ephemeris time, remains in use.

Obviously, the determination in ephemeris time of exactly when something happens cannot be done when the event occurs: this, the most advanced scientific time scale, could not be shown by any clock. Ephemeris time is calculated time, knowable only for past events. I have stressed earlier that each time measurement involves our ideas of how different processes ought to correlate. Ephemeris time makes this explicit: it depends on what we think nature ought to be doing. It is always from memory. "It is in you, my mind, that I measure time," as St. Augustine noted sixteen centuries ago.

Considering that ephemeris time is available only ex post facto, is it ever used for anything "practical"? Most definitely. For instance, it is used worldwide for navigation. Figure 12 shows two pages of the *Nautical Almanac for the Year 1986,* published in 1984, calculated in 1983 and earlier.

G.M.T. (UT)	ARIES G.H.A.	VENUS −4.1 G.H.A.	Dec.	MARS −1.2 G.H.A.	Dec.	JUPITER −2.4 G.H.A.	Dec.	SATURN +0.8 G.H.A.	Dec.
10 00	348 44.0	139 59.7	S15 36.1	59 59.8	S26 48.1	359 11.7	S 6 08.6	106 25.3	S19 16.8
01	3 46.5	155 00.2	37.1	75 01.3	47.9	14 14.4	08.7	121 27.6	16.9
02	18 49.0	170 00.8	38.1	90 02.8	47.7	29 17.2	08.9	136 29.9	16.9
03	33 51.4	185 01.3	·· 39.1	105 04.3	·· 47.4	44 20.0	·· 09.0	151 32.2	·· 16.9
04	48 53.9	200 01.8	40.1	120 05.8	47.2	59 22.7	09.1	166 34.6	17.0
05	63 56.4	215 02.3	41.1	135 07.4	47.0	74 25.5	09.3	181 36.9	17.0
06	78 58.8	230 02.9	S15 42.1	150 08.9	S26 46.8	89 28.3	S 6 09.4	196 39.2	S19 17.0
W 07	94 01.3	245 03.4	43.1	165 10.4	46.5	104 31.1	09.5	211 41.6	17.1
E 08	109 03.7	260 03.9	44.1	180 11.9	46.3	119 33.8	09.7	226 43.9	17.1
D 09	124 06.2	275 04.5	·· 45.1	195 13.4	·· 46.1	134 36.6	·· 09.8	241 46.2	·· 17.1
N 10	139 08.7	290 05.0	46.1	210 14.9	45.9	149 39.4	09.9	256 48.5	17.2
E 11	154 11.1	305 05.5	47.1	225 16.4	45.6	164 42.1	10.0	271 50.9	17.2
S 12	169 13.6	320 06.1	S15 48.1	240 17.9	S26 45.4	179 44.9	S 6 10.2	286 53.2	S19 17.2
D 13	184 16.1	335 06.6	49.1	255 19.4	45.2	194 47.7	10.3	301 55.5	17.3
A 14	199 18.5	350 07.1	50.1	270 20.9	45.0	209 50.5	10.4	316 57.8	17.3
Y 15	214 21.0	5 07.7	·· 51.0	285 22.4	·· 44.7	224 53.2	·· 10.6	332 00.2	·· 17.3
16	229 23.5	20 08.2	52.0	300 23.9	44.5	239 56.0	10.7	347 02.5	17.4
17	244 25.9	35 08.7	53.0	315 25.5	44.3	254 58.8	10.8	2 04.8	17.4
18	259 28.4	50 09.3	S15 54.0	330 27.0	S26 44.0	270 01.5	S 6 11.0	17 07.1	S19 17.4
19	274 30.9	65 09.8	55.0	345 28.5	43.8	285 04.3	11.1	32 09.5	17.5
20	289 33.3	80 10.3	56.0	0 30.0	43.6	300 07.1	11.2	47 11.8	17.5
21	304 35.8	95 10.9	·· 57.0	15 31.5	·· 43.4	315 09.8	·· 11.4	62 14.1	·· 17.5
22	319 38.2	110 11.4	58.0	30 32.9	43.1	330 12.6	11.5	77 16.5	17.6
23	334 40.7	125 12.0	15 59.0	45 34.4	42.9	345 15.4	11.6	92 18.8	17.6
11 00	349 43.2	140 12.5	S16 00.0	60 35.9	S26 42.7	0 18.2	S 6 11.8	107 21.1	S19 17.6
01	4 45.6	155 13.0	00.9	75 37.4	42.4	15 20.9	11.9	122 23.4	17.7
02	19 48.1	170 13.6	01.9	90 38.9	42.2	30 23.7	12.0	137 25.7	17.7
03	34 50.6	185 14.1	·· 02.9	105 40.4	·· 42.0	45 26.5	·· 12.2	152 28.1	·· 17.8
04	49 53.0	200 14.7	03.9	120 41.9	41.8	60 29.2	12.3	167 30.4	17.8
05	64 55.5	215 15.2	04.9	135 43.4	41.5	75 32.0	12.4	182 32.7	17.8
06	79 58.0	230 15.8	S16 05.9	150 44.9	S26 41.3	90 34.8	S 6 12.6	197 35.0	S19 17.9
07	95 00.4	245 16.3	06.8	165 46.4	41.1	105 37.6	12.7	212 37.4	17.9
T 08	110 02.9	260 16.9	07.8	180 47.9	40.8	120 40.3	12.8	227 39.7	17.9
H 09	125 05.4	275 17.4	·· 08.8	195 49.4	·· 40.6	135 43.1	·· 13.0	242 42.0	·· 18.0
U 10	140 07.8	290 18.0	09.8	210 50.8	40.4	150 45.9	13.1	257 44.3	18.0
R 11	155 10.3	305 18.5	10.8	225 52.3	40.1	165 48.6	13.2	272 46.7	18.0
S 12	170 12.7	320 19.1	S16 11.8	240 53.8	S26 39.9	180 51.4	S 6 13.3	287 49.0	S19 18.1
D 13	185 15.2	335 19.6	12.7	255 55.3	39.7	195 54.2	13.5	302 51.3	18.1
A 14	200 17.7	350 20.2	13.7	270 56.8	39.4	210 56.9	13.6	317 53.6	18.1
Y 15	215 20.1	5 20.8	·· 14.7	285 58.3	·· 39.2	225 59.7	·· 13.7	332 55.9	·· 18.2
16	230 22.6	20 21.3	15.7	300 59.7	39.0	241 02.5	13.9	347 58.3	18.2
17	245 25.1	35 21.9	16.6	316 01.2	38.7	256 05.3	14.0	3 00.6	18.2
18	260 27.5	50 22.4	S16 17.6	331 02.7	S26 38.5	271 08.0	S 6 14.1	18 02.9	S19 18.3
19	275 30.0	65 23.0	18.6	346 04.2	38.3	286 10.8	14.3	33 05.2	18.3
20	290 32.5	80 23.6	19.6	1 05.6	38.0	301 13.6	14.4	48 07.6	18.3
21	305 34.9	95 24.1	·· 20.5	16 07.1	·· 37.8	316 16.3	·· 14.5	63 09.9	·· 18.4
22	320 37.4	110 24.7	21.5	31 08.6	37.6	331 19.1	14.7	78 12.2	18.4
23	335 39.8	125 25.2	22.5	46 10.1	37.3	346 21.9	14.8	93 14.5	18.4
12 00	350 42.3	140 25.8	S16 23.5	61 11.5	S26 37.1	1 24.7	S 6 14.9	108 16.9	S19 18.5
01	5 44.8	155 26.4	24.4	76 13.0	36.9	16 27.4	15.1	123 19.2	18.5
02	20 47.2	170 26.9	25.4	91 14.5	36.6	31 30.2	15.2	138 21.5	18.5
03	35 49.7	185 27.5	·· 26.4	106 15.9	·· 36.4	46 33.0	·· 15.3	153 23.8	·· 18.6
04	50 52.2	200 28.1	27.3	121 17.4	36.2	61 35.7	15.5	168 26.1	18.6
05	65 54.6	215 28.7	28.3	136 18.9	35.9	76 38.5	15.6	183 28.5	18.6
06	80 57.1	230 29.2	S16 29.3	151 20.3	S26 35.7	91 41.3	S 6 15.7	198 30.8	S19 18.7
07	95 59.6	245 29.8	30.2	166 21.8	35.4	106 44.0	15.8	213 33.1	18.7
08	111 02.0	260 30.4	31.2	181 23.3	35.2	121 46.8	16.0	228 35.4	18.7
F 09	126 04.5	275 30.9	·· 32.2	196 24.7	·· 35.0	136 49.6	·· 16.1	243 37.7	·· 18.8
R 10	141 07.0	290 31.5	33.1	211 26.2	34.7	151 52.4	16.2	258 40.1	18.8
I 11	156 09.4	305 32.1	34.1	226 27.7	34.5	166 55.1	16.4	273 42.4	18.8
D 12	171 11.9	320 32.7	S16 35.1	241 29.1	S26 34.3	181 57.9	S 6 16.5	288 44.7	S19 18.9
A 13	186 14.3	335 33.3	36.0	256 30.6	34.0	197 00.7	16.6	303 47.0	18.9
Y 14	201 16.8	350 33.8	37.0	271 32.0	33.8	212 03.4	16.8	318 49.3	19.0
15	216 19.3	5 34.4	·· 38.0	286 33.5	·· 33.5	227 06.2	·· 16.9	333 51.6	·· 19.0
16	231 21.7	20 35.0	38.9	301 34.9	33.3	242 09.0	17.0	348 54.0	19.0
17	246 24.2	35 35.6	39.9	316 36.4	33.1	257 11.7	17.2	3 56.3	19.1
18	261 26.6	50 36.2	S16 40.8	331 37.9	S26 32.8	272 14.5	S 6 17.3	18 58.6	S19 19.1
19	276 29.1	65 36.7	41.8	346 39.3	32.6	287 17.3	17.4	34 00.9	19.1
20	291 31.6	80 37.3	42.8	1 40.8	32.3	302 20.1	17.6	49 03.2	19.2
21	306 34.1	95 37.9	·· 43.7	16 42.2	·· 32.1	317 22.8	·· 17.7	64 05.6	·· 19.2
22	321 36.5	110 38.5	44.7	31 43.7	31.9	332 25.6	17.8	79 07.9	19.2
23	336 39.0	125 39.1	45.6	46 45.1	31.6	347 28.4	17.9	94 10.2	19.3
Mer. Pass.	0 41.0	v 0.6	d 1.0	v 1.5	d 0.2	v 2.8	d 0.1	v 2.3	d 0.0

STARS

Name	S.H.A.	Dec.
Acamar	315 33.4	S40 21.1
Achernar	335 41.1	S57 18.0
Acrux	173 33.2	S63 01.5
Adhara	255 28.7	S28 56.8
Aldebaran	291 12.7	N16 29.2
Alioth	166 38.7	N56 02.1
Alkaid	153 15.1	N49 23.0
Al Na'ir	28 08.6	S47 01.6
Alnilam	276 07.1	S 1 12.3
Alphard	218 16.4	S 8 35.8
Alphecca	126 28.3	N26 45.7
Alpheratz	358 04.4	N29 01.1
Altair	62 27.9	N 8 50.0
Ankaa	353 35.1	S42 22.6
Antares	112 51.4	S26 24.3
Arcturus	146 14.5	N19 15.2
Atria	108 11.7	S69 00.6
Avior	234 27.0	S59 27.6
Bellatrix	278 53.9	N 6 20.5
Betelgeuse	271 23.4	N 7 24.6
Canopus	264 05.4	S52 40.8
Capella	281 04.6	N45 59.1
Deneb	49 45.2	N45 14.0
Denebola	182 54.7	N14 39.0
Diphda	349 15.9	S18 03.4
Dubhe	194 16.8	N61 49.5
Elnath	278 38.4	N28 35.9
Eltanin	90 55.5	N51 29.6
Enif	34 06.8	N 9 48.9
Fomalhaut	15 45.9	S29 41.6
Gacrux	172 24.6	S57 02.3
Gienah	176 13.7	S17 28.0
Hadar	149 17.7	S60 18.7
Hamal	328 23.6	N23 24.1
Kaus Aust.	84 10.7	S34 23.7
Kochab	137 19.3	N74 12.8
Markab	13 58.4	N15 08.1
Menkar	314 36.2	N 4 02.5
Menkent	148 32.1	S36 18.4
Miaplacidus	221 45.2	S69 39.4
Mirfak	309 09.5	N49 48.8
Nunki	76 23.4	S26 19.0
Peacock	53 50.6	S56 47.0
Pollux	243 52.8	N28 03.7
Procyon	245 21.2	N 5 15.8
Rasalhague	96 25.3	N12 34.2
Regulus	208 05.4	N12 02.2
Rigel	281 31.6	S 8 12.7
Rigil Kent.	140 20.3	S60 47.0
Sabik	102 35.9	S15 42.6
Schedar	350 03.6	N56 27.8
Shaula	96 49.6	S37 05.9
Sirius	258 51.8	S16 41.5
Spica	158 53.0	S11 05.4
Suhail	223 08.0	S43 22.4
Vega	80 52.6	N38 46.4
Zuben'ubi	137 28.2	S15 59.2

	S.H.A.	Mer. Pass.
Venus	150 29.3	14 39
Mars	70 52.8	19 56
Jupiter	10 35.0	23 54
Saturn	117 37.9	16 48

FIGURE 12   A PRINTED ASTROLABE: PAGES FROM THE NAUTICAL ALMANAC FOR 1986. The object of this publication is to provide data required for navigation at sea.

In both pages, the left column is Greenwich mean time (GMT), also called universal time (UT), although the two are not exactly the same. For our purposes that column may be thought of as ephemeris time (ET). It shows days and hours; values to seconds may be interpolated. In the left-hand page, the second column gives the Greenwich hour angle (GHA) of the first point of Aries (the Equinox). Then follow the GHA and declination of Venus, Mars, Jupiter, and Saturn, with ET as the argument. The same page gives the sidereal hour angle (SHA) and the declination of 56 stars.

The right-hand page tabulates the positions of the sun and the moon. The tables

G.M.T. (UT)	SUN		MOON					Lat.	Twilight		Sunrise	Moonrise			
	G.H.A.	Dec.	G.H.A.	v	Dec.	d	H.P.		Naut.	Civil		10	11	12	13

(Full numerical data of the Nautical Almanac table. The table continues with hourly entries for 10, 11 and 12 September, with columns for SUN G.H.A./Dec., MOON G.H.A./v/Dec./d/H.P., and right-hand panels for Twilight, Sunrise, Moonrise, Sunset, Moonset, and the SUN/MOON summary with Eqn. of Time, Mer. Pass., Age and Phase.)

Lat.	Sunset	Twilight		Moonset			
		Civil	Naut.	10	11	12	13

Day	SUN			MOON		Age	Phase
	Eqn. of Time 00ʰ	12ʰ	Mer. Pass.	Mer. Pass. Upper	Lower		
10	02 46	02 57	11 57	17 12	04 43	06	
11	03 07	03 18	11 57	18 13	05 42	07	◗
12	03 28	03 39	11 56	19 16	06 45	08	

| S.D. | 15.9 | d | 1.0 | S.D. | 16.1 | | 16.2 | | 16.2 |

on the right show the times of morning twilight, sunrise, moonrise, sunset, evening twilight, and moonset, with geographical latitude as the argument. There are also other columns for the adventurous to explore.

Taking any two readings along a horizontal line on the left page (say, Aries and Venus) we have a simultaneity of the *lover & 14* kind. Taking a corresponding reading along a different horizontal line we have a *bark & 8* simultaneity. The separation between the two coincidences, read from the tables, is a time measurement.

Illustration from *Nautical Almanac for the Year 1986* (Washington, D.C.: U.S. Naval Observatory, and London: HMSO. U.S. Government Printing Office, 1984).

Whatever corrections to 1986 ephemeris time will have been found necessary shall be incorporated in the figures for a later year possibly 1990.

In 1884, the selection of Greenwich, England, for the prime meridian was
adopted over French objections; the French would have preferred the
prime meridian to go through Paris. For over a century, millions of people
have visited the Royal Greenwich Observatory and pointed to a white
line, "See, that's it!" We have had Greenwich Mean Time (GMT). For
a number of years, the Greenwich clocks whose averaged readings were
transmitted to Paris have not actually been located in Greenwich but in
Herstmonceux Castle, Sussex, though of course they still show GMT.
That bank of timekeepers in Sussex comprises six atomic clocks. Beginning with 1986, however, Britain is "off the clock." As the vacuum tubes
burn out, they will not be replaced. (Money, you know.) The Observatory will keep operating three cesium clocks and relay their averaged
readings to Paris. GMT will become a local measure and remembered
glory. The world will be approaching what later I shall describe as the
anthill threshold, on Coordinated Universal Time (UTC), maintained at
the Bureau International de l'Heure in Paris.

The history of timekeepers sketched in the first section of this chapter
illustrated the following belief: Each improved clock shows more closely
than do the earlier, less accurate, ones the rate at which time *really* passes.
Increased accuracy meant being closer to the true, ultimate rate of time's
flow. But what we learned about scientific time measurement in the
present section suggests that there is no final standard to which one can
point and state, "This is the rate at which the time of the world flows."
There are only comparisons among clocks, and there are judgments, loose
ones or rigorous scientific ones, about the uniformity of the processes used
as clocks.

Now that we have encountered more time scales than necessary to boil
an egg—how long is a year? That depends on what is to be meant by
a year.

The *tropical year* is the time taken by the earth to make one complete
revolution around the sun, with respect to the sun. In practice it is the
interval between two successive returns of the sun to the vernal equinox;
this is the year of the four seasons. Its length for 1985 was 365.242,191
ephemeris days. The *sidereal year* is the time taken by the earth to make one
complete revolution around the sun with respect to the fixed stars. This is

the year used in calculations where the earth must be regarded as a part of a larger system. Observatories usually have a clock that shows sidereal time. Because of the precession of the equinoxes, the sidereal year of 1985 was 365.256,363 ephemeris days, which made it about 20 minutes longer than the tropical year. The *anomalistic year* is the period between two successive passages of the earth through its perihelion (the point where it is closest to the sun); it is longer than the sidereal year. For 1985 it was 365.259,641 ephemeris days. There are also other ways to measure a year; you name it, the astronomers have it.

How long is a month? That depends on what is to be meant by a month.

The *sidereal month* is one revolution of the moon with respect to the fixed stars; it may be thought of as the period between two conjunctions (meetings) of the moon with the same star, as seen from the center of the earth. This is the month used in astronomy and related sciences; its length is 27.321,661 mean solar days. The *synodic month* is the time between two conjunctions of the moon with the sun; one may think of it as the time from new moon to new moon. This is the month to which our biological and lunar calendars are tuned; its length is 29.530,588 mean solar days.

How long is a day? That depends on what is to be meant by a day.

As do all time measurements, the length of the day also demands another process for comparison. Thus, one mean solar day equals 1.002,737,909,3 mean sidereal days, or 24h 03m 56.555,37s sidereal time. One mean sidereal day equals 0.997,269,566,3 mean solar days, or 23h 50m 4.090,55s solar time.

How long is a second? That depends on what is to be meant by a second.

The simplest, early definition equates it to 1/86,400 of the day. But since days can have different lengths, the International Astronomical Union defined a second of ephemeris time as 1/31,566,925.9747 of the tropical year 1900.

As I already mentioned, the rate of the earth's rotation slowly decreases, hence the length of the day slowly increases. The rate of the earth's revolution around the sun also changes. It is a complicated deal because every time measurement is a comparison of two clocks, each of which continuously changes its rate with respect to yet other clocks. Thus, while in terms of ephemeris days the length of the tropical year steadily decreases, the length of the sidereal year steadily increases. It is for this reason that I had to specify that the lengths of the tropical, sidereal, and anomalistic years given were for 1985

The rate changes are gradual; clocks do not jump on long weekends. If the length of the second remains a fixed part of the year and the year changes, so must the second. To make it permanent, the astronomical second was equated to the second on January 0 of the tropical year 1900, which is the same day as December 31, 1899. And not just any part of that day but at its first instant, which was 12 hours ephemeris time, because in that year the astronomical day still began at noon.

In 1967, the length of the second was redefined in terms of the frequency of a particular type of atomic clock. In 1972 atomic time was adopted as the primary reference for all scientific timing.

Atomic oscillations are recurrent, cyclic phenomena with similarities to planetary motions and pendular swings. But unlike planets or pendulums, whose periods may assume any value, the frequencies of atomic resonances are given by nature and cannot be easily altered. These narrow resonances correspond to well-defined changes in the energy states of the atoms. In 1958, W. Markowitz of the U.S. Naval Observatory determined that one particular oscillatory mode of the cesium atom produced 9,192,631,770 cycles during an ephemeris second. That number was adopted as the definition of the length of an atomic second.

The primary standard atomic clock of the United States, NBS–6, is at the National Bureau of Standards in Boulder, Colorado. Figure 13 is a schematic drawing that shows how it works. Its output is used to control other clocks around the world by radio signals. And for the benefit of all, its cycles are counted. Each time the count reaches 9,192,631,770, a beep is generated, yours for the hearing on station WWV.

Two of the most sophisticated atomic clocks, built to the same specifications and carefully maintained, would get out of step by one second in three million years. But it is not necessary to wait that long to confirm it; it is sufficient to let the two beat against each other for a day and observe by how many cycles they are off by the evening. Whatever way it is measured, their accuracy is so phenomenal that a decision as to which of the two is to be believed cannot be based on astronomical tests. Atomic clocks can only be tested against each other, with a democratic majority having the decisive beep.

Yet why should scientists believe atomic clocks rather than other clocks? That atomic clocks closely agree among themselves is in itself no

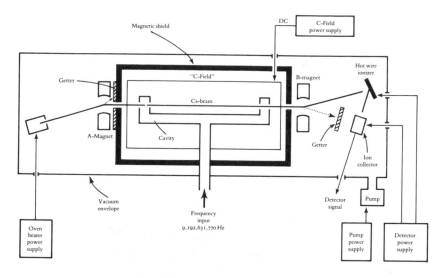

FIGURE 13   UNITED STATES PRIMARY STANDARD ATOMIC CLOCK. This is a schematic of NBS–6, in use at the National Bureau of Standards in Boulder, Colorado. The outermost enclosure is a vacuum chamber, evacuated to below 0.000,000,001 atmospheric pressure. The oven contains cesium metal that is heated to produce cesium gas in the form of an atomic beam that leaves the oven and enters the vacuum. From that beam, the A–Magnet separates out atoms in energy states not used by the clock. These unwanted atoms are absorbed by the getter. The beam then enters the magnetically shielded region and, within it, a microwave cavity.

The magnetic shield isolates the atoms from the earth's magnetic field, reducing it almost to zero. A small, controlled "C-Field" is then applied to produce the necessary separation between atoms of slightly different magnetic properties.

The resonance frequency of the cavity is determined by its size, just as the acoustic frequency of an organ pipe is by its dimensions.

The cavity is supplied with microwave energy that already centers on 9,192,631,770 Hertz (cycles per second) but shifts slightly back and forth. A feedback control system clamped to the narrow absorption band of the atoms is then used to keep the external oscillator narrowly at the atomic resonance frequency. The clock only interrogates the atoms; the signal comes from the controlled external oscillator.

Courtesy, Time and Frequency Division, National Bureau of Standards, Boulder, Colorado.

guarantee that their ticks are uniform. To answer this question, we must consider the nature of the closest competitor among clocks, the rotation of the earth.

What would enable scientists to use the earth's rotation for a clock as accurate as an atomic clock? It would be necessary to have a detailed theory and knowledge of the tides so precise that it could predict what will happen during high tide in our local inlet ten years from now; to have a precise theory and knowledge of the motion of the earth's mantle, of its cores, and of its winds; and even to calculate the effects of manmade changes of the earth's mass distribution. No such theories exist. If they would, the necessary measurements would be impossible to make. For these reasons, short-term intervals are best measured by atomic clocks.

Time kept by atomic clocks and astronomical measurements regularly get out of step; not by very much, but they do. In case of such disagreements, scientists tend to believe the clock whose functions they understand better and declare its time measurements as more reliable. The better understood device happens to be the atomic clock. The preference for atomic clocks has nothing to do with an imagined ideal rate of time's flow.

Let us assume that two identical atomic clocks gave very different figures for the time span between two events. No experimental errors have been made and nothing seems to be wrong with either clock.

The first thing physicists would do is to look for a theoretical explanation for the different readings. If none were found among existing theories, they would seek to formulate a new law of nature. If the new law appeared formally acceptable, and if it could predict future discrepancies correctly, then—agreement or no agreement between the clocks—a valid time measurement has been made.

It is puzzling. What if the difference is substantial? Which clock would be correct? Both may be, provided there is an acceptable theory that connects their readings. Let me repeat verbatim what was said a few pages back:

"To perform a time measurement it is necessary—and sufficient—to have two processes usable as clocks, as well as the conviction that their indications are connectable in a meaningful way. Such a conviction, when carefully stated and expressed in mathematical terms, is known as a scientific law of nature."

My concern with this issue is not an idle spinning of fantasy. I am preparing for an encounter with time in relativity theory.

The noble English art of change ringing consists of ringing a set of bells of different pitches according to a stable rule of permutations. A change is a set of rings, a complete set of changes is a peal. Three bells can ring a peal of 6 changes; five bells, a peal of 120 changes; twelve bells, a peal of 479,001,600 changes. This last one would take forty years of ringing, day and night. I like the Cambridge Surprise Major; it uses eight bells and offers a peal of 40,320 changes.

A peal well rung leaves one with a sense of ceaseless change. The mind attaches itself now to this pattern, now to that pattern in search for a stable melody, but finds none. Yet there is a strict mathematical rule beneath it all.

From the oscillations of atoms to the vast exhalation of the expanding universe, the physical world comprises an infinite number of bells, each ringing its own peal. Each clock of nature works by a strict law of physics, but the cycles are incompatible. Nothing fits into anything else a whole number of times. There is no instant-to-instant coherence among the innumerable physical clocks. The world is made of trillions of Cambridge Surprise Majors rung simultaneously and at different speeds.

Nevertheless, by such means as I have sketched, scientists have been able to construct a time scale that appears reliable to our critical faculties and uniform enough for our needs, for the time being.

The next task, then, is to put some such a time scale or time scales to work in the arrangement of our daily lives. Compared to that challenge, the scientific measurement of time will appear simple.

When Is Easter and Who Are You? Calendars and Chronologies

The first two sections of this chapter spoke about *Homo faber*—man the maker—who learned numbering, invented clocks, and used them to find order in the world and put order in his life. The introduction to the chapter spoke of time reckoning as "metaphysical bookkeeping." This meant having faith in time reckoning as such, without the need for anyone to convince us of its reasonableness. That faith comes partly from our instincts, partly from Plato. It says that the universe has been ordered by time, that the universe is lawful, and that those laws are expressible by number. In sum, you may question whether you really ought to meet him or her at 7:30 P.M., but "7:30 P.M." makes sense. This unquestioned belief in temporal order underlies each act of time reckoning, even if it is noticed only when one gives special thought to time, such as we have been doing in this book.

The present section extends our survey of time reckoning to calendars and chronologies. As it continues, the reader will notice the metaphysical cat getting out of the time-reckoning bag. With calendars and chronologies—and later, even more so, with cosmologies—the measurement of time becomes an exercise in human value judgments, for which the numerical relations serve only as vehicles.

Clocks and calendars belong in the same family of artifacts. Some clocks tick but not all of them; some calendars are printed but not all of them. When it comes to reckoning years, calendars usually take over, but not always. When it comes to the reckoning of centuries, chronologies take over. Always.

Calendars

The man who used to deliver charcoal to my childhood home would regularly appear in early December and pull from his pack a new calendar. It would show a winter scene with finely powdered white glass for snow, a small house, brown horses, and a young woman in a red dress. The powdered glass was pleasantly ticklish to a growing boy's fingers whenever he wanted to reassure himself that the picture of the woman was raised. It was.

The time-reckoning part of that calendar had as long and as distinguished a history as did the mind of the woman in the red dress.

The phases of the moon vary from night to night, its cycle becoming evident in 29 days. The sun's orbit varies from day to day, but it takes a year to notice its cycle. This may be partly the reason why paleolithic calendrical notations were lunar and later lunisolar, but not solar. Early Egyptian calendars were also lunar, though soon they took note of the year. The Egyptians knew that their bright Dog Star, named Sothis by the Greeks, located in the constellation we call Canis Major (the Greater Dog) would always rise with the sun, just at the time of the annual flooding of the Nile. We call the same star Sirius. They judged Sirius to be responsible

for the life-giving waters and described it as the creator of all green, growing things.

To reconcile lunations with the solar year they constructed a calendar of
12 months of 30 days each, plus 5 intercalated days, making for a 365-day
year. But the length of the year was about 365.25 days. As a consequence,
their civil, lunar New Year retrogressed by ¼ day each year with respect
to the religious, Sothic New Year celebrated at the heliacal (near-the-sun)
rising of Soth. The two new years would coincide only once every 1,460
years, a period called the Sothic cycle. It is known that in 139 A.D., the two
new years began on the same day. This led historians, calculating backward, to conclude that the Egyptian calendar was promulgated in
4241 B.C.

I am writing these lines during the 387th year of the Fifth Sothic Cycle,
which is also the 211th Year of our Independence. There are many ways to
count years.

Astronomers of ancient Greece, from the sixth century B.C. on, began
to devise methods for combining solar and lunar cycles into a single
calendrical system. The problem was that the synodic month of 29.53 days
did not fit the 365.25 days of the tropical year a whole number of times.
But the Athenian astronomer Meton did notice that 19 years nearly equaled
235 lunations, making for a period that could serve as a common lunisolar
cycle. Using current data, the lunar and solar clocks at the end of 19 years
differ only by 1.92 hours. The 19-year-period is called the Metonic cycle
and has been built into the Christian calendar, as we shall soon see.

According to legend, the newborn and unwanted twins Romulus and
Remus were floated down the Tiber in a trough and were washed ashore.
They were then suckled by a she-wolf and fed by a woodpecker, after
which they founded Rome and Romulus designed a calendar. It had six
30-day and four 31-day months and an unaccounted-for winter gap of
61 days. As everyone who has lived in Rome knows, lovely as the other
three seasons are, the winters are miserable. Maybe Romulus hoped that if
he neglected them, those days would go away! The year began with
March and ended in December. Until the Gregorian calendar was adopted
in England in 1752 (see below), the English year also began in the spring,
with the 25th of March, the Day of Annunciation. Although our years
start in January, the names of four of our months still carry the enumera-

tion of the ancient Roman calendar. We have *September* (the seventh month), *October* (the eighth), *November* (the ninth), and *December* (the tenth month).

In the sixth century B.C., the Roman Republican Calendar was introduced with months of 28, 29, and 31 days, adding up to 355 days. The announcement of the necessary intercalations was the duty of the Pontifex Minor.[4] But the intercalations did not work, and, since it was a lunar calendar, it became increasingly out of phase with the seasons. By about 50 B.C., the calendrical spring equinox came eight weeks after the astronomical one.

During the first century B.C., Julius Caesar invited an Alexandrian astronomer to advise him on calendar reform. He recommended that they begin with an old form of the Egyptian calendar and improve it by giving up the idea of keeping the months in step with the moon. The Romans could have gone the other way by creating an improved lunar calendar and forgetting about the sun. But their daily lives were sufficiently detached from the cycles of the moon to reject it as their time reckoner. The Julian calendar, as it became known, traveled with the expansion of the Empire to what is now Spain and France, to the Near East, and to North Africa.

The length of the Julian year was 365.25 days. The months followed each other neatly, alternating between 31 and 30 days, adding up to a 366-day year. For three out of every four years February was reduced from 30 to 29 days, which made for three years of 365 days followed by one with 366 days. July, named after Julius Caesar, had 31 days, but August, named after Augustus Caesar, had only 30 days. It is said that Augustus did not like having less days in his month than did Julius and prevailed upon the powers that be to change August so that it would have 31 days. The extra day was taken from February, making it 28 days. The months after August were then rearranged to contain 30, 31, 30, and 31 days. So much for Augustus Caesar's pride, which left us with the messy "thirty days hath September. . . ."

But since the tropical year is 365.242,191 and not 365.25 days, the difference began to accumulate. By the mid-sixteenth century, the vernal equinox of the calendar, used to determine the date of Easter, came ten days after the astronomical event. Had the calendar remained unchanged,

---

4   The Junior Bridge Builder. The Pontifex Maximus (Greatest Bridge Builder) was a name for the emperor and, later, for the supreme pontiff, better known as the pope.

Easter would eventually be celebrated in the summer. To prevent this calamity, the Council of Trent, held in 1545, authorized the pope to take whatever corrective action was necessary. Although work began, it was not until 1572, the first year of the reign of Pope Gregory XIII that the reform was completed and a new calendar promulgated.

The Gregorian year is 365.2425 solar days. To insure this length, the leap years are manipulated. No centennial year is a leap year unless divisible by 400. For instance, the year 1900 was not a leap year, but 2000 will be one. The Gregorian system will remain good without adjustments until 4000 A.D., when the calendar year will be 1.12 days longer than the astronomical year.

The pope's interest in the calendar reform stemmed from a concern for religious issues rather than for scientific accuracy. The calendrical position of Easter determines the dates of the movable feasts and thus sets the rhythm of the ecclesiastical year with respect to the seasons. The steadfastness of Easter in the year symbolizes the certainty that the moral teachings of the Church are stable, that they can guarantee victory over time, and that the position of the Christian in the natural world is a commanding one. It is a feast of rebirth in harmony with nature's own. Its position in the spring is not to be tampered with. But all this is only a coarse tuning. There was also a fine-tuning problem.

All four Gospels concur on the sequence of events that led up to Christ's Passion: the celebration of the Passover; his betrayal and trial; his crucifixion and resurrection. They also agree that he was crucified and died on a Friday, the first day of Passover. "And the third day He rose again according to the Scriptures," which places his resurrection on the first day of the Jewish week, which is Sunday.

What became known as the Easter controversy dates to the early centuries of our era. While the Western Christian churches celebrated the Passion on the day of Resurrection, the Eastern churches did so on the day of Crucifixion. Up to our own epoch, the greatest feast of Greek Orthodox rites is Good Friday and not Easter. But Good Friday, if taken to be the eve of Passover, is thus tied to the Jewish celebration.

Passover commemorates the liberation of the Hebrews from slavery in Egypt. Its day is set in the Jewish calendar as a result of a very intricate system. Passover is celebrated in the month of Nisan. That month begins with the crescent new moon, immediately preceding the ripening of the barley in Judea. The first day of Passover is the 15th of the month, but the feast begins on its eve, the 14th of Nisan. If that day coincides with the solar equinox, then the eve of Passover is to be celebrated the day after.

During the second century there ensued a struggle between the Western and Eastern churches, each insisting on its preference for the day of celebrating Easter. Eventually the Western view prevailed: Easter was to be celebrated on a Sunday, the day of Resurrection, and not on the 14th of Nisan. Those who did not agree with the Sunday celebration became known as the Quatrodecimans (fourteenth-dayers) and were judged heretics. In this scheme, Easter was still tied to the Jewish Passover, albeit to the third day following it.

To make the Christian celebration independent of the Jewish commemoration, the Council of Nicea in 325 agreed to the following calendrical rule. Easter was to be celebrated on the next Sunday after (and not on) the fourteenth day of the Paschal moon, reckoned from the day of the new moon, inclusive. The Paschal moon was defined as the moon whose fourteenth day falls on, or is next following, the vernal equinox, taken as 21 March.

To translate this rule into everyday language: The vernal equinox for the purposes of Easter became a fictitious event independent of its astronomical twin. It became fixed with respect to the Gregorian calendar. The calendrical moon (month) whose fourteenth day fell on or followed closest to the vernal equinox was named the Paschal moon. Again, this moon had no relation to the actual one. The Sunday following the fourteenth day of that month, but never the fourteenth day itself, was Easter.

The new rule assured that Easter would remain in the spring close to the equinox, but would not be tied to the celebration of Passover. It is an example, one of many, of the use of scheduling guided by a desire to establish distinct group identity. Sociologists call this practice temporal segregation. The three great monotheistic religions unify their members and separate them from others by having three different days of rest: Muslims have Fridays, Jews Saturdays, and Christians Sundays.

Temporal segregation through calendars and chronologies (histories) is a powerful means of creating and maintaining communal distinctness. The scientific measurement of time knows nothing of this, neither did the man who used to deliver our charcoal and bring the new calendar. But he knew the difference between "them" who celebrated on Saturdays and "us."

To facilitate the computation of Easter, tables had to be calculated and instructions written. These tables used variables such as the Golden Number (which located the year within the Metonic cycle and thus helped fix the age of the moon), the Dominical Letter (it identified all Sundays of a specific year), and the Epact (a replacement for the Golden Number). Once the tables became available, the calculations of Easter became routine. But it was tedious.

This is how one calculates the day of Easter for 1986. From the table of Dominical (Sunday) letters, that for 1986 is *E.* The Epact, from the Table of Epacts, is *19.* Using the Epact as the argument, the Table of Ecclesiastical New Moons shows March 12 as the beginning of the Paschal lunation. The 14th day of the Paschal moon is therefore March 26. The Table of Perpetual Calendar for the Julian and Gregorian systems, when entered by the Dominical Letter *E,* gives the day of the Paschal full moon as Wednesday. The Sunday following, which is the 30th of March, is Easter.

Every fourth year, in the Table of Dominical Letters the sequence leaps over what would otherwise have been a Sunday Letter for that year. Hence, we have leap years.

For those who have the rest of the tables the *Farmer's Almanac* for 1986, which was available free in our local bank, gave the Dominical Letter and the Epact for that year. It also listed the Golden Number in case the farmer is interested in the position of Easter for years before the Gregorian calendar was introduced in 325 A.D.

Computus, the art of computing Easter, became the first extensive elaboration of medieval science. It called for an increasingly better determination of solar and lunar periods and through that demand it promoted observational and theoretical astronomy and craftsmanship. After the Renaissance, computus made possible the construction of a family of remarkable devices known as monumental astronomical clocks. They may be represented by the astronomical clock of the Strasbourg Cathedral (fig. 14). Finished in 1574—having replaced an earlier version—it was again rebuilt in 1842.

Calendrical science and clock craftsmanship evolved to make artifacts that explained and praised the Christian universe. They were steps in the historical development of the West, forerunners of the later desires to put

FIGURE 14   THE STRASBOURG ASTRONOMICAL CLOCK, an outstanding monumental clock and geared model of the universe. A combination of planetarium, clock, perpetual calendar, and ephemeris.

The clock shows solar, solar mean, and sidereal times, the times of sunrise and sunset, the motion of the moon and displays—of course—the days and months of the year. It takes into account leap years, including the 400-year corrections. It shows the dates of Easter, the movable feasts, the fixed feasts, the names of the saints celebrated each day, and the Dominical Letter of the year. It gives advance notices of lunar and solar eclipses and displays them as they come up, including their extents. It shows the position of the year within the solar cycle of 28 years (after which the days of the months return again to the same days of the week). It indicates the precession of the equinoxes.

A seven-day succession of allegorical figures, pagan deities representing the days of the week, travel on chariots among the clouds, appearing and vanishing in their turn. On a different stage a genie, holding a bell, strikes with his sceptre the first stroke of each quarter hour; the remaining strokes are struck by figures of a child, young man, man, and old man. A figure of death with a bone strikes the number of hours. Just before that photo opportunity, another genie, holding a sandglass, turns it.

After the last stroke at 12 noon, a procession of the twelve apostles appears and passes beneath the figure of Christ; each turns his head toward the Master, bows, and receives his blessing. After they vanish, Christ gives his final benediction. Meanwhile a golden cockerel flaps its wings, moves its head and tail, rifles its neck, and, by all this, reminds the faithful of Christ's prophecy of St. Peter's denial come true: "Amen, amen I say to thee, the cock shall not crow, till you deny me thrice."

The technically most interesting part of the clock is behind its facade, visible only from the back and not normally open to the public. It is the mechanism that governs the astronomical works, an intricate machinery of gears, curved profiles, and differential systems. The clock does more than what I have listed; something ought to be left for the visitor to discover.

Although remarkable for its precision, this monumental clock was not conceived and built to provide time measurement. Rather, it was intended to serve as a model of the universe that works according to a mathematical order and under the tutelage of a moral deity.

Courtesy, Dr. Henry C. King from his private collection, used here with special appreciation by the author.

the skills of scientists and artisans to use for the earthly benefit of people. The much-heralded ascent of man through science and technology was a spin-off of political and religious strivings. In our own age, space science and technology has come about mainly through the energetic American response to the first Russian Sputnik satellite, placed into orbit by the Soviets to advertise their political system.

The age of Gregory XIII was also that of the Reformation and Counter-Reformation, with the pope appropriately threatening with excommunication all those who did not adopt his new calendar. While Catholic towns and countries immediately did, Protestant lands did not. From 1582, the year the calendar was promulgated, until the mid-eighteenth century, Europe was a kaleidoscope of calendars with neighboring towns often using different systems. Great Britain went Gregorian in 1752, Russia in 1918, and Turkey in 1927. In the 1980s there are still at least forty different calendars in use.[5]

The Islamic calendar governs the religious and civil life of half a billion people. Its year contains twelve lunar months of alternately 30 and 29 days, making for a 355-day lunar year. The last month sometimes has an intercalated day that makes the months keep up with the moon, but they regress with respect to the sun by about 11 days a year; the lunar and solar years have a common cycle of 32½ solar years. The calendar originated in 622 A.D., the year of Mohammed's flight from Mecca to Medina, known as the Hegira.

The Chinese calendar is much older. Inscriptions on fortune-telling bones show that as early as the fourteenth century B.C. the Chinese recognized the year as 365¼ days long and lunations as 29½ days, and a century before Meton they knew of the Metonic cycle, by a different name, of course. Their twelve months added up to 354 days, which they completed by an intercalary month from time to time. Calendars were sacred documents because they were necessary for agriculture; they also demonstrated the Emperor's power over the future: he could tell what would happen in the sky. Prepared by court scientists, they were presented in elaborate

---

5   The software of my word processor used for the present manuscript offers a calendar display of the month if one presses the right sequence of keys. The calendar may be moved forward and backward in steps of months or years. When taking backward steps, it does not go to dates before February 1582. It is a learned software: it knows that February 24, 1582, was the date of the Papal Bull, *Kalendarium Gregorianum Perpetuum*.

ceremonies to the Emperor, the Son of Heaven, whose duty was to promulgate it. Astronomical systems, together with the mathematical techniques needed for the making of calendars, were all part and parcel of the Emperor's paraphernalia.

The Chinese calendrical system identifies days by the simultaneous running of two counts: the 10 celestial stems and the 12 terrestrial branches.[6] Like two cogwheels, they mesh and run together, repeating a complete cycle in 60 days. This sexagesimal count has been in continuous use for over three millennia. The combination of the 10 celestial stems with the 12 animals of the zodiac (Rat, Ox, Tiger, Hare . . .) yields a 60-year cycle; within it, each year is identified by two names: a celestial stem and a zodiacal animal. The year itself is divided into 12 months, but in practical use it is the 24 fortnightly units that are important. Each of them bears a meteorological name, for example, Beginning of Spring, The Rains, Awakening of Creatures (from hibernation), and Spring Equinox.

Until recently, Mesoamerican calendars were thought to have formed a world unto themselves. However, research during the past two decades has recognized certain intriguing parallels between them and the Chinese and Hindu calendrical systems, suggesting early cultural contact—"trans-Pacific echoes and resonances," in the words of Joseph Needham.

The fundamental element of the Mesoamerican calendars—the 260-day cycle called a lamat—was probably invented before 400 B.C. It was still the master rhythm of the life of the Aztecs when they were decimated by the Spanish, two thousand years later, and it still regulates the life of large groups of indigenous people in the Central Highlands of Guatemala. Of the Mesoamerican cultures that used the 260-day system, it was the Maya who developed it to an imposing intellectual edifice. They did so during their Classic Era, between 300 and perhaps 1100 A.D.

In the words of the Mexican scholar Miguel Leon-Portilla, the Maya civilization was based on a "chronovision," a total absorption of the individual and collective life in the rhythms of nature, mapped into a mathematical system that had several cyclic counts running simultaneously. The 260-day cycle was that of 13 numbered days meshed with 20

---

6   These are names for any sets of objects countable in terms of tens and twelves. We could do the same, for instance, for the dozen if instead of counting by one, two, three, four, five . . . we would count by *partridge in a pear tree, turtle dove, French hen, calling bird, golden rings.* . . . "How many children do you have?" "We have partridge in a pear tree son and calling bird daughters."

named days, also like the teeth of two gears. In its turn, this unit was meshed with the 365-day calendar year. That year was divided into 18 months of 20 days each, to which 5 unnamed and unlucky intercalary days were added. The 260- and 365-day cycles were themselves meshed to form the Calendar Round of 18,980 days, which is both 52 × 365 and 73 × 260. Counting simultaneously along these four cycles, any one combination would come up only once every 18,980 days.

The Maya also developed a chronology whose reference epoch was some time between 3015 and 3000 B.C. To measure time from that epoch forward and backward, they developed a slightly modified base 20 system. While our decimal system is 10× 10× 10× 10 . . . their system was 20× 18× 20× 20× 20. . . . A unit of 20 days was called a Uinal; 18 Uinals made a Tun, 20 Tuns a Katun, 20 Katuns a Baktun, 20 Baktuns a Pictun, 20 Pictuns a Calabtun, 20 Calabtuns a Kinchiltun, and 20 Kinchiltuns an Alautun. One Alautun is just over 63 million years. These very long periods seem to have been used to weld mythical and historical events into a continuity of cycles.

The various cycles were the units of the gods' and man's journey through eternity. Each cycle was burdened by its own fate, in the Maya version of "sufficient unto the day is the evil thereof." By extending these cycles to inconceivably vast temporal horizons, the calendar made the burden of daily life appear insignificant.

Calendrical records have been found carved on stone slabs and pillars, carved in jade, drawn on pottery, and written in their sacred books, the Codices. These served as aids to the Maya priests in selecting auspicious days for weddings, warfare, and all things in-between. The Codices also contained records of history, instructions in the practice of trades, and medical treatises.

Figure 15 shows the month of September 1986 in a greatly simplified and updated version of the Maya calendar. Figure 16 is a segment of a masterpiece of Maya art, the Dresden Codex.

In a change of venue we spiral back to Europe and also jump ahead in time. In eighteenth-century France there was popular demand for a civil calendar separate from that of the Church. Four years after the storming of the Bastille, the French Republican calendar was introduced and made retroactive to the proclamation of the Republic on September 22, 1792, "The first year of liberty." Year counts began with September 22. A year

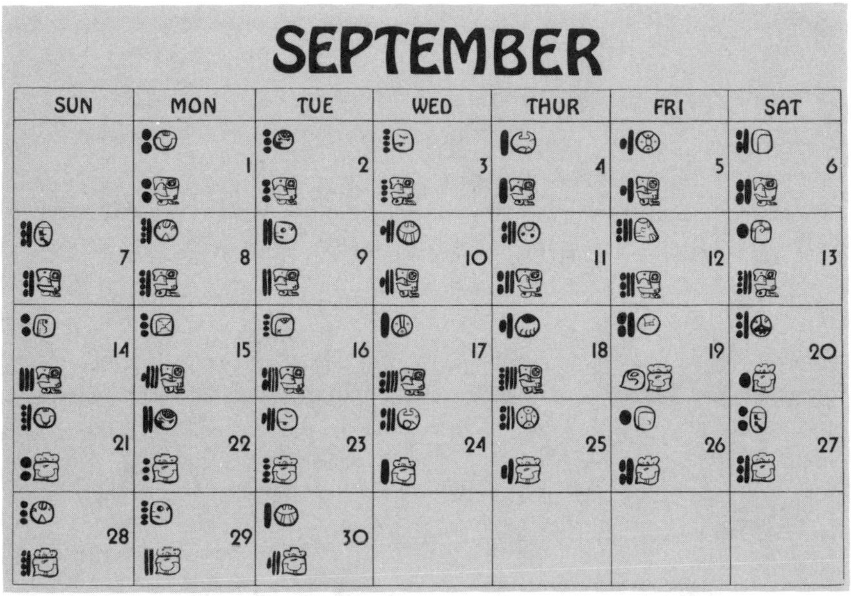

SEPTEMBER

SUN	MON	TUE	WED	THUR	FRI	SAT
	1	2	3	4	5	6
7	8	9	10	11	12	13
14	15	16	17	18	19	20
21	22	23	24	25	26	27
28	29	30				

FIGURE 15   SEPTEMBER 1986 IN THE MAYAN CALENDAR. A simplified and updated schematic representation.

Vertical lines stand for fives, dots for ones. The other symbols are called glyphs. The top row of symbols identifies two endlessly meshed cycles: one of 13 numbered days, one of 20 named days. The 260-day cycle so created is the fundamental unit of the Maya sacred calendar. The bottom row shows the days of a 20-day month that has 19 numbered days; the 20th day, the beginning or "seating" day of the month, is identified by a special glyph (see, for example, September 19). The glyphs that follow the numbered days are the names of the months.

The top symbols for September 10, 1986, identify the day as 11 Ben. It is a good day for children, for sugar cane, other fast-growing reed plants, and also for corn plants. The bottom symbols identify the same day as 11 Kayab. The patron of the month Kayab is Itchell, goddess of medicine, weaving, and childbirth. It is the month for dancing and for pleasure. The next month begins on September 19, with the seating of Cuiku. The patron of that month is the earth dragon.

Correspondence between the Maya and the Gregorian counts, used in the preparation of this calendar, is based on a scheme recently proposed by David Kelley, a scholar of the correlation problem.

From *1986 Maya Calendar,* designed by Philip Chapnick, illustrated by Linda Bea (Glenn Ellen, Calif.: Athanor, 1986). Courtesy, the designer and the illustrator.

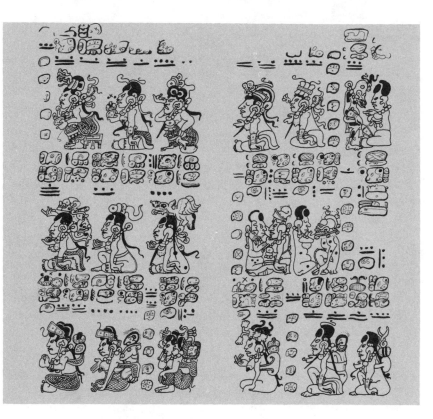

FIGURE 16   A CALENDAR OF NUMBERS, GODS, AND BURDENS OF TIME.
Pages from the Dresden Codex. This codex is believed to have been made during
the first half of the thirteenth century, prepared by and for the use of Mayan
calendar priests. It is assumed to have been part of a gift of treasures sent by Hernan
Cortes early in the sixteenth century to Charles, king of Spain and Holy Roman
Emperor. The first known record of the codex is of its purchase by the Royal
Library in Dresden, hence its name. It suffered water damage during World War II.

The codex treats a variety of topics such as astronomical records, instructions for
farmers, details of new year's ceremonies, and eclipse tables. This illustration is
from the material related to the Moon Goddess.

Note the four vertical "pages" with three rows each; the rows are to be read from
left to right across the "pages." Consider the bottom row. Counting from the left,
between the fourth and fifth figures there is a vertical column of five glyphs,
representing five named days from the 260–day cycle. They are not consecutive
days, but rather the first days of each of five 52–day groups that together account
for the 260–day cycle. Each almanac begins with a vertical column of five glyphs
and continues until the next similar column.

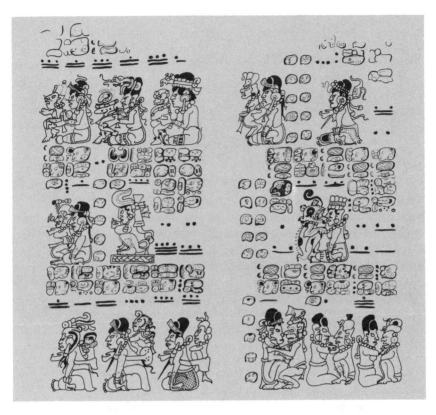

The black-haired figure is the Moon Goddess. In some of the almanacs she carries different stylized creatures, each of which held significance for the ancient Maya. The association of these figures with the Moon Goddess is explained in the hieroglyphic text over her head. In some positions the goddess consorts with figures of evil or good. Sometimes she and others have their bodies covered with black pockmarks, suggesting illness, of which the Moon Goddess was probably both dispenser and healer. Some of her attire may be read in a rebus fashion by those who speak the tongue.

The wealth of the Mayan chronovision and its relation to astronomy, prophecy, history, and cosmology may be best appreciated through the oral tradition, much of which, unfortunately, has been lost. In one known tradition, the Moon Goddess lived on earth before she assumed her elevated position. Some writings refer to her as Our Lady. With the advance of Christianity, she became identified with the Virgin Mary.

Drawing by Linda Bea from a facsimile of the original. Copyright by the artist, reproduced here through her courtesy.

of 365 days was divided into twelve 30-day months, each named according to the season, resembling the Chinese division of the year.[7]

The five intercalary days left over were feasts of Virtue, Genius, Labor, Opinion, and Reward. Ten-day weeks (*les décades*) were introduced; each day was divided into 10 hours, each hour into 100 minutes, each minute into 100 seconds. Sundays were replaced by *décadi* on which people were expected to attend decadal temples. A system of holidays was worked out and people were forbidden to wear their Sunday bests on Sundays. But the precipitous effort got nowhere. The calendar had no more popular acceptance than did Ford's Edsel in its time; on 11 Nivôse XIV, France went back to January 1, 1806.

In 1929 the Soviet Union introduced a plan for a continuous-shift work week, with workers on the job for four days and resting the fifth. They had hoped to increase productivity and also abolish the tradition of the seven-day week, with its Sunday. But technical difficulties, such as having no more than 80 percent of the work force at hand each day, were impossible to solve, and after two years they returned to the long established rhythm of the seven-day week.

Currently, the Vatican opposes any post-Gregorian reform that would require intercalated days and thus disturb the cyclic rhythm of Sundays; the Ecclesiastical calendar is not to be tampered with.

Once a temporal regularity has become well established it acquires a staying power of its own, an example of cultural inertia or momentum, a conservative force comparable to the conservative nature of biological cycles in living organisms. Still, given the right conditions, even very well-established social cycles may be changed. Thus, a calendar reform of the kind the French and the Soviets have tried is now in the making in the United States and the West. As we shall see, it does not stem from government order, but rather is part of the silent time revolution, arising from the grass roots of social change.

To close this subsection on calendars, I want to mention that strict scheduling of daily and yearly activities, which is also a form of calendar,

---

7   In poetic French, the autumn months had a grave sound, winter ones a heavy sound, spring ones a lively sound, and summer ones a sonorous sound. Here they are, for the Francophile: *Vendémiaire*, vintage; *Brumaire*, mist; *Frimaire*, frost; *Nivôse*, snow; *Pluviôse*, rain; *Ventôse*, wind; *Germinal*, seed time; *Floréal*, blossom; *Prairial*, meadow; *Messidor*, harvest; *Thermidor*, heat; and *Fructidor*, fruits. The English words are not translations of the French names but correspondences for what those names stand for.

was born in the Benedictine horary that ordered the lives of the monks. *The Rule of Saint Benedict,* written in the early 500s, told the brothers what to do and when.

> From Easter to Pentecost the brothers shall have dinner at the sixth hour and supper at night. From Pentecost through the summer the monks ought to fast until the ninth hour on Wednesdays and Fridays. . . .
> From September 14th until Lent dinner will be at the ninth hour. . . .
> Immediately upon hearing the signal for the Divine Office, all work must cease. Each monk must hurry to it. . . .
> . . . all things ought to be done at the designated hours. . . .
> Idleness is the enemy of the soul. . . . From Easter to October the brothers shall work at manual labor from Prime until the fourth hour. From then until the sixth hour they should read. After dinner they should rest (in bed) in silence. . . .

We will discuss *The Rule of Saint Benedict* later, in the context of social time, scheduling, and the social present. For now, we will leave the man who used to bring our new calendars with the white snow and the young woman in red, and turn to the learned scholars who count the epochs of the past.

Chronologies

As the reader may recall, the Julian calendar was invented because around 50 B.C. the Roman Republican calendar was out of step with the corresponding astronomical events by about eight weeks. How do we know that it was around 50 B.C.? How may an event be anchored in historical time? Surely, Marcus Antonius could not have said,

> Friends, Romans, countrymen, lend me your ears;
> I come to bury Caesar, 100 to 44 B.C., not to praise him.

Neither is there any mention in the Gospels of the three Wise Men turning to each other and saying, "This is year One, Anno Domini!"

Identifying a suitable reference event to which calendrical counts may be anchored is the task of the science of chronology. It consists of assigning calendrical positions to events with respect to a selected event called the reference epoch or, simply, epoch. The years that follow that epoch form an era. The Declaration of Independence was passed as a proclamation by the Continental Congress in 1776 of the Christian Era. The epoch of the Christian Era is the birth of Christ in year one. Reference epochs are usually more or less fictitious; they gain their authority through convention.

The idea of the Christian Era was invented by a Scythian monk, Dionysius Exiguus (Denis the Little), who lived in Rome. There were several chronologies current in his time; one of them counted years from the founding of Rome, Ab Urbe Condita (A.U.C.). At the request of Pope St. John I, Denis the Little computed the year of Christ's birth as 753 A.U.C. That was to become the epoch of the Christian Era.

With his calculations Denis the Little made himself live around 500 Anno ab Incarnatione (Year of Incarnation). The designation A.D. (Anno Domini) and the use of the century as a chronological unit were not invented until the sixteenth century. The notion of counting years backward as Before Christ (B.C.) first appeared in the seventeenth century; its inventor is unknown. Denis' own writings are dated in a system called the indiction, made up of 15-year cycles. It arose in connection with Roman tax laws and meant something like a fiscal cycle. The same *Farmer's Almanac* that brought us *E* for the Sunday Letter and the Epact of 19 also tells us that 1986 is year 9 in the current Roman indiction, a count that began during Emperor Diocletian. Knowing the Roman indiction is useful for farmers who pay taxes to the Emperor.

The year of Christ's birth is 1 A.D., the year preceding it is 1 B.C., there is no year zero. As it also happens, Little Denis made a little mistake; modern scholarship places the birth of Christ in 4 B.C. Mistake or not, Christendom has had for some time a good chronology in terms of which other epochs may be arranged.

But no time reckoner in itself can reckon time; one needs at least two of them. Where does the year of Christ's birth, whether 1 A.D. or 4 B.C., fit into the broader scheme of other eras?

The epoch of the Jewish Era is the creation of the world some time between 3762 and 3758 B.C., depending on how it is calculated. The number in use is 3761 B.C. This figure—the number of years between the Fall and the Redemption—remained a fact of religious knowledge until the advance of scientific cosmology. This is reflected in an early fifteenth-century poem.

Adam lay ibounden,
Bounden in a bond,
Foure thousand winter
Thoght he not too long. . . .

The reference epoch of Islamic chronology is 622 A.D., the year of the Hegira, Mohammed's flight from Mecca to Medina, previously mentioned. Although the Japanese have been using the Gregorian calendar for over a century, the epoch of their Era is 660 B.C., the year of enthronement of the first Japanese Emperor. India uses the Gregorian calendar and the Christian count for official purposes, but years are preferentially reckoned from 78 A.D., the epoch of the Saka Era. Saka is the name of a Scythian tribe, founded by a semimythical king whose reign is guessed to have begun in 78 A.D. Unlike in the Christian count, the name of a Saka year is the number of the year passed. This is no different from the way we count our age or name the hours. In this mode of naming, Columbus landed in the West Indies on October 12, 1491.

Putting it all together, today falls in the year 5747 A.M. (Anno Mundi, as calculated in the Jewish and traditional Christian chronology), 2739 A.U.C., 1407 A.H. (Anno Hegira), 2646 of the Japanese era, and 1908 of the Saka era. In Chinese chronology I am writing in the year Ping-yin. Ping is the 3rd celestial stem of 10, yin is Lion, the 3rd animal of the zodiac of 12 animals, which makes this the 3rd year of the 60-year cycle of 10 celestial stems and 12 celestial branches.[8]

Astronomy has its own reference epoch and its own era, hence its own chronology. It is called the Julian period, but it has nothing to do with the Roman Julius Caesar, for whom the Julian calendar was named. The Julian period used in the astronomy is the creation of Joseph Scaliger, a sixteenth-century French literary scholar who named it after his father, Julius Caesar Scaliger. The Julian chronology is not a linear scale, as is the Christian one, in the spirit of salvation history. It is a cycle that repeats itself, but it is long enough for astronomers to take it to be linear for purposes of bookkeeping.

Scaliger calculated the least common multiple of the 28-year solar cycle, the 19-year Metonic cycle, and the Roman indiction; it is 7,980 Julian years

8   An elderly couple interviewed in the 1985 television series on China, *The Heart of the Dragon*, gave their wedding day in terms of this sexagesimal count.

(after the original Caesar) of 365.25 days each. Scaliger judged this a long enough time to embrace human history. In his desire to establish a long cycle he shared the strivings of the Mayans, the Buddhists, the Chinese, and all civilizations that do not hold history to be linear. The epoch of the Julian Era is the year when the beginnings of the four cycles coincided: 4713 B.C. Astronomers specify each day by its Julian Day (J.D.) number, having assigned 0 to the day that began in Greenwich, England, at noon Greenwich Mean Time on January 0, 4713 B.C.

That there was no Greenwich, England, sixty-seven centuries ago or that the paleolithic inhabitants of that region, if any, knew little of mean noons or of January 0's presents no hindrance. People can extend the reaches of their temporal horizons beyond the confines of their lives and, with that, expand the boundaries of human reality. Plato did so when he wrote of the "perfect year," the period after which "all the eight revolutions, having their relative degrees of swiftness" reassemble themselves in a configuration they have already held at least once.[9] The perfect year of modern astronomy is the Julian (Scaliger) cycle of 7,980 Julian (Caesar) years.

The *Astronomical Almanac* helps a person find each minute of his life in the Julian system; it tabulates Julian dates and tells you how to interpolate for geographical location and part of the day. Today's astronomical day began at noon Greenwich Mean Time (because Scaliger lived before astronomers switched to midnight). Taking everything into account, I am writing this line at 2,446,684.4166 J.D.

The Julian count will commence again on January 0, 3267. The ephemeris for that year, prepared jointly by the Nautical Almanac Offices of the United States of America and the United Kingdom, will note the change.

The desire to provide temporal ordering in our lives, relate history to the rhythm of the stars, and express it all in numbers is ancient and universal. The Psalmist spoke for everyone, everywhere, and in all ages: "Teach us to number our days, that we may arrive at wisdom of heart." But how to number the days and what constitutes wisdom varies greatly from place to place and age to age.

Because of these variations, the best authority on calendar making and

---

9   The eight revolutions were those of the sphere of the fixed stars (it rotates once a day) and those of the Greek planets: the moon, the sun, Mercury, Venus, Mars, Jupiter, and Saturn.

the writing of chronologies is *Alice in Wonderland*. The issue before the court was, Who stole the tarts? The Hatter addressed the jury.

> "Fourteenth of March, I think it was," he said.
> "Fifteenth," said March Hare.
> "Sixteenth," said the Dormouse.
> "Write that down," the King said to the jury; and the jury eagerly wrote down all the three dates on their slates, and then added them up, and reduced the answer to shillings and pence.

They seem to have been debating the correct date of Easter.

Cosmic Birth and Death Judged in Terms of Human Time

As chronologies were extended to increasingly longer periods, the question inevitably arose: Did they have any boundaries to their validity? Did the universe and/or time have a beginning and will it have an ending?

Ideas about the origins of the universe are called cosmogonies. From legends and religious teachings to philosophical speculations and scientific theories, each cosmogony is appropriate to the age, place, and temperament of the people who composed it. Ideas about the end of the universe have also been the subject of much thought. In religious and philosophical reflections they are often called eschatologies (from the Greek for "last" or "farthest"). Scientific work on cosmogonies is rich; that on eschatologies, rather meager. People are much more interested in the first few hours than in the last ones.

The belief that the universe had a beginning could not have come from observing nature at large: never, during the life of our species, was the world anything but cyclic. It could not have come from observing havoc: havoc is not a precondition of new life. The belief that the universe will end could not have come from observing calamities: out of all past calamities retained in human memory, life has always and again arisen.

Scholars hold that beliefs in cosmic creation and destruction came from observing births and deaths of animals and people, and that it is a recently acquired idea. Paleolithic art shows little or no evidence of belief in cosmic beginnings or endings.

Between the earliest records made 26,000 to 30,000 years ago and the first narrative cosmogonies recorded 4,500 years ago, some 10,000 generations of people passed. During these eons of humanization, notions of births and deaths entered the inner landscape of the mind, there to give rise to feelings and ideas of creativity and destructiveness. Language and the arts made possible the application of these notions to the largest imaginable stage, the universe.

By "universe" one ought not mean the cosmos of our age, with its tiny, insignificant earth, but rather an awed sense of an infinite earth, surrounded by a mysterious sky that was populated by unknowable powers. The "big everything" was usually judged as being alive. To contemplate its birth and possible death was an obvious exercise of the mind in search of temporal order.

The desire to identify in the passing world the stable patterns of a life cycle—birth, growth, decay, and death—are as intense today as they must have been when the earliest creation legends were born. Only the terms in which the features of that life cycle are expressed have changed.

The oldest stories identify the universe with local geographical regions. The history of cosmologies is one of expanding geographical and temporal horizons. During this expansion of space and time, the preferred modes of perception have also changed. At first and until recently, the world was seen as a concrete structure and process, a friend, an enemy, or both, a fellow-creature of sorts. In its modern form, to be considered in Chapter 4 in the discussion of general relativity theory, the universe became an abstract geometrical construct quite alien in its nature to the daily concerns of people.

The earliest known creation legends are parts of the myths and liturgies of the third and second millennia Egypt, Mesopotamia, China, and India. In Egypt, the cosmos was thought to have come from the dry land that emerged from a primordial ocean. Egypt was the center of that universe, a country where the act of creation was repeated daily with the rebirth of the sun and yearly with the rise and fall of the life-giving Nile River. The creation event was modeled on the features of a stable and predictable environment. In contrast, the natural environment of Mesopotamia (present-day Iran and Iraq) was unstable: the Tigris and Euphrates Rivers were unpredictable, as was the climate with its torrential rains and scorching winds. For the Babylonians, the inhabitants of Mesopotamia, the world was born of violent conflict.

The *Enuma Elish* ("When above . . .") is a Babylonian creation legend with origins in a Sumerian story of around 3000 B.C. This is how it begins.

When above the heaven had not yet been named
And below the earth had not yet been called a name,
When Apsu primeval, their begetter,
Mummu and Tiamat, she who gave birth to them all,
Still mingled their waters together,
And no pasture had been formed and not even a reed marsh was to be
    seen;
When none of the other gods had been brought into being,
When they have not yet been called by their names and their destinies
    had not yet been fixed,
At that time were the gods created within them.[10]

"Them" refers to Apsu and Mummu. Apsu was the primeval father, a personification of fresh water; Tiamat, the primeval mother, personified sea water. Mummu is believed to have been the deified form of mist.[11] From the mingling of fresh and sea water came new, deified forms of silt and, after appropriate copulation, some other gods and still more others.

With that many gods, as with too many lawyers around, havoc followed. It ended only when Marduk, king of Babylon, killed Tiamat, sliced her into two, and from the two halves made heaven and earth. Then he designed a calendar, the duty of Babylonian kings (and of Chinese emperors). He

bade the moon come forth; entrusted night to her;
Made her a creature of the dark, to measure time;
And every month, unfailingly, adorned her with a crown.
"At the beginning of the month, when rising over the land,
Thy shining horns six days shall measure;
On the seventh day let half [thy] crown [appear].
At the full moon thou shalt face the sun. . . ."

---

10   This and the following citation from *Enuma Elish* are from A. Heidel, *The Babylonian Genesis* (Chicago: University of Chicago Press, 1951). The bracketed words were supplied by the translator, being best guesses of words obliterated on the tablets.

11   With its ambiguous position between water and air, mist fascinates the mind, which fears and reveres the pull of chaos. Mist has served as the home of deities: the gods of Nordic mythology live in mist homes, as do the godlike characters of that delightful modern epic, Tolkien's *Lord of the Rings*.

From where did Marduk appear? This question did not seem to have occurred to the tellers of the story, because for them the social organization of Babylon was a fact of nature. In their oral and written traditions Marduk—whose name meant "the calf of the sun" and "spring"—was as much an eternal fact of the world as were fresh water, sea water, or silt. Did time, to be measured by the moon, already exist? As concrete cycles of durations in need of counting, it did. It was another fact of nature so obvious that the question, "Did it have an origin?" had no meaning. As a universal framework, the idea of time had not yet been invented.

Around the globe from Sumer, the Chinese had their own creation stories, with images as characteristic of China as the silt was for Egypt or violence for Mesopotamia. One of the Chinese legends is told in the framework of stone masonry, which, along with architecture and sculpting, is an ancient Chinese art.

The creator of the world was P'an Ku, the architect of the universe. He is portrayed as a dwarf clothed in leaves and is usually shown with chisel and mallet in hand, working on granite (fig. 17). The setting suggests P'an Ku's origins as that of a nature god in a region of stony cliffs. Upon his death, his body gave birth to features of the earth: his head became mountains; his breath, wind; his voice, thunder; his blood, rivers; his teeth and bones, minerals; and the vermins on his body, people. With the last feature, suffering arrived on earth.

Unlike the Babylonians, the Chinese also had an abstract cosmogony of great refinement. From all eternity was Tao, the Cause, the Reason, the Principle, the Way, the Unknowable. In the beginning was Wu, the Nonexistence, the No-limit. From it came Chaos, the Great Ultimate. This gave rise to Yin and Yang, terms originally meaning the dark and the sunny sides of a hill. Yin stood for the earth, moon, darkness, quiescence, female, duality, and numbers divisible by two. Yang stood for heaven, sunlight, vigor, male, penetration, and odd numbers. Through several steps, Yin and Yang together created all things, including man.

Neither the concrete creation narrative of P'an Ku nor the philosophical cosmogony that traces the coming about of the world to the principle of Tao addresses the issue of the origins of cosmic time. It is probable that the question of a possible beginning of time would have appeared meaningless to the creators of these cosmogonies.

The natural setting and cultural atmosphere of India was different from those of the Middle East or China and gave rise to creation legends of yet

FIGURE 17 P'AN KU CHISELING THE STARS AND PLANETS FROM THE BOULDERS OF PRIMEVAL CHAOS. The sun, the moon, and the stars are visible through the opening he has made. He labored for 18,000 years, making the heavens rise and the earth thicken. Then he died for the benefit of his creation. While alive, the dragon, the phoenix, and the tortoise kept him company. They were three of the four intelligent creatures of the world (the fourth one was the unicorn). From where did the stones come? That was not asked in this cosmogony. That kind of a question can only be considered in abstract cosmogonies, those that are able to handle the idea that the universe has come about from conditions that were substantially different from the world as we now know it. Illustration from C. A. S. Williams, *Outlines of Chinese Symbolism and Art Motives,* 2d ed. (Shanghai: Kelly and Walsh, 1933).

a different tenor. The following is from the Rig Veda, one of the Vedas mentioned earlier.

> Then even nothingness was not, nor existence.
>   There was no air then, nor the heavens beyond it.
> What covered it? Where was it? In whose keeping?
>   Was there then cosmic water, in depth unfathomable?
>
> Then there was neither death nor immortality,
>   Nor was there then the torch of night and day.
> The One breathed windlessly and self-sustaining.
>   There was that One then, and there was no other.
>
> At first there was only darkness wrapped around darkness . . .
> In the beginning desire descended on it,
>   That was the primal seed, born of the mind . . .
> Seminal powers made fertile mighty forces . . .
>
> But after all, who knows, and who can say
>   Whence it all came, and how creation happened?
> The gods themselves are later than creation,
>   So who knows truly whence it has arisen?[12]

The hymn senses the paradoxical issues involved in the creation of the world, as does the Tao in its own way, both quite unlike the cosmogonies of the Middle East.

Let the Egyptian, Mesopotamian, Chinese, and Indian cosmogonies represent the rich variety of creation legends, all of them being responses to the need of seeing in the universe a fellow being that, as do animals and people, had at one time to be born. Now, let us turn closer to home.

In pre-Socratic Greece, let us listen to the poet Hesiod, who lived around 800 B.C. His *Theogony* (Creation of the Gods) related "how at first gods and earth came to be, and rivers and the boundless sea with its raging swell and glowing stars."

> Verily first of all did Chaos come into being, and then broad-bosomed Gaia [earth], a firm seat of all things for ever, and misty Tartaros [a bottomless abyss], and Eros, who is fairest among

12   A. L. Basham, *The Wonder That Was India* (New York: Grove Press, 1954), pp. 247–48.

immortal gods, looser of limbs, who subdues in their breasts the
minds and thoughtful counsel of all gods and men. Out of Chaos
Erebos [darkness] and black Night came into being [see fig. 2] and
from night again came Aither [ether, a notion we will reencounter in
premodern physics] and Day. . . . And earth first of all brought forth
starry Ouranos [sky], equal to herself, to cover her completely round
about. . . . Then she brought forth tall Mountains, lovely haunts of
the Nymphs. . . .[13]

It is a long and delightful story, the Greek mind seeking the orderly
beginning of the world in the Greek countryside. It is human-centered,
sensual, and colorful. Three centuries after Hesiod came the thinkers
I have spoken of in Chapter 1: Heraclitus, Parmenides, Zeno, and after
them a quantum jump to Socrates, Plato, and Aristotle.

The preeminent character of the universe, as Plato saw it, was its ordered
beauty. It was a rational universe that we could comprehend because our
souls were also rational: both spoke the language of the eternal, timeless
forms of geometry. The early meaning of *kosmos* was the order of a well-
ordered society. Only later did it come to mean order in the universe,
which for Plato was "by god's providence, in very truth a living creature
with soul and reason" and not what we regard today as physical. *Kosmos*
also meant good behavior, decency, ornament, decoration, and, in
women, the art of being skilled in adornment, called *kosmetikos*. Order
meant order in social, political, scientific, and personal contexts, all of
which was included in that single idea. Order remained an aspect of a one
and single reality until modern science began to peel the orderliness of the
physical world from the idea of orderliness in general.

The English "universe" comes from the Latin *universum,* literally,
"turning into unity." It suggests that the multiplicity of our changing
impressions has been turned into a single concept that can stand for something permanent in time. Its opposite is chaos, which in its original Greek
meant abyss, an image that survives in "chasm."

Hesiod, just quoted, assumed that chaos preceded order. So does the
Book of Genesis: "The earth was without form and void; and darkness
was upon the face of the deep." The Biblical story is not concerned with
what was in the deep, but only with the fact that it was formless, without

13   From G. S. Kirk and J. E. Raven, *The Presocratic Philosophers* (Cambridge: Cambridge University Press, 1975), p. 24.

organization, like the black holes of modern physics. No credit for black holes need go to the Greeks or Moses, though. They should be credited instead with the discovery of something more basic: the formulation of the twin concepts that define each other: order and chaos.

For Plato, the only things that existed before Creation were the geometrical rules in the mind of the Craftsman (the Demiurge), the maker of the universe. For Christians, it was the articulated will of God: "In the beginning was the Word, and the Word was with God, and the Word was God." These beliefs are variations on the same theme: before the world of space, time, and matter, there was only the timeless Idea, Form, Law, or, in Christian terms, the Word.

Early along the trajectory between the Word and the Processed Word of our age, during the merger of Greek, Hebrew, and Christian thought, the notion of Creation left the domain of mixing waters, murderous heroes, and copulating gods and became an event describable only in abstract terms.

It was thus that St. Augustine could separate the world from time and decree that the two, though they could have been created separately, were created together. Before Creation there was only timeless eternity and God's high regard for number. Wisdom, wrote St. Augustine in his *On Free Choice of the Will,* may or may not be the same as number, but it is "clear that both are true and immutably true." Twelve centuries later, Johannes Kepler remarked in his Preface to his *Mysterium Cosmographicum* that "Quantities are identical with God, therefore they are present in all minds created in the image of God." The latest child of this long love affair with the timelessness of number, as we shall see, is the creation story of physical cosmology.

This section began with the question: do chronologies kept by clocks and calendars have any boundaries to their validity? So far we have considered speculations about a beginning of the universe. Is the universe going to have an end? Will there be an epoch fundamentally different from ours, in that it will be followed by eternal timelessness, whatever "followed by" and "eternal timelessness" might mean?

If time's passage from the instant of Creation to the presumed end of the world is thought to be representable by a straight line, then the beginning and ending of time are symmetrical. Flip the line over and it will look the same as before, because it has two symmetrical, indistinguishable endings. But cosmogonies and world endings (eschatologies, to refresh the reader's memory) are anything but symmetrical and indistinguishable. A geometrical line is an awfully poor image of time, even if useful as a time-line to graph increased sales and decreased profits.

Cosmogonies educate, entertain, and account for the world and society as we find them to be. They appeal to the intellect by offering explanations. Eschatologies restrict, frighten, moralize; at best, they promise a continuation of life in some form, provided the tithes have been paid. The asymmetry is a projection of our mind's unequal concerns with the beginning and ending of our lives, as previously discussed. Hardly anyone worries about what it was like being himself or herself before he or she was conceived. If anyone does, it is an intellectual game. But probably everyone is concerned with what, if anything, will happen to him or her at the instant of and after death.

According to the Vishnu Purana of ancient India, the end of the world will take place in three large steps, each of which will contain many smaller steps. The scheme resembles the cyclicity of Hindu ideas of time, related earlier. Each eon will be progressively less blissful and also shorter. At the final ending there will appear seven suns, firestorms, deluges, vast elephants, and darkness.

> At the end of a thousand periods of four ages the earth is for the most part exhausted. . . . The eternal Vishnu . . . enters the seven rays of the sun, drinks up all the waters of the globe, and causes all moisture whatever, in living bodies and the soil, to evaporate. . . . The destroyer of all things, Hari, in the form of Rudra, who is the flame of time. . . . proceeds to the earth and consumes it. A vast whirlpool of eddying flame then spreads to the region of the atmosphere, and the sphere of the gods, and wraps them in ruin.
>
> [Heavy clouds like vast elephants spread over the sky. Some are black as the blue lotus, white as the water lily, some deep blue, azure, or bright red, and some like the painted jay. The world is enveloped in darkness, all things perish, and the clouds pour down their water.]
>
> When the universal spirit wakes, the world revives. . . . Awaking at the end of the night, the unborn Vishnu, in the character of Brahma, creates the universe anew. . . .[14]

The last ending was, after all, not a final ending. Let us consider another story of world ending.

---

14   *Vishnu Purana*, trans. and illus. H. H. Wilson (Calcutta: Punthi Postak, 1964), p. 493.

In Scandinavian mythology, Ragnarok is the name of the end of the world of gods and man. The Ferris Wolf, the World Serpent, and the hell-hound Garmr slay Odin, father of all gods; Thor, god of thunder (as in Thor's Day); and Loki, the mischief maker. Instead of huge elephants and fire, appropriate to the Indian subcontinent, there is a monstrous winter fitting for a Scandinavian nightmare. The stars vanish and the earth sinks into the sea, but then rises again. Balder the Innocent, God of Light (killed earlier by Loki), returns at the helm of the host of the just. A new life begins in a Scandinavian spring that resembles an Ingmar Bergman movie with Liv Ullmann. The earth is

> most lovely and verdant, with pleasant fields where the grain shall grow unsawn. . . . From a woman named Lif and a man named Lifthasir the sun shall have brought forth a daughter more lovely than herself [Lif] who shall go in the same track formerly trodden by her mother.[15]

Although the setting of these two world endings are very different, the messages are the same: after suffering and horror, there is continuity. Similar postmortem universes may be found in other eschatologies. In all such stories the problems that are associated with an ending and/or beginning of time cannot come up for consideration because time is taken to go on; history is assumed to be cyclic and infinite. Each ending is only a provisional one, each beginning only a milestone.

The idea of a final, absolute ending of the universe of time followed by a timeless existence of the soul is a unique feature of Judeo-Christian cosmology, built on the idea of salvation history. There was a beginning, followed later by the act of redemption, and there will be a final ending called Apocalypse. In Greek it means "revelation," which explains why what in Catholic bibles is the Book of Apocalypse, in Protestant ones is the Book of Revelations.

The staging of the Apocalypse resembles other world endings, but differs from them in its finality and in its consequences for the individual. Good people this way, bad ones that way. "No, sir, you must go to the Other Place, and for good."

15   *The Elder Edda and the Younger Edda of Saemund Sigfusson,* trans. I. A. Blackwell (London: Norroena Society, 1907), p. 327.

Concerns with the finality of heaven and hell are not among the most pressing worries of today's men and women, at least not in its religious form. But the idea of an end of time has been an essential part of Christianity since the ministry of Christ. Reinforced by complex historical forces, it helped create that undisguised urgency which has become the hallmark of Western civilization.

The expected–unexpected character of the Christian universe, explicit in its religious form during the Middle Ages, has changed with Protestantism into a keen consciousness of temporal passage. In that new form it came to infuse modern Western life and, through it, the lives of all people on earth. Men and women of the rat race, people who seek immediate solutions to problems that need long periods of reflection, and others who bargain away the future because that future may never come seldom realize that they are acting out the legacy of the Apocalypse.

Homo faber put his intuitive knowledge of time in the service of daily life by creating clocks, calendars, chronologies and cosmologies: the ways of time reckoning. What did we find out about them?

We learned that no single clock in itself can measure time. It is always necessary to compare at least two clocklike processes, each with countable events, and then express the result of comparison in number. In daily life, one of the processes is usually masked by convention, but a second process is always involved. Scientific time reckoning is no exception to this rule. It differs from nonscientific time reckoning only by its demand that the relationship between the clock readings be well defined and unambiguously expressed.

Calendars and chronologies help us time individual and social behavior by the heavenly clocks. In the history of man and the universe, they extend the reach of all events orderable by time. As the boundaries of time are pushed out to ever-lengthening periods, the influence of political, philosophical, and religious views on time reckoning becomes increasingly evident. Narrative cosmologies further extend the limits of human time by projecting our concerns with birth and death upon the universe.

At first thought, time reckoning seemed to have been mainly a problem of skills and techniques. But as the temporal horizons expanded, technical details became irrelevant, human values relevant.

THREE

To whom there belong?

THE TIMES OF LIFE, MIND, AND SOCIETY

The two preceding chapters gave ample evidence that "time that takes survey of all the world" has been much debated and cleverly reckoned. This and the following chapter take a new survey of time in terms appropriate to the science and thought of our age.

Based on what has been said thus far, learned men, women, and clever clock and chronology makers seem to agree only on this much: that mature people, in their waking state, are aware of a "now" or "present," with respect to which they recognize a future and a past.

Let me again represent this view by a metaphor, that of the cable of the San Francisco cable car. It is invisible, inaudible, and moved by distant, irresistible, and unseen powers. The cars move because they attach themselves to a point along the moving cable.

Likewise, one may imagine a cosmic present—a cosmic now—moving through the universe. Sticks and stones and dry bones, frogs, horses, people, and societies attach themselves to that cosmic now of the physical world, and together, we all travel through time.

"Alas, poor Yorick! I knew him, Horatio," said Hamlet, contemplating Yorick's skull, "A fellow of infinite jest, of most excellent fancy."

Alas, poor Time, we knew him for millennia. But the physical sciences of this century have revealed that there is nothing in the physical world to which our idea and experience of a present could correspond. The physical world is "now-less." Since future and past make sense only with reference to a present, the physical world—quite consistently—is also futureless and pastless. It is not at all timeless in some mystical sense; it is temporal. But its time is qualitatively different from what we usually mean by time: the metaphor of flow cannot be applied to it. Or if you wish, it can be said to flow in both directions simultaneously.

This chapter leads the reader from the intuitive notions of time—that fellow of infinite jest and most excellent fancy—to the bare bones of non-flowing, undirected, now-less time that physical science has revealed to be the reality of the physical world.

A reasonable sequence of presentation should perhaps begin with time in the physical world. To jump ahead of ourselves a bit, there are three kinds of physical temporalities corresponding to the organizational levels of light, particle, and massive matter. Then we could go on to discuss bio-temporality, nootemporality, and sociotemporality.[1] But the nature of the physical temporalities is very far from our intuitive notions of the nature of time. So far, in fact, that of all temporalities, the physical ones are the most difficult to appreciate.

For that reason, this chapter will address the nature of the time of life, of the mind, and of society and leave the time of the inanimate world for the following chapter.

When my children were small, someone brought us a small carved figure of a European peasant woman. She wears a babushka, her hands are folded over her round stomach, and she may be holding a flower. She is painted with many colors. At the level of her equator, the lady can be opened. Inside her there is a smaller copy of herself, and inside that, a yet smaller copy, and so on, all of them being hand-painted figures. As they become smaller, the details disappear. The last little woman looks rather drab.

The set of these women form a hierarchically nested system: each one contains those beneath it; each adds some refinements to what it contains.

The organizational levels of nature also form a hierarchically nested system. Beginning with society, the most complex one, then going on to the world of the individual mind, of life, and of massive matter, particle, and radiation, each contains—and is built on—the structures and func-

---

1   We are already familiar with nootemporality or noetic time. The concepts of bio-temporality (the temporal world of living organisms) and sociotemporality (the temporal world of a time-compact globe) will be introduced later in this chapter.

tions of the integrative levels beneath it, and each adds some new freedoms to what it subsumes. Thus, societies, acting as communities, have more freedom than do any of their individual members. Humans, while biological creatures, are free in ways that species without possessing a mind are not. Living organisms, while made of matter, have degrees of freedom that matter has not.

This chapter will demonstrate that an understanding of time based entirely on intuition—assumed as a valid basis for reasoning in the first two chapters of this book—is totally inadequate for an appreciation of the intricate structure of time revealed by contemporary scientific thought. There is nothing intuitive at all about the existence of a hierarchy of temporalities. To wit, of qualitatively different forms of time appropriate to the world of light, atomic particles, massive matter (that is, the world of stars and galaxies), living organisms, the mind of man, and cultural systems, respectively.

Yet it is precisely for a universal understanding of time that we ought to strive, because, as humans, we belong to all the organizational levels of nature. We are made of matter, hence we are kin to photons, atomic particles, and the stars and galaxies; we are alive, hence kin to the bacteria, the dragonfly, and the grizzly bear; we have minds, hence are kin and belong with all other men and women; we are also members of societies and hence kin to other people with whom we can share social values.

To be able to master the counterintuitive aspects of the nature of time that we shall encounter as our voyage continues, it will be useful to become familiar with a scientific idea of reality. It was first suggested early in this century by the German theoretical biologist, Jakob von Uexküll. He drew attention to the fact that an animal's receptors and effectors determine its world of possible stimuli and actions. Through them, they determine the nature and extent of the animal's universe. He called such a species–specific universe the umwelt of the species. What is not in that umwelt must be taken as nonexistent for the members of that species (fig. 18). For instance, ultraviolet patterns on certain butterflies exist for other butterflies but not for vertebrates; vertebrates have no sense organs through which they can read those patterns. What an earthworm cannot know might kill it but it still won't know what hit it.

The word *umwelt* has been naturalized into English. In modern psychology it is defined as "the circumscribed portion of the environment

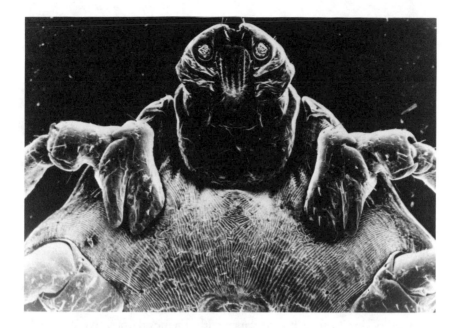

FIGURE 18 WHAT IS REALITY? HELP FROM THE TICK, VIA JAKOB
JOHANN BARON VON UEXKÜLL. This is a portrait of a dog tick, magnified
60 times by scanning electron microscopy. Shown are parts of two of the eight
crablike legs and the swollen body.

The female of the tick hangs on a bush or tree until a victim passes by. It then
drops on it, seeks out a hairless spot, bores through the skin with its beak and
gorges itself with blood. From the size of a flea or pea, depending on the species, it
can become the size of a large acorn. After her meal the tick drops to the ground,
lays her eggs, and dies. The larvae, called seed ticks, move up on blades of grass and
wait. When a suitable victim walks by they do what their mother did: they drop,
gorge, drop. Then they mate. The males, their mission accomplished, die. The
females crawl up on a bush or tree, there to hang and . . . wait.

Most of the tick's life is spent waiting for the stimuli that can trigger its feeding
and reproducing regimen. Some ticks have been known to be able to wait for their
meal for over twenty years. The list of stimuli that can act as releasers is a very short
one.

Some species respond to butyric acid present in the furs of warm-blooded
animals. (This is the stuff that gives sweat and rancid butter their characteristic
odor.) Once on their host they respond to temperature, seek out a hairless spot,
insert their beaks through the skin, and fill themselves with blood. They are as
ready to drop on a balloon that smells right as they are on a deer; they are as ready to
drink poison as blood if it has the right temperature. Some ticks use their first pair
of legs to sense carbon dioxide exhaled by the animals whose blood they seek.

Some have eyes that sense the difference between light and shadow.

What comprises the universe of a tick? Light and shadow. The presence or absence of carbon dioxide. The presence or absence of butyric acid. What comprises the universe of man? All things and events visible and invisible, possible and impossible, present, past, and future.

All species have their species-specific realities, including *Homo sapiens*. Human time is one aspect of that species-specific reality. But whereas the temporal horizons of all other species are limited to what they may immediately sense or to what extends only very slightly beyond it, our temporal horizons are unbounded. While a tick can only wait for butyric acid or the like, we can wait for Godot, that is, for symbolic realities that may be freely moved in an imagined future and past.

Illustration, courtesy the Smithsonian Institution, National Museum of Natural History, and the Hoogstraal Center for Tick Research.

---

which is meaningful and effective for a given animal species and that changes its significance in accordance with the mood operative at the moment."[2] Note that the "environment" of which the animal's umwelt is a portion is our own, human umwelt.

By means of instruments capable of translating the signs and signals of other umwelts into those of our own, we can expand our umwelt. We know of the ultraviolet patterns on butterflies because we can make photographic plates sensitive to ultraviolet rays and thus translate the patterns into the visible spectrum of light. "Visible" means visible for us. We know about the time of animals because we experiment with them. We know about the time of atoms or of the universe because we write equations that depict their behavior, and those equations tells us about the temporalities of the atomic and stellar worlds. We are thus authorized to talk about the temporal umwelts of matter, animal, and man, and to assume that they are really real, so to speak.

When the time-related teachings of the different sciences are systematically surveyed and arranged to correspond to the hierarchically nested organization of nature, then five distinct temporalities may be identified, themselves also ordered in a hierarchically nested fashion, like the little carved peasant women.

---

2   H. B. English and A. C. English, *A Comprehensive Dictionary of Psychological and Psychoanalytic Terms* (New York: McKay, 1958).

These temporalities are aspects of the different umwelts and must be taken, as I have stressed, to be real, complete, and sufficient in themselves for the integrative levels where they appear, even if they seem strange and incomplete when compared with the nootemporal reality of the human mind.

Let me represent them by means of a visual metaphor.

Imagine a long, well-defined arrow drawn on a piece of paper: head, shaft, and feather (tail). The picture stands for the kind of time that is unique to the human mind. It is a time informed of a sharp division between future and past, of long-term expectation and memory, and of a *mental present* with continuously changing boundaries. These are the hallmarks of *nootemporal* reality.

Let the head and the tail of the arrow become ill-defined, amounting to ambiguous but still distinct limits to the shaft. The picture is a visual metaphor for *biotemporality*. This is the temporal reality of all living organisms, including man, as far as his biological functions are concerned. Living species display a very broad spectrum in the degree to which they distinguish among future, past, and present, but the biotemporal arrow is quite different from the noetic one. Here the mental present of the nootemporal world reduces to the *organic present* of the life process.

The head and tail of the metaphorical arrow may even be absent. What remains is a shaft of the arrow, a line, an image of *eotemporality,* so named after Eos, the goddess of dawn. This is a temporality that does not have a present, and it consistently has neither a future nor a past. The metaphor of flow cannot be applied to it. This is the time of the astronomical world of galaxies and stars, of the macroscopic, physical universe.

The shaft of the arrow may now disintegrate into disconnected fragments. This image stands for a world in which it is impossible to distinguish between "now" and "then," even in principle. This is the universe of elementary particles. Such a world is called *prototemporal* for proto-, the first of a series.

Finally, even the remaining fragments of the arrow may vanish. We are left with a blank sheet of paper. Its blankness stands for the *atemporal* world of electromagnetic radiation. It does not stand for nothingness, but rather for a condition in which no meaning whatsoever may be given to any notion that we associate with time. This atemporal world is the most primitive level of the universe, that of radiative chaos.

From the atemporal world let us climb back to the nootemporal—and beyond. A good case can be made for *sociotemporality* being the time appropriate to a time-compact globe. But the difficulties of delineating the features of sociotemporality are great. Namely, it is necessary to use a new language whose vocabulary and syntax derive entirely from social rather than individual experience. Being quite ignorant of the main characteristics of this postulated sociotemporality, it is not possible to suggest a visual metaphor that could join the arrow. We will return to the issue of sociotemporality in the last sections of this chapter and toward the end of this book.

The sequence: matter, life, mind, and society also represents steps in the history of inorganic and organic evolution. We all came from the big bang, having taken an immense journey. Each of us shares some functional and structural aspects of matter and living matter, each of us exists in all the different realities that life and our species internalized along its evolutionary journey. We can study the hierarchically nested components of time because we have minds that can explore the world through symbols. Like Tennyson's Ulysses, each of us can say of himself or herself,

> I am part of all that I have met;
> Yet all experience is an arch wherethrough
> Gleams that untravelled world, whose margin fades
> For ever and for ever when I move.

Biotemporality: The Cyclic Order of Life

Living systems are distinguished from nonliving systems by being able to secure instant-by-instant synchronization among the multitudes of their clocks. This is not a minor task because the frequencies of those clocks are spread across a spectrum of seventy-eight orders of magnitude. This section fills in the details.

Birds and bees do it, chipmunks and tulips do it, men and women do it. Each of the millions of cells in all living organisms does it. They all oscillate. Single oscillating living systems, such as cells, are said to have *biological clocks*. Cooperative oscillations involving many biological clocks are said to manifest *biological* or *physiological rhythms*.

In historical times, the harmony between the cycles of nature and the behavior of living things, including people, have often served as a source of inspiration, a reassurance of continuity, such as in the Song of Solomon.

> For winter is now past,
> The rain is over and gone,
> The flowers have appeared in our land . . .
> The fig tree hath put forth her green figs . . .
> Arise my love, my beautiful one, and come.

Of more recent vintage are these lines:

> Summer is icumen in!
> Lhude sing cuccu!
> Groweth sed and bloeth med!
> And springth the wde nu.
> Sing cuccu![3]

The authors of these lines, written twenty-four and seven centuries ago, respectively, did not ask whether there was a clock in the fig tree, the lover, or the cuckoo. Tree, man, bird, and the heavens formed a single cyclic world. The significance of universal cyclicity began receiving attention during the Renaissance. Then, at the end of the nineteenth century, biological science rediscovered living cycles and began to supply an increasing torrent of marvelous details.

The biological day

Carolus Linnaeus, inventor of the binomial system for naming species, observed in 1727 that a hawk's-beard and a hawkbit (two species of daisies) opened and closed with an accuracy of one-half hour each day. Later he suggested the planting of a floral clock "by which one could tell time, even in cloudy weather" (fig. 19).

It is probable that all living organisms down to the molecular level display circadian (approximately 24 hour) oscillations; scientific literature on the biological day is already vast and is increasing rapidly. Circadian

---

3   Now the summer's come again! / Loud sing, cuckoo! / Seed a-growing, mead a-blowing! / Green grow the woods too. / Sing cuckoo!

FIGURE 19   RECKONING THE SOLAR DAY BY BIOLOGICAL CLOCKS IN
PLANTS. This is a modern rendering of Linnaeus' idea mentioned in the text. The
layout shows red hawk's–beard and white water lily opening between 6 and 7 A.M.,
the lily closing between 5 and 6 P.M. St. Bernard's lily and St. John's wort open
between 7 and 8 A.M. and close between 3 and 4 P.M.; the scarlet pimpernel opens
between 8 and 9 A.M. and closes between 1 and 2 P.M. The marigold opens between
8 and 9 A.M. and closes between noon and 1 P.M. I suspect that the rhythm of these
flowers has been modified for the purpose of the drawing because the real ones
behave differently, as we know from Perdita in Shakespeare's *Winter's Tale*:

> Hot lavendar, mints, savory, marjoram
> The marigold that goes to bed with the sun,
> And with him rises weeping.

The dial–like arrangement is not a necessary feature of the flower clock; it is used
only to help illustrate the idea. The drawing is by the German artist Ursula
Schleicher–Benz. From *Lindauer Bilderbogen* 8, courtesy, Jan Thorbecke Verlag,
Sigmaringen.

behavioral rhythms have been studied for hundreds of invertebrate species from algae to diatoms, dinoflagellates, corals, sea anemones, jellyfish, parasites that live in the blood, aquatic and terrestrial worms, molluscs, starfish, sea urchins, crustaceans, millipedes, centipedes, spiders, scorpions, mites, ticks, and, of course, insects. Equally extensive work has established the presence of circadian rhythm in mammals from Norway rats and deermouse to mice, squirrels, beavers, muskrats, shrews, weasels, kangaroos, sea otters, dogs, cats, apes, and humans.

There is also an internal biological day that can only be inferred from experiments rather than observed directly in behavior. At least sixty internal circadian cycles have been identified in the mouse alone. These rhythms are present in the functions of the hypothalamus; those of the pituitary and adrenal glands; in the chemical composition of the bone marrow and blood; in the susceptibility of organisms to harm by X-rays, by noise, and by chemicals. You annoy the poor mouse; a chronobiologist will tell you about the expected circadian variation in the intensity of its reaction.

Our fellow humans have been studied for circadian responses from sexual performance to solving abstract problems; from hand coordination to reactions to more chemicals than those that line the pharmacist's shelf.

The daily variation in the readiness to respond to environmental stimuli has not escaped the attention of jailers, dictators, and military leaders. Judicial executions, unless public events, have usually taken place very early in the morning while people still had sleep in their systems; it is more traumatic that way to all who witness it and even to those who only hear about it. Military campaigns often begin a few hours after midnight; if we were nocturnal beings, they would start in the early afternoon. Taking advantage of the biological day is often combined with taking advantage of social rhythms. The attack on Pearl Harbor in 1941 took place on a Sunday morning; the Polish government crushed Solidarity in 1981 on a Saturday night. In modern totalitarian systems, it is in the wee hours of the morning when the policeman is most likely to knock at the door.

Advanced organisms contain many cells; the human body is made up of some 10^{14} (a hundred trillion) of them.[4] They are all kept cycling in

---

4   *A note on the use of shorthand.* While no mathematics is used in this book, it is impossible to keep away from very large and very small numbers. If I wished to tell the reader at what

constructive cooperation by local, semi-independent regional offices known as pacemakers. For a healthy organism, the cooperation among pacemakers may be so complete that the organism appears to have only a single master clock.

The sleep–wake cycle is the most obvious circadian rhythm. In some birds, such as the house sparrow, the pineal gland serves as the sleep–wake pacemaker. That gland is believed to be a vestigial third eye that was developed in some now extinct reptiles. In one living reptile it still functions as a sensory organ, called the pineal eye. It is an aperture in the vault of the skull right beneath the scales; some pineal eyes even have a lens and a cornea. In mammals the sleep–wake cycle is controlled by a pair of millimeter-sized organs in the hypothalamus—the oldest part of the brain—and are located just above where the right and left optic bundles cross. They are called suprachiasmatic (above-the-crossing) nuclei.

The closeness between the nerves used for sensing light and the pacemaker that controls sleep and wakefulness is significant. It suggests that light and darkness have been the primary synchronizing signals of the circadian clock. Upon some thought, this is not at all surprising.

There is ample evidence that many rhythms of the biological day remain functional even when the organism is isolated from external cues. Those rhythms are said to be endogenous (originating from within). But since the free-running periods of circadian clocks are seldom exactly as long as the solar day, to remain in phase with its environment, the biological day must be continuously pulled into synchrony with the solar day.

Human infants are born with a readiness to adapt to day and night, but how this rhythm is passed from parents to offspring is not known. In rats and squirrel monkeys, the circadian clock already oscillates during fetal

---

frequency the molecules of the skin oscillate while responding to ultraviolet light, I could write it out as 10,000,000,000,000,000 Hz (Hertz, meaning cycles per second, named in honor of Heinrich Hertz, the inventor of the radio). But this many zeroes get boring. I could go to the dictionary, look it up, and say that they oscillate ten quadrillion times a second. But such a name is eyewash, because hardly anyone knows what it means. Instead, I will use the mathematical shorthand for the powers of 10. In this case, it is 10^{16}, which stands for the numeral 1 followed by 16 zeroes. 10^{1776} means the numeral 1 followed by 1776 zeroes. Very small numbers are represented by negative powers. For instance, 0.000,000,001 is 10^{-9}, the numeral 1 being at the 9th place after the decimal point.

life, having become synchronized in utero with the external light-dark cycle experienced by the mother.

Not so in humans. The human embryo and the fetus are both subject to the mother's circadian system, yet in the human infant, circadian behavior is acquired after birth (fig. 20). How soon circadian activity appears depends on maturity: preemies develop it later than do full-term babies. Once developed, it remains with each of us for life and is regularly synchronized with the sun, directly or indirectly. In the absence of such synchronization, the daily rhythm becomes free running (fig. 21).

Seventy-eight octaves of biomusic

It was a custom of traditional China for the concubines of the Emperor to present themselves for the sharing of the imperial couch, beginning with the highest-ranking woman, at full moon. It was believed that the Yin (female) influence was then the highest. That, combined with the powerful Yang influence ever-present in the Son of Heaven, would give the highest virtue to the child then conceived. The custom neatly ties together biology, astronomy, and social guidance. It uses the moon as a time reckoner to help schedule an activity that, if not regulated, could easily lead to rivalry. There is no evidence of moon-dependent virtues in infants, but the ubiquity of monthly rhythms in people and animals makes the belief a rational one. These thoughts lead us to considering the *biological month*.

The most obvious lunar timers are the tidal rhythms and the cyclic variation of the intensity of moonlight. It is from these timing signals that organic evolution has constructed the biological month.

High and low tides follow by a fixed amount the moon's crossing of the meridian in both the visible and the invisible hemisphere. Since the lunar day is 24.8 hours long, the basic cycle of the tides is 12.4 hours. But the amplitude of the tides is not constant. There are lunar, solar, and lunisolar variations of both the diurnal and the semidiurnal species. There are also fortnightly variations, a rhythm of 27½ days and even a semiannual rhythm. Figure 22 shows the spring tides (at new and full moon), the neap tides (at first and third quarters), and a semidiurnal inequality due to the slightly different lengths of the solar and lunar half days.

Animals in the intertidal zones have been living with the rise and fall of water levels ever since life first appeared on the seashores. To different

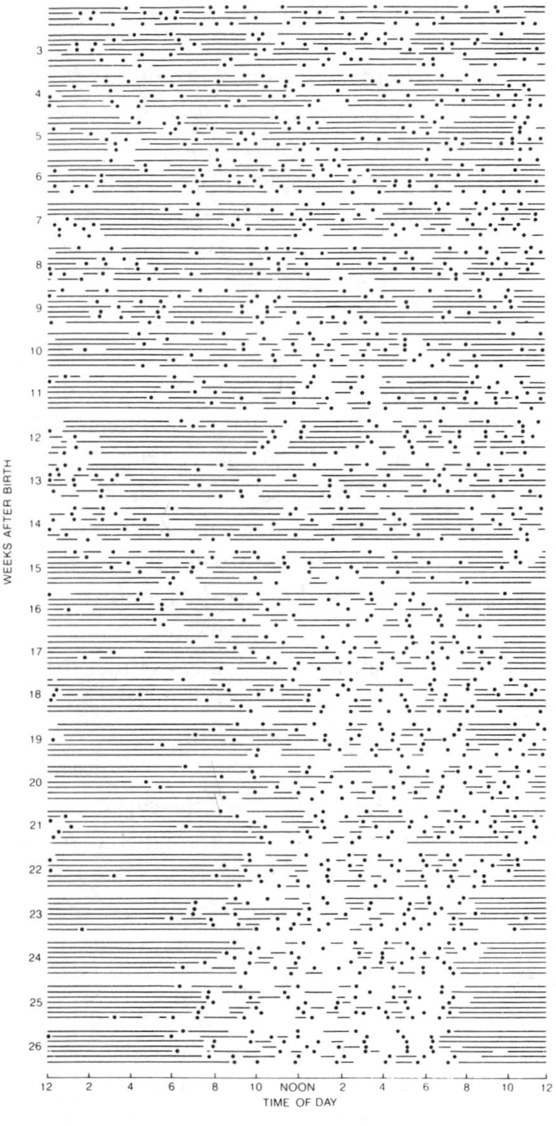

FIGURE 20 ENTRAIN-
MENT OF AN INFANT'S
CIRCADIAN RHYTHM.
The graph shows the development of the sleep–wake
cycle in a human infant and
the way it becomes en-
trained (pulled into phase)
with the day–night cycle.
The horizontal axis shows
the time of day; the vertical
axis, the days after birth,
grouped by week. Solid
lines represent sleep; open
spaces, wakefulness; dots,
nursing. Although the in-
fant was fed on demand, the
mother succeeded in training it to daytime feeding.
Random feedings, though
possibly with a circadian
component, may be ob-
served until about the 15th
week. By the 26th week, the
night–day, sleep–wake cy-
cle has been established,
more or less, and the parents
will be able to get their sleep.
From: N. Kleitman and
T. G. Egelman, "Sleep
Characteristics of Infants,"
Journal of Applied Psychology
6 (1953): 169. Courtesy, The
American Physiological
Society.

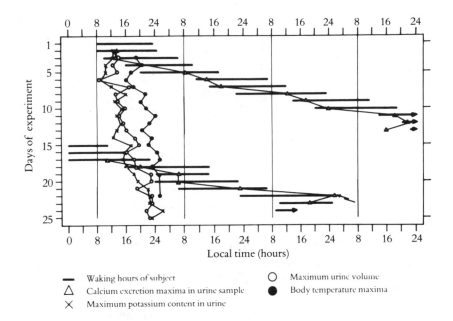

FIGURE 21 FREE-RUNNING CIRCADIAN RHYTHMS IN AN ISOLATED
HUMAN SUBJECT. The ordinate (vertical coordinate) shows the days of the experiment; the abscissa (horizontal coordinate), the hours of the day. On the first
day, the subject rose at 8 A.M. and retired just before midnight. On the second day,
he again rose at 8 A.M. and retired soon after midnight. On the third day, he rose
around 10 A.M. and retired around 6 A.M. the next day (outsiders' time). He then
rose around 4 P.M. (16 o'clock) the same day and went to bed at 10 A.M. the next
morning.

Through the 24-day test, the body temperature, urine volume, and potassium
content in the urine kept a circadian period of 24.5 hours, their maxima coming on
the average only half an hour later each day. But the subject's sleep–wake cycle
averaged 31.2 hours. This is an unusually long free-running cycle. Most peoples'
free-running biological days differ from the solar one by no more than ±2 hours.

The internal desynchronization between the sleep–wake cycle and the other
cycles resembles that associated with jet lag.

Illustration, courtesy Jürgen Aschoff, Max Planck Institut für Verhaltens-
physiologie, Andechs, German Federal Republic.

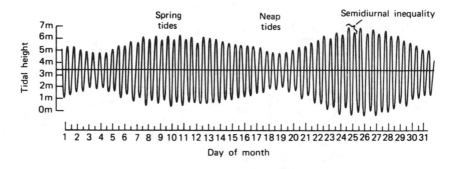

FIGURE 22   TIDAL VARIATIONS about an assumed average tide of 3.5 meters. After J. D. Palmer, *Biological Clocks in Marine Organisms* (New York: Wiley, 1974).

extents, these variations have been programmed into the behavior of marine organisms and are manifest in their locomotor activities, in their migrating and mating patterns, and their reproductive cycles and color changes.

Lunar rhythms that are synchronized with the variations of lunar brightness have been identified in many terrestrial species. The flight activities of many insects vary with the biological month, as does the behavior of monkeys, bats, and kangaroo rats. Some night-active species synchronize their lives with the 24.8-hour lunar day and even with specific phases of the moon. Lunar clocks are as inborn as solar ones, and like solar clocks, they become free running in the absence of external synchronizing cues.

Both the median and the mean lengths of women's menstrual periods are 29.5 days. The closeness of this figure to the 29.53 days of the synodic month—the period between successive new moons—cannot be accidental. There is no well-researched and established reason why and how menstruation became lunar; there are only plausible guesses. It may be imagined, for instance, that as the rhythm of being in heat once a year was replaced by being in heat continuously, the menses of females became synchronized with the moon—perhaps with the new moon—because this offered certain advantages, such as easier collective protection from predators. Later, the decreasing danger from predators made synchronous menstruation unnecessary, but by then the biological month was entrained.

The lunar month manifests itself in the prey-predator rhythms among people, even in our own days. In his sensitive book on the sociology of time, Michael Young reports on the night-time activities of the Los Angeles Vice Squad, as he observed it.

> But on this night of the lunar month an arrest [for prostitution] is unusual. It is nearly full moon and cloudless. [The police officer] told me before we went out that, unfortunately, there would for this reason probably be little activity in the streets that night. Clients are put off and so are the prostitutes. In the backs of cars or on the ground in the dark alleys where they sometimes go they are more likely to be seen, and, if they do risk it, they are more likely to be arrested. So full moon is nearly always a slack-time for [the Vice Squad].
>
> *(The Metronomic Society,* p. 224)

There are also annual rhythms, manifestations of the *biological year*. Leaves in the autumn change their color, insects become dormant, many species begin to hibernate, and I get a gathering-in feeling. In the spring new leaves come out of silent buds, insects buzz, and starlings celebrate their annual testicle growth. Tennyson, having been a poet and not a biologist, did not measure the testicle size of starlings but spoke instead of the change in the color of a dove's throat during the mating season.

> In the spring a livelier iris changes on the burnished dove,
> In the spring a young man's fancy lightly turns to thoughts of love

Scarlet gilia, a flowering herb of the West, changes from pink to white from July to early September in order to attract two different pollinators: hummingbirds and hawkmoths. Hummingbirds are attracted to pink flowers. They arrive, do their task, then move on, giving room to the next group of traveling pollinators, who prefer the color white. The scarlet gilia internalized the annual rhythms of two animal species because it was useful for maintaining the continuity of its own species.

In mammals, yearly cycles have been identified in aggressive behavior, eating and drinking habits, food selection, self-stimulation, sexual behavior, and vocalization. In birds, there are annual cycles in courtship, copulation, drumming, egg-laying, feeding, the storing of food, nest building, the amount and manner of preening, song patterns, and territorial defense.

In September, garden warblers in Germany begin their migration. First, they fly southwest to Spain. In November they cross over to Africa at Gibraltar, then take a southeastern course toward their wintering grounds. In experimental tests they have been kept in cages in Germany, where their motions were monitored. In August they began hopping in the direction of Spain in a display of migratory restlessness. During November they changed their direction of hopping from southwest to southeast, as if they were over Gibraltar. Obviously, they had a way of telling directions, a program for what to do, and a yearly clock that controlled timing. If locked in jail, humans do something similar: they travel in their minds.

The regularity of bird migration, such as the departure and return of the stork, has been woven into the fabric of fairy tales and into the reality of country life, constituting a rite of passage (fig. 23).

Some genetic rhythms have periods longer than the lifetime of an individual. For instance, a caterpillar is a metamorphic insect whose life includes two different forms with a division of labor. Caterpillars feed and grow, but cannot reproduce; butterflies feed on nectar and reproduce, but do not grow. The genetic program of the organism addresses two different forms of the same individual. It says: crawl, eat, grow, change into a pupa, emerge, flap wings, fly, eat but do not grow, mate, die. Repeat cycle, a capo dal fine.

If the genetic cycle is longer than a form of life, is it the cycle of an individual? Lewis Carroll must have wondered about this, for he made the Caterpillar ask Alice, "Who are you?" Alice could not answer because, as she said, "being so many different sizes a day is very confusing."

> "It isn't," said the Caterpillar.
> "Well, perhaps you haven't found it so yet," said Alice, "but when you have to turn into a chrysalis—you will some day, you know—and then after that into a butterfly, I should think you'll feel it a little queer, won't you?"
> "Not a bit," said the Caterpillar.

The biological day, month, and year are internalized rhythms that correspond to astronomical cycles. But living organisms show many other cycles whose external correspondence is not in astronomy but in the behavioral cycles of other species. The scarlet gilia I described previously

FIGURE 23   RETURN OF THE NATIVE: BIOLOGICAL CLOCKS AND
SOCIETY. Time reckoning by bird migration is an ancient custom. It was still very
much alive in the Europe of the 1930s, when this drawing was made, but began to
disappear as industrial and commercial rhythms took over the task of timekeeping.
It is still possible to feel in one's spine the primordial call of such a rhythm by
watching Canada geese as they fly south, wending their way from the James Bay
region to their wintering grounds on the Gulf of Mexico. Their characteristic
honking comes from a distant evolutionary past and carries with it the timelessness
of eternal return. "Our village stork returned, spring arrived!" From Jacques
Waltz Hansi, *L'Alsace Merveilleuse* (1935), ©S.P.A.D.E.M., 1935,
Paris/V.A.G.A., New York, 1985. Courtesy of VAGA.

is a good example. By what means have these many and different cycles been communicated to the species, thereby permitting natural selection to select for certain rhythms and create the biological clocks?

There are many means. Taped bird songs can entrain circadian motor activities in the house sparrow, suggesting that some time in the distant past birdsongs and circadian rhythms evolved together. Songs can be pulled into synchrony by sound signals each time two or more people sing together. Temperature cycles can produce entrainments in plants and lizards. Menstrual periods were pulled into involuntary synchrony in a test that used olfactory communication.[5] One may speculate that olfactory signals were important in the primordial eons when our female ancestors menstruated at the same time. Similar involuntary synchrony was produced among college women sharing a dorm: they talked about their lives for extended periods of time, using the most advanced means of communication, that of human language. But the most universal means for synchronizing biological clocks with the environment, as I stressed earlier, is light.

Rhythmic patterns in disease are often due to the cyclic behavior of the infecting organism. Malaria is a good example. When it manifests a 48-hour rhythm, it is called tertian malaria (coming every third day), when it displays a 72-hour rhythm, it is called quartan malaria. The periodic chills and fevers of the disease are caused by the life cycle of a single-celled parasite in the blood, called the plasmodium, that reproduces every third or every fourth day.

The administration of drugs in medicine is often designed to correspond to the underlying cyclicity of the disease. A good example is the use of hormonal preparations called corticosteroids that are used to compensate for deficiencies of natural hormones in disorders such as Addison's disease, a malfunction of the adrenal cortex. In administering the drug, physicians try to simulate the pattern of hormonal release that would take place normally, which means that a larger quantity is given in the morning.

The recognition that the patients' responses to medication can and often do vary cyclically gave rise to the field of chronopharmacology. Its task is the optimization of the desired effects and reduction of the undesired ones

---

5   Underarm sweat, collected from a woman serving as the mistress clock, was regularly applied to the lips of the other women.

by proper timing of medication. The taking of sleeping potions in the evening, diuretics a week before the onset of menstrual flow, or anti-histamines according to a person's circadian sensitivity are three examples. Figure 24 shows a traditional form of the cyclic administration of medication, represented by the labels placed on medicine bottles used to guide non-English-speaking Navajo Indians.

We have learned about many biological cycles that manifest themselves in the behavior of organisms. There is yet another group of organic clocks whose frequencies need not correspond to any external cycle. These are the internal household rhythms necessary to maintain life. When all these rhythms are included, the spectrum of physiological clocks turns out to be very wide.

The fastest ticking clocks are the atoms and molecules of the body, such as those of the human skin that respond to ultraviolet light and change a sunning beauty into a suntanned one. They belong in our organic clock complement, for they are a part of the life process. These clocks oscillate around 10^{16} Hz, making the length of a single cycle 10^{-16} sec.[6] Working our way toward lower frequencies, retinal cells respond to light that oscil-lates between 10^{15} and 10^{14} Hz. Chemical reactions involved in photo-synthesis may be measured in picoseconds (10^{-12} second). In the metabolic domain, the periods of oscillations stretch from a few seconds to a few minutes. The fastest growing bacteria reproduce every ten minutes, that is, they have a life cycle of 600 seconds. Cultured mammalian cells divide at rates between half a day and a day, therefore their periods are between 10^4 and 10^5 seconds. The period of lunar clocks is around 10^6 seconds, those of annual clocks close to 10^7 seconds. There are bamboos that flower every 7 years and cicadas that emerge every 13 or 17 years. These long periods are of the order of 10^8 seconds (fig. 25).

The ratio of the fastest to the slowest oscillations in biological clocks is 10^{24}:1, an immense range. Musicians call a frequency range of 2:1 an octave. The instruments of the living orchestra extend across 78 octaves.

Life comprises biological rhythms of cycles upon cycles, a biomusic by nature's own Muzak that cannot be turned off without harm to life.

---

6   Within the length of a second there are 10^{16} cycles, or 10,000,000,000,000,000 waves. It follows that the length of each cycle is 1/10,000,000,000,000,000 second or 0.000,000,000,000,000,1 second, which, in the mathematical shorthand notation introduced earlier, is 10^{-16} seconds.

FIGURE 24 EVERYDAY CHRONOPHARMACOLOGY. The top label says, "At sunrise (top) two pills (center band) for the young woman (bottom)." The bottom label says "Two teaspoonful for the baby at night." The labels are used for Navajo Reservation Indians. Courtesy, Indian Health Service, Health Resources and Services Administration, Rockville, Maryland.

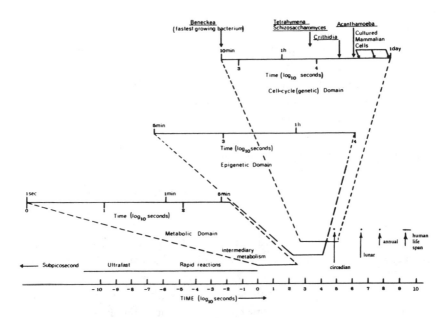

FIGURE 25   THE SPECTRUM OF BIOLOGICAL CYCLES. The horizontal axis orders the lengths of the periods; the numbers along it are the exponents of 10. (For example, −3 stands for 10^{-3}, which is 0.001 second; 0 stands for 10^0, which is 1 second; and 5 stands for 10^5 or 100,000 seconds.) The lowest horizontal plot represents the domain of oscillations from light reception (subpicoseconds) to rapid reactions at around 1 second. The horizontal line above it stands for the domain of metabolic reactions. The same line is also shown magnified between 1 second and 5.5 minutes. The third line up (the epigenetic domain) is the range of periods needed for the development of entirely new biological structures from undifferentiated ones during embryonic growth. The genetic domain is that of cellular oscillations. The circadian, circalunar, and circannual rhythms are shown by upward-pointing arrows. The ratio between the fastest and slowest biological cycles is 10^{24}, a range that a musician would describe as 78 octaves.

Adapted from Leland N. Edmund, Jr., "Chronobiology at the cellular and molecular level: models and mechanisms for circadian timekeeping," *American Journal of Anatomy* 168 (1983): 391.

The organic present

For an organism to remain alive, it is necessary that the multitudes of its inner clocks be kept cycling according to their intricate and complex demands of mutual dependence. Biochemical events that should happen simultaneously must, and those that should not ought not, or else the organism will come to harm.

The instant–by–instant synchronization that assures the necessary collective viability is the *organic present* of all life forms. The maintenance of its integrity is a necessary and sufficient condition for life. When it is not properly maintained, illness results; a total internal lack of coordination leads to death.

To say that a living organism is an orchestra with trillions of instruments that are kept playing in a coordinated fashion from instant to instant is, of course, a metaphor. But it is a useful and appropriate one, because it reminds us of the organic solidarity that constitutes the life process.

Jet lag is an example of a mild interference with the integrity of the organic present. Imagine someone flying from Louisville, Kentucky, to London, England. Some of her biological cycles will reset themselves to the five hours' difference quite rapidly; with a bit of skill she can adjust to the sleep–wake cycle in three days. But there are metabolic processes that take much longer to get in phase with the new environment. For a while, therefore, her internal clocks will be out of phase with respect to each other, making her experience a washed-out feeling, an effect of internal desynchronization.

In some cases, such as in juvenile insulin-dependent diabetes, the organic present must be maintained by conscious action, substituting for biological coordination. A healthy body takes care of the production of the hormone insulin, and gets it to where and when it is needed, by means of a timed delivery process of which we are normally unaware. When this process does not work, a person must inject the right amount of insulin one or more times a day. It must be done at the right time, in a continuous balancing act between a fast slip into life-threatening hypoglycemia and the coma of hyperglycemia. The mind must take over the responsibility for maintaining the instant-by-instant coordination among the biological clocks involved.

Instead of the rhythm of insulin production, consider the autonomous oscillations of the eye. The human eye has a high-frequency, small amplitude, irregular movement with a rate of between 30 and 70 per second. Also, about once a second, the gaze is transferred slightly from one point to another so as to allow the contours of an image to fall upon a new set of rods and cones. This motion, called saccadic movement or flick, permits the just-used rods and cones to regenerate. The eye also has slow, irregular drifts.

What makes the eyes look alive is their many, coordinated oscillations of which we are not consciously aware, but whose absence we notice. As death approaches, a person's eyes stop their complex dance, which accounts for the stare of the dying, as it is sometimes said, into eternity. Since gazing in one fixed direction overloads the retina and is known to make images disappear, it is likely that images of the world may vanish from the sight of the dying before self-awareness does. The end of the eyes' tremor signifies the approaching end of the organic present.

In the sections on time measurement I related how timepieces developed from machines that modeled the motions of the sun, the moon, the planets, and the stars: Plato's moving images of eternity, our heavenly environment. An improved clock meant a device that imitated the regularity of stellar or planetary movements more closely than the devices before it. To achieve the desired improvements it was always necessary to add more gears and/or some other parts that moved or controlled motion. In their turn, the new parts always necessitated the rearrangement of the insides of the devices. Clocks became increasingly more complicated. The process remained open ended because each improvement only revealed that what was earlier thought to have been a uniform process was, in fact, not uniform. Each discovery called for the creation of even more complex devices and for a program of testing them against all other clocks, including those of the sky. Each clock was, and each new clock remains, an imperfect model of the cyclic order of nature.

The history of biological clock making through evolution by natural selection followed a similar path. It began by copying astronomical and other environmental periods, then, having found the new model wanting, kept on refining and retuning it by adding more and more sophisticated parts. Organic evolution perpetuated the organic present by making physiological clocks of increasingly greater complexity.

The literature of biological clocks regularly asserts that the reason they evolved was to help the organism survive. This is as inadequate a view as if

I claimed that the musicians of an orchestra help the orchestra make music. Musicians do not help the orchestra, they *are* the orchestra. Likewise, biological clocks do not help a living organism survive, they *are* the organism.

The organic present signifies life's need to maintain *simultaneities of necessity* through uninterrupted internal coordination. Necessity, that is, for the survival of the organism. This need sets life apart from nonliving matter. The physical world can manifest only *simultaneities of chance*. These are the kind of simultaneities that relativity theory speaks of, as we shall see. Certain events in the physical world might happen at the same place and at the same time but if they do not, no autonomy would be lost, because nonliving structures have no autonomy to be maintained through internal synchronization. A physical system never needs anything, because it cannot have a self-directed purpose or plan.

Summing it up: the creation and the maintenance of the integrity of the organic present is a necessary and sufficient condition of life. It is also the phenomenon through which future and past—that is, the flow of time—are introduced into the flowless time of the physical world.

Biotemporality: The Aging Order of Life

As living things, we belong not only with the cyclic order of life but also with its aging order. We not only oscillate so many glorious ways but also age, gloriously or otherwise.

The idea of time's passage, so I reasoned earlier, had its roots in the dawning realization by our distant ancestors that aging and death were the inevitable destiny of creatures of their kind. Life, aging, death, and time have formed a closely related constellation of ideas ever since our species established its identity as a group of persons who could use their memory to prepare for future contingencies. Yet neither aging nor death by aging are necessary corollaries of the process of life.

Those life forms that replicate asexually are potentially immortal. If a mother cell splits, we have two or more daughter cells but no body is left to be buried, because nothing died.

This is the case in unicellular organisms that replicate by fission, such as protozoans and bacteria, and in multicellular invertebrates that propagate by budding, such as sponges and yeasts, or by fragmentation, such as jellyfish. These organisms can and do die from innumerable causes, but if kept in the right environment, they show no senescence and can replicate indefinitely.

Death by aging was an evolutionary development that took place long after life began, and was a necessary corollary to sexual reproduction. This section explores that fateful evolutionary event, important to the study of time, because three billion years after it happened it begat the constellation of issues that correlate, in our minds, the idea of time, death by aging, and issues of sexual behavior. To be able to conduct this exploration, we begin with a refresher course on organic evolution by natural selection.

It is estimated that there are between 3 and 4 million species alive today and that this is only a small portion of the species that have ever existed. How did this great variety of living forms come about?

The machinery responsible for the origin of species was first recognized by Charles Darwin a century and a half ago. His ideas were patterned after the ancient know-how of selective breeding, practiced in animal husbandry from prehistoric times. But unlike in artificial breeding, where people do the selecting for specific purposes, the selecting of preferred individuals and groups in natural selection is done by the changing fortunes of the environment, for no evident purpose.

The direction of evolutionary change is determined by the competition for food and mate, the presence of predators, geological changes, and vagaries of the climate. But before these and similar forces may become effective in selection, two conditions must be fulfilled. First, there must be sufficient variation among members of a species so that those better suited for the altered conditions may preferentially survive and reproduce. Second, the particular advantages for survival that these individuals have must be genetically based so that they may be passed on to the offspring.

Organic evolution is a myriad-sided process. It involves replication and reproduction, with all their related genetic machineries. It involves ontogeny (the development of the individual from the fertilized egg) and phylogeny (the evolutionary development of the group). It involves interaction among individuals and among the species. As is the case in our life experience, each improvement achieved by a species appears as deterioration of the environment for another, competing species that shares the same ecological niche. If changes in the biology of one species make it possible for its members to catch more fish, there will be less fish left for another species on the same diet. This aspect of natural selection will be of special interest later when discussing the time-compact globe.

The process of natural selection works on the individual members of the species, but it is the species that evolves. The individual animal has no power to change his biological makeup. A short-haired dog cannot grow long hair to keep himself warm if the climate becomes colder. Rather, the colder climate will select, for preferential survival, those dogs that because of the inaccuracies and randomness of genetic reproduction have longer hair.

Organic evolution by natural selection has been creative in the strongest meaning of the word: it has brought about an untold number of new living forms and processes, unpredictable from those forms and processes that existed before the new organisms in fact appeared.

Organic evolution has no plan—its steps are unpredictable even in principle—but it does tend toward increasing complexity. It has no goal, for there is no such thing as a perfect, final form of life. Neither do new life forms create more peace on earth, as it were. Each evolutionary change that has come about in response to some unbalance led to new unbalances, which then served as new selection pressures. Along such a tortuous and creative path has the kingdom of life become larger, more sophisticated, more complex.

We are now ready to inquire into the evolutionary origins of death by aging, that annoying detail which has given rise to and remained central to the human sense of time.

All living organisms used to be thought of as belonging to one of the two great kingdoms: animals and plants. But taxonomists (specialists in classifying species) have come to recognize sufficient differences among groups of species to justify dividing life into five kingdoms.

Our interest here is in the kingdom of Monera, which encompasses bacteria, blue-green algae, and viruses. The oldest known fossilized life is bacteria dating back to two and a half billion years ago. Even at that early age they were much more complex than one might imagine the very earliest crystal-like life forms to have been. There is no such thing as a simple life form except by comparison with more advanced forms. Today some 20,000 species of bacteria are known; blue-green algae have about 1,500 named species.

Members of the kingdom Monera reproduce asexually: they split into two and the two daughter cells continue living, as good as new. As I noted earlier, there are "no bodies left to be buried." All descendants of an asexually reproducing individual are identical with their parent; further offsprings remain so from generation to generation unless mutations occur. But if a species is to evolve, it must produce different individuals. The only way asexual species can do so is by random mutation. When a mutation does occur, the old and new cells become competitors. If the mutation was unfavorable, as chance mutations usually are, the new forms are likely to die out. If it was favorable, then the old form either vanishes or decreases substantially in number compared with the new form.

The speed at which a species evolves depends upon the rates at which favorable mutations can spread and get established, and unfavorable ones spread and be eliminated. In asexual reproduction, a mutation spreads from the original individual to her offspring; if favorable, it may become dominant after many generations, while the nonmutated forms are eliminated.

How does this work out for two mutations? How may they spread through an asexually reproducing species? One way would be for the two to occur simultaneously in the same individual. But since probabilities multiply when joined, the probability of two mutations occurring in the same individual at the same time is very much less than that for the two mutations to occur in two separate individuals. Another way would be for the second mutation to arise in a direct descendant of the first mutation.

Primordial life is known to have reproduced asexually. But somewhere along evolutionary history, there appeared certain unicellular organisms that reproduced in ways quite opposite to the asexual method. Whereas asexual reproduction involves the splitting of one cell into two, sexual reproduction involves the fusion of two cells into one. Sexual reproduction, therefore, demands special conditions: the simultaneous presence of two cells. Why should such a more demanding system evolve? The answer must lie in whatever advantages the sexual life cycle has over the asexual one.

In sexual reproduction different mutations present in two separate individuals can combine in each of their offsprings. Mutations can spread much more rapidly than in the asexual mode because there is no need to wait for the one-after-another sequence. If the combination is un-favorable, it will die out faster; if it is favorable, it will get established more rapidly. Calculations have shown that for a population of one billion, sexual reproduction is almost six times as rapid in spreading mutations as asexual reproduction. As the population increases, the relative speed goes up rapidly.

Many evolutionary biologists believe that sexual reproduction became established because it was able to produce adaptive changes faster than asexual reproduction could. Sexual species outran their asexual competitors along the road of increasing complexity. It seems that ever since natural selection first commenced, it favored time economy: the faster actor had the stage.

Also, sexual pairing from generation to generation was able to create unique individuals. Each sexually produced organism is a biologically distinct system with a mixed bag of genetic endowments. Upon the different individuals natural selection could advantageously work and create, in due course, the great variety of life forms that populate the earth.

The scheme I presented, though broadly accepted, is not without its problems because there are also some disadvantages to sexual reproduction. Far too little is known of the actual origins of sex. That it involves the swallowing of the sperm by the egg has suggested to some biologists that its origins might have been in the advantages conferred upon a cell whenever it ate the right kind of other cell. We will leave the debate over the origins of sexual reproduction on this appetizing note.

The task of reproduction was relegated to the germ cells that replicate themselves asexually and therefore remain potentially immortal. The task of carrying and caring for the germ cells was delegated to the soma, a system of much greater complexity and flexibility than the germ cell. But somatic cells age, and when they become incapable of maintaining their collective, organic present any longer, the soma dies.

Death by aging and sexual reproduction have thus come about as complementary aspects of the same crucial evolutionary step. It was an ingenious division of life's labor into the unaging germ cell and the aging, mortal soma. It was the birth of the aging order of life.

Time, life, aging, and death have formed a single constellation of thought in the human mind ever since time and the inevitability of death were jointly discovered. Yet, as we have seen, death by aging is not a true corollary to time because not all living things die by aging. Time relates to life not via aging and death but via the organic present, which is a necessary corollary of all forms of life.

None of this makes aging of lesser importance to a person. Members of our species, as members of all sexual species, have an upper limit to their lives that may be approached but not surpassed; within that span, aging is a disease, a necessity for letting biological novelties have their way with the species. Meanwhile, each person marches along the seven ages of man in Shakespeare's *As You Like It,* approaching the

> Last scene of all,
> That ends this strange eventful history,
> Is second childishness and mere oblivion,
> Sans teeth, sans eyes, sans taste, sans everything.

Before reaching mere oblivion, most men and women have reproduced. They have created other humans who, at the time of their conception, looked rather different from what they look like at the age of thirty-two. Their sizes at the time of their conception were immense in terms of molecular dimensions. So were their weights—about 10^{-13} gram—which is some 10^{10} times the weight of a water molecule.

Our epoch is so deeply immersed in science that the reader has surely no difficulty in understanding what was said about our original sizes and weights. But it is difficult to imagine that once upon a time the person you are now was a wiggling, conservative molecule. In that world the passage of time made little sense. As the zygote grew into the embryo, fetus, infant, toddler, and young man or woman, the reality of time changed its character.

Since we are members of a sexually reproducing species and therefore each of us ages, we begin serving our death sentence the very instant we begin our life sentence. But for psychological and cultural reasons, living and dying are kept by our minds as separate and opposite categories of the experience of passing time. But the common origins of sex, and of death by aging, survives among our feelings in a hazy, primitive fashion and is ready to haunt the mind.

Eons after it happened, in the mere yesterday of history, transmuted by the cultural labor of societies, those feelings created the intimate bonds among the themes of age, death, love, and noetic time, whose uncountable combinations in kind and intensity form the fabric of all civilized concerns.

The Origins of Life and of Biotemporality

Hans Christian Andersen's Ugly Duckling was snapped at by the ducks, pecked by the hens, and spurned by the henwives because she was different from them. One day she saw a dignity of swans and began to swim toward them for she knew she belonged with them.[7] She realized that it did not matter in the least "having been born in a duckyard, if only you come out of a swan's egg." It does not matter in what yard a person was born. If—and only if—she came out of a human egg, she will be able to experience and have an idea about noetic time. But human eggs did not always exist. Even life did not always exist. How did it come about?

The subject of this section is biogenesis, the coming about of living systems from nonliving matter. In the study of time biogenesis signifies the emergence of the now—the organic present—from the presentless world of physics.

Some 3.5 billion years ago, as perceived and measured by us clock watchers, self-organizing systems characterized by a know-how for defining an organic present came into being. Biogenesis created a new kind of time, one that was more advanced, more evolved, than the time of the physical world: expectation and memory created new categories of time in terms of the organisms' self-interests. Nothing in the physical world ever remembers anything, not even a memory circuit does. Nothing in the physical world ever expects anything, not even an alarm system. Only for living organisms do future, past, and present constitute reality.

Our most distant ancestors

What is to be meant by the birth or emergence of life?

That new little humans may be created through the cooperation of a man and a woman is demonstrated each time a child is born. But this is only an example of sexual reproduction and the development of the zygote; it is not biogenesis. The same goes for test-tube babies. They are made by using the accumulated know-how of living bodies from which the egg and the sperm were stolen. We would be somewhat closer to replicating biogenesis if scientists could construct, from chemicals on the shelf, the fertilized egg of a rabbit so that the grown rabbit would be indistinguishable from the one that ate our lettuce in the garden last night.

---

7   Dignity is to swans as pod is to seals and pride to lions.

The rabbit makers would make the headlines, but they would only have copied what organic evolution had already done.

Perhaps someone could produce the fertilized egg of a Jubjub bird. After the bird had grown to full size, its maker would have to convince other biologists that his creature is not a Disney World Audio-Animatronix figure but a member of a new species of birds with all the main features of the class Aves: warm-blooded, feathered, winged, and able to sing the Jubjub bird song. The bird maker would still only have modeled the principles of life from living organisms created by organic evolution.

The examples could be multiplied because by the creation of life people always meant the coming about of structures that behaved the way living things were supposed to behave, as understood at that place and in that epoch. Whenever that understanding changed, so did the specifications of what the creation of life must create.

In our epoch, the secret of life is preferentially understood through biochemistry and sought in the nature of certain large molecules. Accordingly, research on biogenesis has been directed toward the identification of conditions under which certain molecules could have spontaneously gathered into self-replicating molecular systems that could mutate and interact with the environment.

Since in all cells that task is performed by the nuclear genetic material, it is to the presumed history of those molecules that we now turn.

Once in his or her lifetime every man and woman resided in the world of those asexually reproducing giant molecules. He or she was a combination of DNA strands containing a mixed bag of genes. But my lover and I were never bare genes. We were even then surrounded by a cell that contained proteins, left over from prior life cycles.

As imagination is passed backward along the remarkable career of organic evolution, it sees living systems that, on the average, get smaller. At very great distances they may resemble viruses, in that they were no more than replicating units of molecular assemblies or almost bare genes. But even then, they were already of very great complexity compared with any nonliving structure. There the trail gets lost.

That we do not now witness life coming about from no life is not an argument against life having come about from no life. Certain historic configurations happen only once and never repeat themselves. Also, systems between the living and nonliving forms of matter are believed to be metastable, like a cone standing on its point. The transient systems, whatever they were, had to evolve into stable living systems very rapidly (on the evolutionary time scale) or else they would have perished.

We will later learn of other examples of evolutionary metastability and will note that they are always accompanied by the emergence of a new kind of temporality. With biogenesis came biotemporality; nootemporality emerged with man; and, in our own age, we face a crisis of the emerging sociotemporality of the time-compact globe—to be encountered later in the book.

Darwin was aware of the difficulties implicit in the disappearing act played by biogenesis. He remarked in 1871 that if

we could conceive in some warm little pond, with all sorts of ammonia and phosphic salts, light, heat, electricity, etc., present, that a protein compound was chemically formed ready to undergo still more complete changes, at the present day such matter would be instantly devoured or absorbed, which would not have been the case before living creatures were formed.[8]

The environment of the earth during the ages when life was born was not user friendly for people. The days were 18 hours short or shorter. The atmosphere was mostly hydrogen. The small amount of oxygen, perhaps 0.07 percent of the atmosphere, existed only in bound forms such as in carbon dioxide. The first organisms, therefore, had to be anaerobic (not in need of oxygen), as are some prokaryotes of our own days. The 20 percent oxygen of our present atmosphere is the accumulated product of plant life of the last two billion years, during which plants have been absorbing carbon dioxide and exhaling oxygen.

Ultraviolet radiation, deadly to all present forms of life, then bathed the surface of the earth. Today the layer of ozone in the atmosphere, located between 20 to 50 km high, filters out ultraviolet radiation, but the early earth had no ozone layer because there was no free oxygen from which it could be made. Natural radioactivity was much stronger than it is today. There was continuous thunder and lightning as in a Frankenstein movie; the ground shook, boulders rolled and cracked, the earth vibrated at many frequencies. There were electric discharges other than lightning; there

---

8   *The Life and Letters of Charles Darwin,* ed. Francis Darwin (London: Murray, 1887), vol. 3, p. 18.

were shock waves and large, rapid variations of temperature. Such an environment would kill most organisms alive today, but in that epoch, it made possible the coming about of life in its primordial form.

Similarly rough conditions would appear friendly to certain forms of life even today. Recently in the depths of the Pacific Ocean bacteria were found living next to hot volcanic springs where water temperatures rose to 300°C (572°F). In experiments that created conditions corresponding to those that prevail beneath 2,500 m of water and at a temperature of 250°C (482°F), vent bacteria became more productive than they were at lower temperatures. These bacteria convert inorganic chemicals into forms of energy useful for them, which is their way of feeding themselves and enjoying life. To each his own, they say, such as the blue-green algae that live in the Dry Valley lakes of the Antarctic, where winter temperatures drop to −60°C (−75°F). These living fossils, useful oxymorons, that thrive under inhospitable conditions represent the prokaryotic form of life dominant on earth until about 600 million years ago.

Geographically closer to home, algae, yeasts, bacteria, and fungi have been reproducing in the contaminated vessels of the Three Mile Island nuclear power plant so rapidly that they are impairing efforts to clean up the stricken plant. They thrive in an environment that would be deadly to all higher forms of life.

Life is a hardy process, constituting one of the stable integrative levels of nature. It did not need to originate under conditions appropriate to current life forms and their chemistry.

In trying to identify the origins of life, scientists have been following Darwin down to his "warm little pond." It is better known as the primordial soup, full of prebiotic compounds ready to become living molecular aggregates. The literature of biogenesis is extensive and concerned professional organizations, such as the Society for the Study of the Origins of Life, meet regularly.

Recently a Scottish biochemist, A. G. Cairns-Smith, pointed out that to assume the biochemistry of the newly born life as having been the same as that of today's life amounts to assuming a "high-tech" biogenesis. It is true that even the oldest known fossilized bacteria were high-tech. But eons earlier, the birth of life itself had to involve much simpler, low-tech structures. For such low-tech structures, so he reasoned in detail, we ought to look for the presence on the biogenetic earth of certain crystalline solids, certain minerals, kept wet.

The idea that the earliest forms of life were crystal-like is not in itself new. It has been known that between the realms of the living and the non-living, crystals represent the highest degree of stable organization. Inorganic matter is not able to create more ordered, stable systems than those found in crystals. Organic matter can and always does.

Cairns-Smith, a specialist in clay chemistry, postulated that early life inherited at its core a solid-state crystalline structure but replaced its chemistry with what later became the DNA–RNA–protein system of "modern" life. Two and a half billion years modern. During the last fifteen years, he and NASA scientists at the Ames Research Center in Moffet Field, California, have carried these ideas further and, through exploratory tests of the principles involved, began to paint an increasingly plausible picture of life's crystalline origins.

Along the shores of primordial seas, clays were plentiful. Energy from any number of sources was available to feed the molecules and make the molecular clockshops prosper. There was radioactive decay, electrical discharge, and shock waves. By "feeding" is meant the absorbing of noise, radioactive radiation, and electricity. By prospering is meant the maintenance of the unit's collective coherence manifest, for instance, by cyclic expansion and contraction with adaptive advantages.

If the diet I described appears unappetizing, think of the prokaryotes of today, such as bacteria and blue-green algae. Food for these organisms is a functional designation without any human meaning. Some of them feed on sulphur (they are surely the flowers of brimstoned hell), others on iron (it's good for the blood), again others on inorganic compounds (you've got to eat a pailful of dirt before you die). For dessert, they take carbonate of ammonia and thrive on it.

We may now start imagining the landscape of the biogenetic earth. The miniature clockshops would be lodged in the cracks of clays or in some other matrix both suitable and available. In the Precambrian molecular Eden, the protogenes absorbed energy from some of the many sources available, stored it, and then used it to fuel their collective oscillations. These crystals could reproduce by splitting and growing again, which reminds one of the asexual replication of cells by splitting. Since they surely did not replicate without faults, they had to be different from individual to individual. The differences allowed the environment to begin the process of natural selection.

Let us assume that during the Precambrian Era, some three and a half billion years ago, certain tiny clockshops, certain molecular aggregates of matter, have come about and were able to model some of their environmental cycles. Perhaps they cyclically changed their shapes as if breathing. Perhaps they varied their chemical composition or their ability to absorb energy in a fashion that was useful for maintaining themselves as semi-autonomous, organized units.

By copying some of the environmental rhythms, these groups could acquire advantages over other, similar molecular aggregates that could not retain their functional and structural integrities in the face of environmental change: having a collective beat became an advantage.

The miniature clockshops had built into them the need for becoming increasingly complex for reasons similar to the reasons why people construct increasingly more complicated clocks. To wit, for the better modeling of environmental cycles. Imagine, for instance, that some of them were able to internalize the tidal rhythm (see fig. 22). The struggle for available energy—for sunlight, for food from the sea water—had already begun. Those systems that could anticipate the flow and ebb thereby acquired certain privileges of foresight. Those among the privileged who could anticipate even the semidiurnal inequality—or whatever corresponded to it three billion years ago—gained yet added advantages. Foreknowledge meant a better chance for survival.

The successive steps that mapped various component cycles into the systems were improvements, but none represented an arrival at perfection, for a more accurate system could only reveal the need for yet more refined tuning. For instance, after the semidiurnal inequality of the tides, first their fortnightly variation and then their monthly changes came into sight. In the chronic inadequacy that was sustained by the process of improvements I see the sources of tension characteristic of all forms of life, but totally absent in nonliving matter.

Each added step in refinement, each step in evolutionary adaptation, made the system more complicated. In the course of time, many different clocks evolved with many and necessarily different morphologies (structural forms). A biological clock that could oscillate a million times a second had to have very different machinery from one that oscillated 78 times a minute (as the heartbeat) or once a year (as the hibernation clock). The increased complexity also demanded increasingly refined inner control systems, so that the integrity of the organic present could be maintained. As a consequence, a biological division of labor evolved. Increasing complexity and a finely tuned division of labor became, and remained, corollary fundamental features of life.

It is easy to imagine a biochemistry different from ours, such as one using silicon instead of carbon. But it is impossible to think of life without the need for maintaining the organic present and employing the skills of foreknowledge and memory.

Exploring the mini-giant clockshop of the DNA

Perhaps our most distant ancestors were molecular noodles floating in a primordial broth, perhaps crystals rhythmically shivering in wet clay. Regardless of those origins, in all known cells the carriers from generation to generation of the know-how that is necessary to maintain the organic present are the molecules of deoxyribonucleic acid, or DNA. Let us see what a DNA molecule is like and how it helps perpetuate the organic present.

Inside the nucleus of each cell there are threadlike organelles called chromosomes. All somatic cells (that is, all cells other than germ cells) have a duplicate set of chromosomes, whereas germ cells have only a single set; they are haploids, from the Greek *haploeides,* "single." Each human somatic cell has 46 chromosomes, those of a dog 78, a radish 18, and bluegrass from 56 to 249, depending on species. Kentucky bluegrass has 69. There is no direct relation between the number of chromosomes and the sophistication of the mature organism. The instructions that help build the elementary chemical components of life into structures, each with specific functions, come from a large molecule deep in the chromosome. It is the DNA molecule, with its double helix.

What does this famous double helix weigh?

Molecular weights could be expressed in tons, stones, grams, or even carats, but the numbers so obtained would be inconveniently small. Therefore, molecular weights are measured in more convenient units, called daltons. One dalton is $1/12$ of the weight of a carbon atom. An oxygen atom weighs 16 daltons, a water molecule 18 daltons, and a mercury atom 200 daltons. Then things get heavy. An insulin molecule weighs 5,700 daltons, hemoglobin 65,000 daltons, and polystyrene 50,000,000 daltons. The weight of the DNA molecule depends on the manner in which it was obtained, but 10^9 daltons is a representative figure.

What dimensions are we talking about?

When stood upright, a molecule of human DNA fills a cylinder 2 nanometers (nm) across (1 nm = 10^{-9} m). The large-component molecules join in base pairs stacked flat upon each other at intervals of 0.34 nm. Each pair is rotated 36° with respect to the one below it, making the staircase rise 3.4 nm at each full turn. The total length of the molecule is an incredible 2.7 meters. With 23 pairs of chromosomes, this makes the average DNA length per chromosome almost 6 cm.

The molecular structure of the DNA is the logo of modern biology. The double helix, with its slowly rising backbones of phosphate and sugar molecules, and with its interlocking pairs of bases serving as the steps of the spiral staircase, is unmistakable.

Models of the DNA molecule can be seen as computer-generated images in scientific journals and as structures displayed in museums. The one at the Boston Museum of Science is made of metal rods and wooden balls, painted in bright colors. It is about a meter wide. Had it been built to full size in length, it would stretch a million kilometers straight up.

Everyone knows that the DNA is not made of wooden balls or colored dots printed on paper and that models can show only a small part of the whole molecule. But these models still leave the viewer with a misleading impression. Namely, all wood-and-wire structures and printed pictures are necessarily static. Not so the real thing.

The DNA molecule is a dynamic system of millions of atoms. It continuously wiggles, vibrates, and oscillates, moving as if it were breathing. Its rates of vibrations span the electromagnetic spectrum from radio waves to infrared. Groups of oscillating patterns, quantized vibrations called phonons, wander around the molecule looking for a place to settle but they never do, as long as the DNA is an integral, functional unit. If the DNA falls apart, for whatever reason, the phonons vanish like water waves do when the water vanishes. Because of a certain kinship between crystals and the DNA molecule—increasingly, the DNA itself is being thought of as a very thin and very long solid—these vibrations were studied by formal means that come from lattice dynamics, first developed in the science of crystals.

A good way to think of the real DNA molecule is by replacing in the mind the connecting rods of the museum model with springs. Then imagine taking the balls (molecules) apart and connecting their atoms by other springs. Again, each atom is to have its nuclear structure and cloud of electrons held together by yet other springs. Then imagine giving a good shake to that zany contraption which extends from Boston to a million kilometers up, stepping back, and watching what happens.

If the springs could accurately duplicate the intermolecular, inter-atomic, and intra-atomic forces, then one could observe vibrations that travel thousands of base pairs before petering out, having given rise to other traveling waves. There would be longitudinal waves moving along each backbone–necklace, bunching and thinning rhythmically like compressional waves that travel along a line of bumper-to-bumper traffic. The twists of the staircase would be seen vibrating about an equilibrium position, all of which would make the rungs of the ladder oscillate about their horizontal positions in addition to having some vibrations of their own along the rungs themselves.

This whirring, purring, breathing, dancing system remains a single mini–giant clockshop, carrying its message only as long as its innumerable oscillations remain coherent, coordinated from instant to instant in an organic present. It contains hereditary information not by what it is, but by what it does. The information does not reside in the arrangement of the structure in space but in the song sung and the dance danced by that struc-ture. How do the song and dance work?

Imagine a tune heard as you hurry upward along the spiral staircase. It has only four tones but it is a unique composition. Next, transcribe the tune upon a messenger RNA (mRNA) molecule that serves as a "gofer" for the DNA, because the original recording must remain in place. Then let the mRNA, whistling to itself, carry the tune to the organelles called ribosomes, the sites of protein synthesis. In the next step let the ribosome move along the mRNA and, having learned the recorded tune as well as it can, begin the making of one or another of twenty amino acids. These, in their turn, having elaborated the original tune, may begin the building of proteins.

From weeds to hippos, the genetic code of all organisms alive today are the same. Since DNA has also been identified in bacteria over two billion years old, the genetic code has evidently been stable for eons. But this means only that its words (notes of music) remained the same. Its sentences (the tune) have changed greatly. The enrichment of its language through new sentences may be witnessed in the multiplicity of species, all of which evolved from earlier, simpler forms of life.

The variations among members of a species derive from random mutations and a number of combinatorial tricks, as already mentioned. It is impossible to foretell when and at what gene location the next spontaneous mutation will arise. But geneticists have shown that the rate of spontaneous mutations tends to remain more or less constant through time: so many mutations per so many DNA bases per each million years. This is a statistical constancy and not a precise one, but one good enough to serve as the ticks of a DNA clock that helps researchers determine when certain evolutionary changes have occurred.

Here are some interesting readings of the dancing DNA clock, taken in 1986. My hairy ancestor, the one with the weapon for the frumious Bandersnatch, diverged from his closest relative the chimpanzee 7 million years ago. Walking back along the family tree, their common ancestors parted company with the gorilla 9 million years ago, from the orangutan 15.5 million years ago, from the gibbon 20 million years ago, and from the Old World monkeys 30 million years ago. These figures do not exactly match the ones obtained from paleontology and used in Chapter 1, but there are no glaring discrepancies.

What kind of programs do the DNA molecules carry?

Contrary to the belief that this or that feature of the body is in the genes, these marvelous dancing things do not "Pale brows, still hands and dim hair" make. Nowhere along the copying and flow of the original tune was there anything said on how to make a cell, let alone brows, hands, or hair. The original song, greatly modified, is conveyed only as a timetable, teaching the ribosomes how to teach the amino acids about picking up components of an existing environment and in what sequence, so as to be able to manufacture proteins.

But proteins still do not carry instructions on how to make black skin that is beautiful. Only genes do. A gene is the smallest hereditary unit capable of controlling a specific function. They are located on chromosomes but should not be confused with them, because genes are operational units, whereas chromosomes are structures. If and only if a very large number of genes work together, and if and only if they find themselves in an environment with all the necessary building material, then and only then can the development of the individual begin and progress. As the very end product of such a cooperative enterprise, an elephant or a pine tree may then come about with characteristics described as genetic.

The DNA was able to initiate the process and impart its message to the

right fellow-molecules because its dance followed an intricate and necessary choreography and because it sang a prescribed and very long tune.
The song and the dance both originated over three billion years ago, when
biotemporality was born.

The origins of life, as interpreted in this section, should be sought in the
coming about of systems that could use available forms of energy to
produce cyclic behavior, which was advantageous for the maintenance of
their autonomy. Any such system is chronically incomplete because each
adaptive step only makes new and more refined improvements necessary.
From a different perspective, organic evolution increases the variety of
potential selection pressures, which then make possible the coming about
of increasingly more complex forms of life. What probably began as a
single circadian oscillator evolved into the uncountable variety of
biological clocks whose frequencies today extend across a band of 24
orders of magnitude. The broadening spectrum of physiological clocks
also amounted to the development of different morphologies, in an increasing division of life's labor.

   The life process created the possibility of self-referential needs of
organisms, both in their collective and individual dimensions. Goal-
directedness, in response to need, gave a definition to futurity with
reference to the organic present. Memory, in its rigid instinctual form and,
in the most advanced species in flexible forms, gave a definition to pastness
also with reference to the organic present. It is thus that with life, time's
arrow was born.

   Stated in the language of physics (to be introduced in Chapter 4) it is
thus that biogenesis broke the eotemporal symmetry of the physical
world.

   In the course of organic evolution, sometime between 1.6 and 0.5
million years B.P. nature took a quantum jump along its path of increasing
complexity. Out of the living matrix of life, a new species arose with a
new kind of temporal reality, a new species-specific universe or umwelt:
that of the noetic.

Nootemporality: The Brain's Way of Minding the Body

The subject of this section is what has sometimes been called mental time,
sometimes human time. In the study of time it is called nootemporality

and is identified with the umwelt, the integrative level of the mind.[9]

If someone were to prepare an inventory of all the differences between a human and one of the higher primates in terms of their chemical compositions, anatomies, and behavior, the list would contain a large domain of residual, differential qualities that had ill-defined boundaries. On balance, these qualities would favor the survival of humans and their societies to such a great degree that an essential difference between our species and even our closest animal kin would become undeniable. Most of the advantages would be identifiable as behavior that relates to having a mind.

To come to grips with the idea of the mind, we will begin by learning about the biological structure whose functions include the mental, that is, about the brain and the central nervous system. Then we will ask not what the mind is, but what it does. A convenient way to give an answer to that question is to focus on the behavior of a fictitious character, an image on the stage of the mind which, or whom, we know as the self.

With the self in the bag, or rather in our thought, we will turn to the micro- and macrostructure of the mental present. It is in the mental present that we spend the time of our lives: we are never in the future or in the past. The remarkable micro- and macrostructuring of the mental present together with selfhood made possible—and was made possible by—the development of language. It is the different languages that guide their speakers, along so many different paths, to their views on the nature of time.

The section concludes with a discussion of the experienced speed of time, an issue close to the heart of the classic question, "What is time?"

The nervous system and the brain: our organ of time sense

This subsection sums up what we need to know about the nervous system and the human brain so that we may take "the step to mind" in the following subsection.

THE NERVOUS SYSTEM

We sense light with our eyes, heat and cold with our skins, scent with our noses, and sound with our ears. Where is the organ of the sense of

---

9   The concept of umwelt was introduced early in this chapter. A definition can also be found in the Glossary.

time? It is everywhere in the body: it is our nervous system. A good way to understand its task is to have an idea how the nervous system evolved. Therefore, back to the clockshop!

The evolution of increasingly more complicated living clockshops demanded the coevolution of internal control systems: they were needed to maintain the integrity of the organic present. The earliest means available for that chore were electrical and electromagnetic forces that held large molecules together. In today's viruses, made entirely of DNA or RNA, this is still the case. The evolution of cellular and later of multicellular life demanded a different means of inner communication.

Thus the Metazoa, a subkingdom of multicellular animals whose bodies are composed of specialized cell groups, possess a nervous system. The earliest such systems were nerve nets that connected adjacent portions of the body, such as may still be found in the tiny aquatic animal called the hydra. It has no brain but there is a concentration of neurons (nerve cells) around the mouth. The earliest organisms to develop a centralized nervous system were primitive invertebrates, the flatworms. They have nerve cords, brains, and well-articulated digestive and reproductive systems.

The appearance of vertebrates—animals with backbones—was the next step. Inside the backbone ran nerve cords, parts of a complex network called the central nervous system. Its task was to pick up and transmit signals from the sensory organs and cause the appropriate responses to be sent to the muscles and other organs: Observe mouse, process info, instruct paw, catch mouse. The information-processing part determines the nature and extent of the animal's umwelt. The central nervous system also supervises the autonomic nervous system, which governs involuntary action: heartbeat, digestion, micromovements of the eye, the pulling away of the hand from something hot even before we know that we touched it.

Essential responses to token stimuli, found in primitive instinct–reflex organisms, represent the action of a central nerve cord of hundreds of thousands of neurons. An organism with a fully elaborate central nervous system that contains 10^6 to 10^8 neurons can learn to handle the particulars of its environment. In organisms with 10^9 to 10^{10} neurons, behavior ceases to be rigidly programmed, and socialization, including the communal education of the young, becomes a prolonged process. In addition to cyclic, instinctual preparations, such as those for the changing seasons, animals with that many neurons have a degree of foresight and memory. They are able to plan. Humans have about 10^{10} neurons and therefore, while rather advanced compared with the flatworm, they are still in the same league with the chimpanzee and the baboon, as far as the number of neurons in the brain is concerned.

But humans evolved a brain that in some ways is different from other brains, and, most importantly, they learned how to use it.

THE HUMAN BRAIN: BOUNDARY TO BIOLOGICAL COMPLEXITY

At present postal rates, one kind of it may be mailed from here to Louisville, Kentucky, for $3.75; the other kind, back to Connecticut for $3.16. Each package would carry in it a molecular device that, when it was still in its proper environment, could do miracles. It could make a body speak with a Radcliffe English or with a Dutch accent. It possessed factual knowledge more extensive than but not identical with the thirty volumes of the *Encyclopedia Britannica*. It could think up enough new ideas to build and ruin civilizations or colonize a planet. It was able to make a person be aware of love, hatred, and fear; display a sense of tragedy; and enjoy comedy. It was also able to assess the world in terms of distant futures and pasts.

One package would carry the brain of a human male, the other that of a female, average weights 49.5 and 43.5 ounces, respectively. The right environment for each was the body in which each had developed from a mini-giant clockshop. American and Japanese scientists are now working on the hard and software of fifth-generation computers. The brain is a fifty-millionth-generation computer whose hardware and software evolved simultaneously as a structure in space and the creator of new programs in time.

Studies of casts made of the cranial cavities of fossilized mammals show that as long as 65 million years ago, those brains began to increase in volume at higher rates than would be expected from the growth of their owners' body sizes. About 50 million years ago, fissurization began to evolve, probably as a means of increasing the surface available for the cerebral cortex. This is the surface layer of the two cerebral hemispheres (*cortex* is Latin for "bark") and is responsible for all higher brain activities, such as language, memory, and expectation. In the evolution of the genus *Homo* the first enlargements of the brain began perhaps two and a half million years ago from a size of about 400 cc, reaching 1400 cc as recently as 200,000 years ago. The size of the human brain has not changed since.

Among human adults, brain sizes differ slightly, but these variations bear no known relationship to intelligence. Neither does it have any recognizable relation to the intelligence of the higher primates. For instance, the gorilla's brain is 450 cc, about one-third the size of the human brain. Comparisons between animal and human intelligence have been made by such means as comparing degrees of intentionality and methods of cognitive communication. Whatever the common scale of comparison may be, a ratio of 3 : 1 does not do justice to the difference. The uniqueness of the human brain cannot be attributed to its size alone. It is more likely to come from a difference so well put by Andersen's Ugly Duckling, interviewed earlier: it does not matter where you hatched, as long as you came out of the right egg. In this case, the human egg. Does this mean that the human brain is genetically programmed?

The unabridged *Webster's Dictionary* has over 550,000 entries. A literate person could recognize perhaps 450,000 of them as words that could be used in English, even if he or she did not know what many of them mean. The same person could probably assign the right meaning to about 80,000 of them. In his work, Shakespeare used about 34,000 words. Scientists have calculated that to possess an inborn vocabulary of 10,000 words would require 10^{16} kg of DNA. Obviously, the brain could not be genetically programmed in its detailed functions.

What makes the brain educable? How does it do what it does? Explanations have always been given in terms appropriate to the epoch, place, preferred ways of thought, and prevailing values. Today the brain is often described as a computer; hence it is a computer model that I want to consider.

The model employs the idea of brain states and takes advantage of the fact that a brain cell, the neuron, has only two states: It is either on or off. The model assumes that each distinct configuration of neurons corresponds to a distinct brain state. One may think of each different brain state as associated with an element of thought. Just what an element of thought may be is, again, not clear, but it is necessary to begin somewhere.

The world's largest computer system is the American telephone network, with 152,000,000 telephone sets. Each of these sets may be connected to any one and any number of the other sets. At each instant each of these sets may be on or off the hook. If every set plays the on-or-off game, the system may have $2^{152,000,000}$ different configurations. Changing this number to a power of 10, we get $10^{43,700,000}$.

Instead of telephones on or off the hook, think of each neuron as being either on or off and assume that each may be connected to any one and any number of other neurons. (This is not true, but good enough for the game.) The number of different brain states of 10^{10} neurons then equals $2^{10^{10}}$ or 10^{10^9}. This shorthand stands for the numeral 1 followed by 10^9 zeroes. As self-organizing system, the brain has a potential for $10^{956,300,000}$ times as many different states as does the American telephone network.

Large numbers in themselves do not say much, because it is always possible to think of an even larger number. They acquire meaning by comparison. For instance, a cubic centimeter of water has some 10^{22} molecules. How many different configurations can that cc of water have? Let us call it infinitely many. But from the point of view of organization they are all the same: chaotic, mixed up. That small sample of water has only one state: a random one. None can claim to be distinct from the others, none has a distinct and different identity.

Not so for self-organizing systems, such as the living brain. That most couth organ maintains its identity from instant to instant. Its different states are distinguishable from each other by the meaning of thought with which, so we have assumed, they are associated.

What makes the organization of the brain so remarkable is not the large number of its component parts but the immensity of different ways in which those parts may be meaningfully interconnected into a single, functioning unit. Systems of this kind are called complex systems and are of special interest in social and economic theory and in biology. They cannot be analyzed by methods suitable for less complex structures, but only by means of statistics and hierarchy theory. There are reasons to believe that the peculiar powers of the brain reside in its complexity and that the potential, maximal degree of interconnectedness of the human brain represents the boundaries of complexity achievable by biological means.

The natural sciences have revealed a number of boundaries to the physical world. There is the absolute zero temperature, there are the smallest possible objects (particles) and the largest possible object (the universe), and there is the largest possible speed (that of light). All these boundary conditions are approachable along a continuous path from human-sized conditions, such as by making an object colder and colder or by making it move faster and faster.

But in each case, as the boundaries are approached, certain laws of nature, not noticeable for human-sized, garden-variety conditions, become evident. An example is the famous time dilation of special relativity theory, an effect present even for a person interested in knowing the rate at which a moving snail's clock changes because of its motion, but so small that it is unmeasurable. It is not surprising, then, that understanding the functions of the brain—the boundary to biological complexity—has been so difficult. The human brain still appears to be a world unto itself.

In the technical lingo of our epoch, the brain is a molecular computer. Scientists have been thinking about molecular computers independently of any interest in the brain. They write about molecular wires, molecular gates, molecular generators, and even special molecules that would be useful in their work, because to make a molecular computer one needs molecular machines.

Such brain-making molecular machines already exist. They have memories; they can read and be read out; they can print. They may be found in small packages called gametes. The memory bank is called the DNA molecule. When its double helix unwinds to make two thin and very long wiggly solids out of one, it spins at a rate of 15,000 R.P.M. or 250 R.P.S. The separation of the chain and its duplication by copying take place simultaneously. If we assume that each base pair may determine no more than a single letter of a four-letter alphabet, we obtain a copying rate of 10,000 characters per second. This is slower than the fastest printer, which prints 18,000 characters per second. But the tiny device is much lighter than the printer (it weighs only 6×10^{-12} grams), consumes practically no energy compared with its heavy and noisy competitor, and we know how to produce it. But most importantly, it has all the information and know-how for making human brains. It has been coded into it in a top-of-the-line experiment that began four billion years ago.

The molecular system that nature developed through organic evolution will remain for a while the only machine that can make anything that performs like the human brain. Meanwhile, thinking about the brain as a computer is useful, but it is no closer to the mark than the circle drawn by a compass is to the shiny, sad, or happy face of a baby.

THE HIERARCHICAL STRUCTURING OF THE HUMAN BRAIN

Unlike the body, the brain retained its evolutionary history, and each of its archaic layers assesses differently the nature of time. Let us examine the details.

In the early stages of organic evolution, rigid, programmed responses to Food, Friend, and Foe were satisfactory because the three Fs themselves behaved according to rigid, predictable patterns. With the enlarging clock complement of life and the development of increasingly refined inner controls, the behavior of the three Fs became unpredictable, to different degrees.

In the economy of nature, each individual of every species needs food, has friends and foes, and is also all of these itself. Life is a cooperative, coevolutionary enterprise. The increased unpredictability demanded increasingly flexible responses. In evolving them, all species with nervous systems began to internalize—to record in their brains—the permanent features of what their senses told them about the world. These dynamic charts of their umwelts became their external realities. Our species also made such inner records.

There is no one-to-one correspondence between what members of a species can be said to recognize as their world and what may be observed in their brains. The most beautiful or horrifying experiences correspond, in our brains, to the mere passing around of biochemical and electrical signals. Same for the crocodile. But the neural organizations responsible for crocodile and human experience are different.

Curiously, however, the difference is not exclusive but hierarchical. Reptiles have reptile brains. Humans have a human *as well as* a reptile brain because reptiles have been our evolutionary ancestors.

Mammals and birds evolved from mammal-like reptiles that populated the earth 200 to 250 million years ago. The remains of these, the therapsid reptiles, may be found on all continents, but no present-day reptile is in the lineage of that now-extinct order. They were the main reptile groups until the appearance of the dinosaurs. From the early mammals came the late mammals, including the primates; a number of primates evolved into manlike creatures; one of those evolved into our species.

Comparative neuroanatomical findings show that the human brain has retained three distinct neural assemblies that resemble—have been

inherited from—our reptilian, early mammalian, and late mammalian
forebears. Retaining ancestral structures and functions in this manner is
unique to our brain; it is not shared by the rest of the body. In the evolution
of the body, new structures generally replace the old ones. For instance,
our bones resemble closely those of our ancestors. But our leg bones are
14 inches long, theirs were 8 inches long. It is not possible to have a leg
bone that is both 14 and 8 inches long. Not so for the brain, which evolved
by growing new structures around and about the old ones. Human behavior is directed, therefore, by three brains, as it were, each according
to the behavioral patterns appropriate to those respective ancestors: the
reptilian, early mammalian, and late mammalian (fig. 26). These three
brains form a single coordinated unit within which they retain, more or
less, their archaic responsibilities.

The functions of the reptilian brain include the signaling of hunger,
pursuit of foe, and possession of mate; its responses to stimuli are rigid and
ritualized. Its temporal horizons are limited, its demands must be met
immediately or else they are judged unmet. Its umwelt is biotemporal; in
it, distant futures and pasts do not exist.

The paleomammalian brain is also concerned with the preservation
of the self and the species, but it has a greater flexibility of response than
its resident roommate and ancestor. What in figure 26 is labeled neomammalian corresponds roughly to the cerebral cortex and is responsible
for hearing, seeing, smell, touch, and control of the muscles. It contains
special nerve tissues, those of the neocortex or neopallium, whose net-
works in man are responsible for language, long-term anticipation, and
memory.

Our ancestral and newer ways of perceiving the world, including time,
survive in the human brain and are in continuous communication with
each other. The older temporal umwelts become dominant whenever the
younger modes of perception and value judgment are bypassed: they
determine the temporal settings of our dreams, they surface in mental illness, and express themselves in mob behavior.

With more room at the top of the brain, in both the physical and the
metaphorical sense, as the brain evolved, the boundaries of human reality
began to expand. A hallmark of that expansion has been the ability to
imagine things and events far in the future (if anywhere), in the past
(where they did or could have happened), and in the present (here, elsewhere, nowhere). Our inner world became filled with scenes, props,

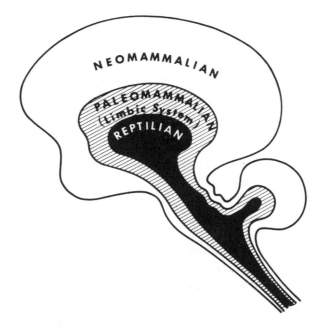

FIGURE 26 HIERARCHICAL ORGANIZATION OF THE PRIMATE BRAIN, a schematic representation. Each of the three evolutionary formations retained its basic structure, neural organization, and chemistry, corresponding to its origins; each is in the lineage of the brain of *Homo sapiens*. The three layers share the cranial cavity and are in constant communication. Each does for the human body what it would have done for its archaic owner before the later layers evolved. Each retained its peculiar way of assessing reality, including that of time.

For instance, the world of the reptile brain is the biotemporal umwelt. That brain can handle only the immediate future and past. The nootemporal umwelt is a creation of the much younger neocortex or neopallium, a type of nerve tissue that constitutes most of the neomammalian brain. The world as perceived by the neocortex possesses unlimited temporal horizons. These different realities, each with its peculiar temporality, are simultaneously present among our mental functions and simultaneously interpret our sense impressions. The older control systems can become dominant whenever the evolutionarily younger systems are short-circuited, such as in sleep, illness, danger, or feelings of elation.

From P. D. MacLean, "The brain in relation to empathy and medical education," *Journal of Nervous Mental Disorders* 144 (1967): 374–82. Courtesy of the author.

creatures, people, and events both possible and impossible. So rich and varied is that world that poets, artists, and scientists have hardly begun to tell us about it. They will never exhaust the store, because its furnishings are being continuously created. From among the infinite variety of those creations, one character stands out. It is called the self, and we shall soon meet it.

The step to mind

Do not ask what the mind is. Ask what it does, and how.

I like walking in the mountains, riding when opportunity arises, writing whether or not it is opportune, and listening to the birds in the morning. There are other people who like to do the same things. Yet no part of anyone's anatomy may be called the walk, the ride, the write, or the listen. My brain has been minding my body and my affairs for a number of years and most of the time I like minding, that is, using my mental faculties. So do many other people. Yet no part of anyone's anatomy may be properly called his mind any more than his mope, his brood, or his celebrate. Nevertheless, for historical reasons that have to do with earlier judgments about man's position in the world, each of us is said to have a mind.

What is to be understood by the mind has been answered in many ways, depending on the changing winds of social, philosophical, religious, and linguistic preferences. The idea of the mind has often been interchangeable or overlapping with the notion of the spirit or soul. The meaning of these words depends heavily on the culture. For instance, it is impossible to translate the English "mind" into any other language without giving examples of how English-speaking people use it in the formulation of their thought. Dictionary translations of "mind" always carry connotations not present in English usage and miss some that are present, because the other language also has its own assumptions about minding.[10]

The problem is that "mind" is a noun and therefore it stands for a permanent structure, as do the words "cigar," "fossil," or "hardware."

---

10   For instance, *Geist,* the most frequent translation of "mind" into German, is conceived of by that language much more in the social setting than is the English "mind." Thus, in idiomatic German, *Geist* refers to the spiritual nobility and cultural refinement of Western European life. The word is easily combined with other words to form compound words describing the furnishings of that civilization. For instance, *Geisteswissenschaften* ("sciences of the mind or spirit") is what we call the humanities.

Instead of thinking of the mind as a thing located in the brain like Georgie the Ghost in the attic, it is better to think of it in terms of the present participle, *minding,* which describes a function and not an object. Trying to locate the mind in the brain or 20,000 leagues from it is like trying to locate a melody in or around the brass of a French horn. Minding, like singing, is not a structure in space, but a process in time. Nevertheless, I will use the word "mind" because it is the common way of speaking about minding.

What did our ancestors' brains learn when they learned minding?

To be able to separate from the barage of their sensory inputs what appeared to be permanent features of the world.

To be able to construct a symbol called the self. It stood for whatever was judged permanent among the features of that body which felt hunger, anger, or satisfaction.

To be able to observe the actions of that self from within, by feeling, and from without, by imagining.

To be able to speak a human language. Only through language could those communal relationships that were necessary to construct the image of a self be established.

The humanoid features of the brain developed with enormous speed; on the scale of organic evolution, the step to mind took only an instant. The four learning processes just listed should be thought of as having taken place simultaneously, not one after the other. They were mutually reinforcing aspects of change, none imaginable without the others. Together they created a new stable integrative level of nature, that of the human mind. It was an evolutionary step in the same category as biogenesis. We saw how, from the point of view of time, the birth of life amounted to the definition of the organic present and how the organic present gave meaning to futurity and pastness. The emergence of man amounted to the definition of the mental present. The mental present alone gives meaning to distant futures and pasts, to the flow of history, and to all the other features of nootemporality that we shall encounter.

The self—the "I," the "me"—is a resident of the external world, which it shares with all other objects that are judged to have sustained identities: stray dogs, rolling stones, fig trees. But unlike these and other objects, the self can only be partially surveyed by the senses. I can walk around the fig tree if it lives near me, but I cannot walk around myself. If you want to see whether your new suit fits, you must look into one or three mirrors; it depends on the store. You cannot explore yourself by touch with the same ease you may explore a willing mate and you could not map your scent chart even if you had Fido's nostrils.

The self is also a resident of the inner world, where it shares its existence with all feelings that cannot be assigned a location in the external world, such as hopes, fears, futurity, and pastness.

In its simplest form the self, the "I," seems to be identical with the body, such as in the sentence addressed to a member of the opposite sex, "I would like to practice birth control with you." It's you and it's me, and that's that. It is equally easy to identify the self in the cry, "I don't want to die," coming from someone standing in front of a firing squad. The "I" stands for the prey; the shout is addressed to the predators.

But what if the same cry comes from someone dying of cancer? The body is hell-bent to go, therefore the "I" cannot stand for the body. What about the young and healthy Shiite Moslem bent on a suicide mission? His body is able and does its biological best to remain alive, but the "I" will not have it. In senility, the body may look very much as it did a few years earlier, but the person is "not all there." Whatever is "not there," is the person's identity.

Consider the curious idea of "my death" that has puzzled psychologists and philosophers. My body is laid out, relatives come and cry, competitors come just to make sure I am really dead, and they all sign the visitors' book. But there will be no living body around in whose brain the laid-out body could be identified with the self. Who, then, is the observer?

The self is an image, an idea so highly placed in our perception of reality that its wishes are our commands. If it says, "Live!" the person does everything to keep alive, regardless of what his body is bent on doing. If it says, "Die!" the person will want to die. The "I" has it, so ordered.

Even though one cannot see, smell, or touch one's own body with the same ease as another person can, one can hear one's self the same way that others do. This immediacy helps us understand the intimate relationship between language and selfhood or, more specifically, between name and the self. Rodgers and Hammerstein knew all about it.

> Do a deer, a female deer,
> Re a drop of golden sun,

> Mi a name I call myself,
> Fa a long, long way to run. . . .

Since the average lifetime of a cell is seven years, by the age of seventy-seven a person's body has been totally replaced eleven times. The way he looks will have changed greatly, as will have his behavior. But what he calls his self will have remained permanent. The self is a symbol and has the same purpose that all symbols do: It serves as a weapon against passing.

When speaking of postmortem existence, it is the self that is assumed to survive or not to survive. Selfhood and concern with death are complementary aspects of our command of noetic time. Even though many animals show mortal terror in the face of an immediate threat to their life, no animal ever behaves as if concerned with its later, inevitable death. It is not that animals have succeeded in overcoming the demands of their selfhood for survival after death, but that they have never had a well-enough defined identity to demand a negation of death. Animals are not wiser, having integrated life with death, but rather they have never separated the two, as humans did when they began thinking in terms of human time.

Belief in the permanence of selfhood is demonstrated by language. According to common usage, a person seldom dies a natural death. He may be killed by external forces, by illness, by accident, or by murder. She may be killed by old age, taken by God, put to death against her will; she may succumb to the fatal blows of fate, or be carried away by the Reaper. In all great religions, death is a step along a path of continued existence, a sea change from one condition here to another one elsewhere.

The hidden wealth of the mental present

This section describes the great wealth of the mental present. Normally we are quite unaware of that wealth, like people in a pleasure boat unaware of the teeming life and struggle, the topography, and the sunken galleons in the sea beneath them.

From cradle to grave, our lives are conducted in the present: we are never in the future or in the past. The time of the present feels simple; its structure is anything but simple. The "now" of our experience is made up of three hierarchically nested components. They are the *organic present* (already encountered at length), the *mental present* (subject of this sub-section), and the *social present* (a subject of the next section). Each of these components derives from the need of systems of differing complexities to maintain an instant-by-instant coordination among its peculiar functions. Coherence among the oscillations of biological clocks is necessary to maintain life; coordination among the neural functions of the brain is necessary to maintain the integrity of the mind; synchronization of social functions is necessary to maintain the viability of a society.

If the biochemical processes of the body are not coordinated, there is still matter but there can be no life; if the neural functions of the brain become incoherent, there can still be a living body but it will be incapable of doing mental work; if social coordination vanishes, there still may be some people around but they will not form a society.

Our exploration of the structure of the mental present will recognize its micro- and its macrostructure.

THE MICROSTRUCTURE OF THE MENTAL PRESENT

In everyday life, even as attention focuses on a task at hand, a large store of fears and hopes about the future, gladness and guilt about the past, and an intricate flow of sense impressions about the present mix in the mind. It is to the microstructure of present sense impression that we shall turn first. A good way to represent that microstructure is to turn once again to the cyclic order of life and ask, What happens when we hear music? Quite a lot.

The frequency band of human hearing is from about 15 Hz to 20,000 Hz. The boundaries of this range do not make us world champions: some insects are able to detect sound at 100,000 Hz. But the range of sound energy (loudness) to which human ears can respond is impressive: the difference between the squeak of a mouse still audible and the roar of the jet engine that does not yet cause permanent damage is a factor of 10^{12} (one trillion). And when it comes to the dexterity of analyzing the content of audio signals, the mind's performance is phenomenal, taking apart and reassembling each instant of time.

The top line of figure 27 shows a short portion of the sound of a single organ pipe. By means of harmonic analysis the components of that sound may be identified. What does that mean? If I were to generate the twelve sine waves of the illustration, with relative amplitudes and phases as shown, then listen to them at the same time, I would hear the original organ sound. This is just what music synthesizers do: they construct complex sounds from their sine wave components. A harmonic analyzer is the opposite of a synthesizer. Composite sound goes in, sine wave components come out.

The mind can do something very similar, but infinitely more sophisticated, as may be illustrated in four steps.

Figure 28 is the first page of Edison's patent on his "Phonograph or Speaking Machine," known to us as the record player. The wiggly lines of its *Fig.3* and *Fig.4* are the draftsman's way of representing what the patent describes as the sound vibrations.

Figure 29 shows sound vibrations as actually recorded by a "Speaking Machine." The four traces are segments of an opera. The visual resolution of the recording, as printed, is insufficient to show that each of the four bands is one single continuous wave of varying amplitude. Each wave is the mathematical sum of its component sounds that are heard from instant to instant, just as the composite signal on the top of figure 27 is the mathematical sum of the twelve component sounds.

Figure 30 has sufficient resolution to show what is meant by a single, continuous wave. The figure is the soundtrack of a single frame from a motion picture about a 107-piece orchestra playing Ravel's "Bolero."

This is how that soundtrack—and the soundtracks of all motion pictures—are made and work. A microphone picks up the amplitude variations of air pressure generated by the sound of horses' hoofs, rumbling volcanoes, lovers' whispers . . . and changes them to electrical vibrations. These, in their turn, are printed and become the wave of the soundtrack. The projector and sound system changes that wave back into air-pressure variations and our ears change those pressure variations to biochemical and electrical signals, which the nerves then conduct to the brain. Then comes the real trick.

The single, continuous, irregular wave is decoded by the mind, which thus recovers the individual sources of the music, such as the different musical instruments and voices. Listening to such wiggles of air pressure, a trained musician might remark that the oboe d'amore should have been a bit louder or that Peter, Paul, and Mary are aging. The complex envelope of the soundtrack of figure 30 has not twelve components, as in the top line of figure 27, but innumerable ones.

Figures 27 to 30 help demonstrate the stunning analytical skill and precision with which our mind takes apart, interprets, and then reassembles

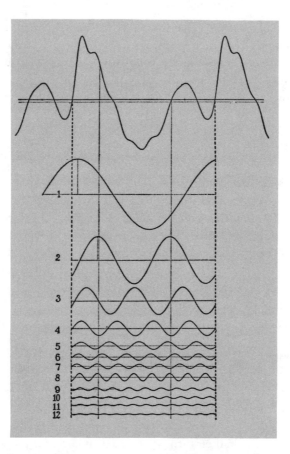

FIGURE 27   THE SOUND OF AN ORGAN PIPE (top line), reduced to its simplest harmonic components using a harmonic analyzer. The graph shows the first twelve components of the sample organ sound.

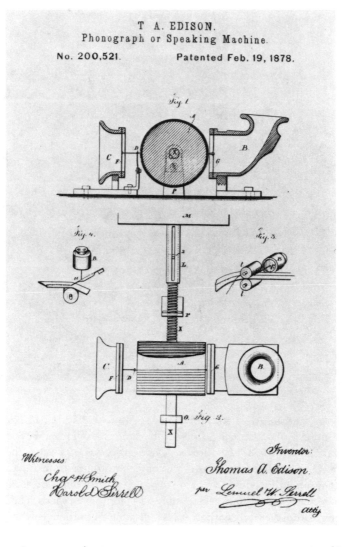

FIGURE 28 **EDISON'S PATENT FOR THE RECORD PLAYER**, granted in 1878. From air vibrations that correspond to the wiggles in *figs. 3* and *4,* the listener can recognize a spoken message, understand it, and even identify the speaker. This remarkable feat of decoding, usually taken for granted, is made possible by the mind's ability to slice a second into thousands of parts, then integrate the string of impressions into a meaningful message. Courtesy, The Burndy Library, Norwalk, Connecticut.

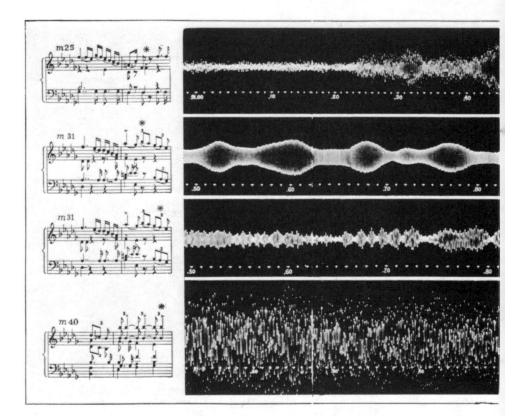

FIGURE 29   PEN RECORDING FROM AN OPERA reproduced by a talking machine in 1920. This is a small portion of the famous sextet from the Second Act of Donizetti's opera *Lucia di Lammermoor*. We hear six voices as well as the orchestra. The fine divisions are ¹/₁₀₀ of a second; the top line is from 91.00 to 92.00 seconds; the second line, from 117.49 to 118.34 seconds; the third line, from 120.50 to 121.29 seconds; and the fourth line, from 160.13 to 161.14 seconds. The excerpts on the left are from the score, with asterisks indicating the portions corresponding to the waves.

The top line is the voice of Enrico Caruso, the legendary Italian tenor; the second is that of the soprano Luisa Tetrazzini, singing a high B flat. In the third line, the

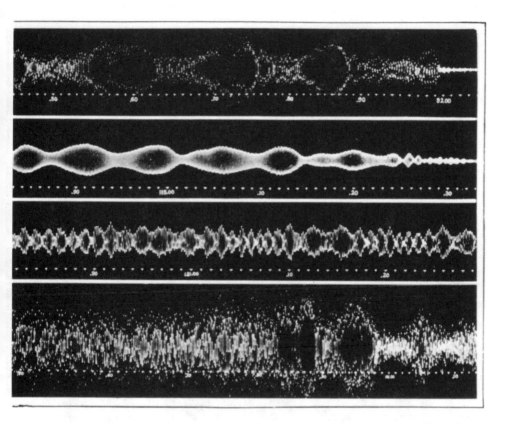

voices of Tetrazzini and that of the baritone Pasquale Amato are seen (heard) together. The fourth line shows all the six voices singing together. Each line is one single, continuous curve, but the picture is not sufficiently magnified to show it. The finely structured irregular curves are changed into air vibrations, picked up by the ear, and transmitted to the brain, where our minds can pick out the individual voices. The disassembling and reassembling is done within the boundaries of the mental present.

Illustration after Dayton Clarence Miller, *The Science of Musical Sound* (New York: Macmillan, 1934).

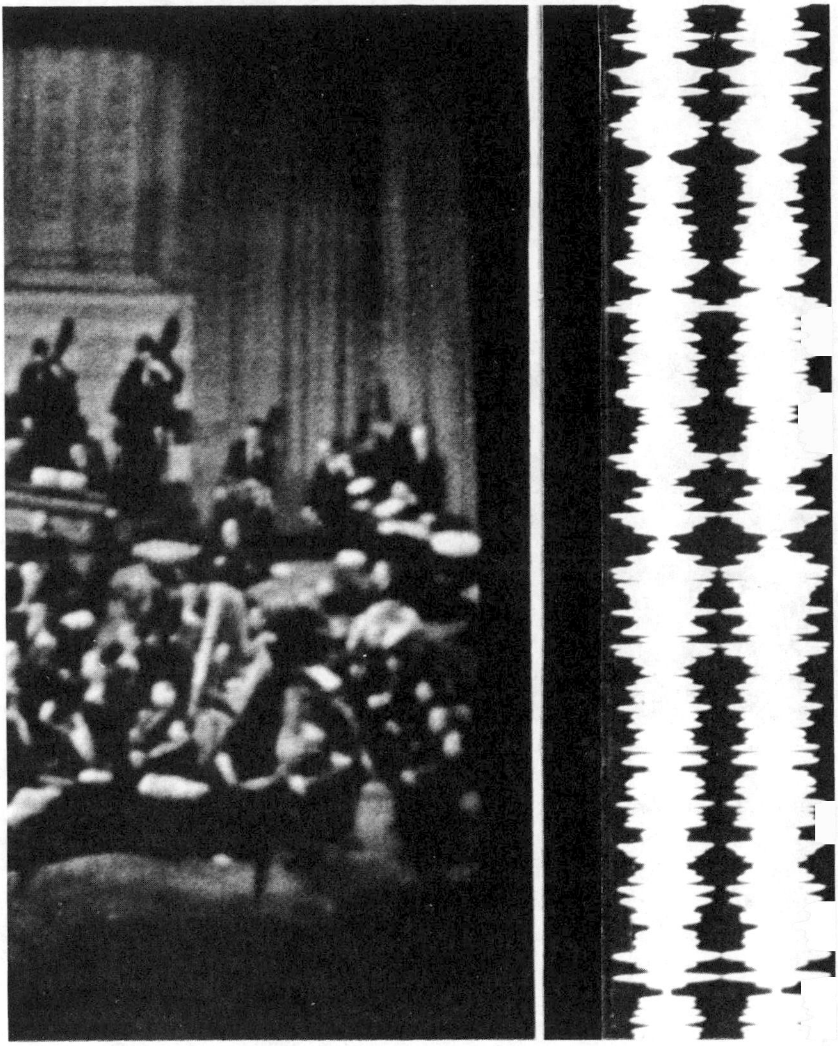

To understand what this illustration tells us about the mind's capacity to structure
the mental present, it is necessary to attend to technical details.

The soundtrack is moved in front of a narrow horizontal slit. The amount of
light that gets through is proportional to the transparent part of that narrow horizontal segment. (For the connoisseur: a dual bilateral track.) As the film travels, the
amplitude variation of the light is picked up as a signal, amplified, and used to drive
a speaker. When the pressure variations created by the loudspeaker reach the ear, a
sound is heard. The mind recognizes the components of the composite sound, such
as those of the individual instruments, if that is what the person wants to hear. Or
the listener may note the rhythm, the melody, the words, or all of these. Each
frame is projected for $1/24$ of a second. The visual resolution of this illustration is
sufficient to show frequencies up to about 30,000 Hz, but motion picture sound
systems seldom have that high fidelity.

A composer with unusual gifts could write music by drawing the wavy lines
directly on a piece of paper. If those lines were then copied as a soundtrack and
played, the listener would hear whatever the composer had intended him to hear,
such as a band playing square dance music and a caller calling, "Do-si-do your left
hand lady, do-si-do your partner."

Or the composer might compose directly for a digital laser disc by writing the
binary code for each small segment of the composite wave: 00100 01110 11001
11010 10101 111. . . . After his 1s and 0s have been changed into the microscopic
dark-light signals of the disc and the disc read out by a laser beam, the listener may
hear "A bicycle built for two," sung by Placido Domingo, if that is what the
composer wanted him to hear. Such a composer would have put into practice the
dream of Pythagoras: he would have reduced all sound to number. We do not have
composers who could do so directly. But it is being done through the technology
of laser discs—and the marvelous gift of the mind to analyze and synthesize the
contents of the mental present.

The illustration is a frame from "What Does Orchestration Mean?" with
a young Leonard Bernstein conducting. Columbia Broadcasting System, Educational Films, 1954.

audio signals. While this is being done, our other senses are also at work. They also feed their inputs to the brain, there to be processed and added to the content of the mental present. Our olfactory stimuli, for instance, are also continuously analyzed, making us aware of pleasant or unpleasant scents. While listening and smelling, we might also have our finger on a red button to ring an alarm or on a belly button to signal laughter.

All this processing takes place without our conscious awareness, involving both the younger and the archaic levels of the brain. What emerges into awareness are only the conclusions or judgments of the higher brain functions, provided sufficient time has been allowed for information processing.

For instance, two very sharp audio stimuli separated by less than 2 milliseconds (0.002 second) will appear to human subjects as simultaneous. That amount of time is not sufficient for the nervous system to tell us that the sounds were distinct. If two sharp signals are separated by more than about 2 milliseconds, they will appear distinct. However, unless they are separated by more than about 20 milliseconds, the subject will not be able to tell which one came first. In other words, the simplest processing step of which we may become aware is the separation of what first appears to be one signal into two. Being able to tell which signal came first—placing them in temporal order—is a higher processing step that takes more mental labor and requires more time. But these perceptual tasks are only the surface phenomena of the analysis, synthesis, and interpretation that goes on beneath them, within the confines of the mental present.

THE MACROSTRUCTURE OF THE MENTAL PRESENT

The process just described pertains only to the temporal organization of current sense impressions and draws mainly upon the cyclic order of life. But the mental present always contains expectations and memories, and through them, it incorporates features of the aging order of life as well. For instance, the sources of the air-pressure variations that carried the audio information from source to the mind could have been many: voices from a fireman's party, the music of *The Music Man,* tanks rolling across a pontoon bridge. Identifying their origins would not be possible were it not for the listener's memories and anticipations concerning them. The smell of a swannery, the sight of an emu, or a touch of the flu are all perceived and judged in terms of memory and anticipation. They and current sense impressions all share the mental present, wherein they are continuously correlated, integrated, elaborated, and placed in the context of nootemporal reality.

How does this remarkable integrative capacity of the mind arise during the development of the individual?

After having been giant molecules, each of us became embryos and, three months after conception, fetuses. The fetal brain gets to work as soon as it begins to develop and keeps on functioning as if it were a new and different organ each day. For each of the developmental stages of the body, it does the appropriate thing. But what the world appears to be for the fetus in its stages of development can only be guessed at through plausible inference, in spite of the fact that each of us has been there.

Psychologists have speculated that the fetus must gather from its existence a feeling of omnipotence because of its biologically privileged position. It has little chance to learn of cause and effect, since everything it needs is supplied, ready-made.

From just before to just after birth, the fetus experiences the greatest change of umwelt a person is able to experience and survive. One may fantasize the change as it occurs. Imagine being placed in a pleasantly heated straightjacket, with food being pumped into you through a tube; the noises are constant and monotonous, and motion is all but impossible. Under such conditions most people would surely go mad. Fetuses love it.

Leaving their "everything's-going-my-way" feeling plunges them into a world of cause and effect. They learn that to satisfy a need they must first do something, such as cry; crying brings food and comfort. With this experience, the feeling of magic causation is born. Later it gives way to the realization that certain actions usually, but not always, bring predictable responses and often not immediately. The experience of delay is thus born.

Slowly, but quite perceptibly, the young child learns the rhythm of life outside the womb. With each sunrise and sunset, he adapts himself more intimately to the enduring and the changing aspects of his environment, illustrated, for instance, by the entrainment shown in figure 20. Space, which first appeared as a chaotic assemblage of changing shapes, becomes filled with objects that retain their identities. To state it differently, as the infant's sense of delay develops, he or she learns to recognize permanence.

Intentionality, which the infant first learned from his needs, becomes assigned to objects: "My bear wants to play." The spoken word itself gets endowed with the power that had earlier been attributed to wishing and imagination. The nootemporal umwelt begins to emerge as the child's mental present becomes defined, maintained, and refined by language.

The first time-related words the child can verbally manipulate and later comprehend refer to the present: "I need it!" This, however, is not the mental present of the adult, but more like the biotemporal present, ruled by demands for immediate satisfaction. In the control of behavior, anticipation appears before memory does: a toddler can follow a plan of attack for long minutes, but seems to forget what he or she was doing in a few seconds. Between twenty-four and thirty-six months of age, the child learns to use time-related words appropriately. This is followed by an increase in his vocabulary, with usually a greater variety of expressions about the future than about the past.

In this learning sequence, the continuity with simpler life forms is clear. Except for rare examples, such as a call indicating just-discovered food, animals cannot communicate about the past. As I have already remarked, there are many ways to tell a dog, "I will feed you," or, "I will take you for a walk," but there is no way one can tell it, "You have already been fed," or, "You already took a walk." In the development of the child, the earliest category of time is that of the future, then that of the past, and finally the mental present.

Somewhere along the line, the child discovers passing: first it is "growing up," then it is "getting older," and then it is aging. On the inner landscape of his mind, the expanding temporal horizons combine with the realization of passing, with selfhood and the recognition of inevitable personal death. These experiences and ideas together create the mature human's world: the nootemporal umwelt.

To what neurophysiological processes in the brain do the experiences of future, past, and present correspond?

A great deal of work has been done on the psychology of memory. There are theories on the psychology of recall, recognition, retention, short- and long-term memory, retrieval, and forgetting. There are also neurological theories of memory. Much is known about details, but no powerful, general principles have been proposed and demonstrated as valid. Neurologists used to speak of engrams, hypothetical loci of altered brain tissues that were presumably the elementary traces of memory. But more recent work shows that memory is retained in the brain in a distributed rather than filamental manner, and hence neurologists now prefer to consider memory and cerebral association centers instead of engrams.

Memory, as distinct from recall by rote, is not comparable to the consecutive frames of a motion picture. It is more like a family album filled in a helter-skelter fashion, in need of someone to put the snapshots in temporal order. For instance, I remember Kaaren running downhill toward three stately copper beech trees, remember the grace of her motion through the grass and the wind blowing her hair. I also remember that she reached the trees, stopped in their summer shade, and awaited her lover. Soon she embraced a man younger than I am. But each of these pictures is only a static, or slightly vibrating, tableau; none of them moves. If I want to remember Kaaren's motion, I must make the pictures move by attaching a clock to them, figuratively speaking. Or else I must make the series of events that whizz by in my mind, slow down. The ambiguity ("I must make it move," "I must make it slow down") demonstrates the inadequacy of language when it comes to conveying inner temporal experience. In either case, the images must be forced to come alive, to partake of the flow of time as I experience it in the mental present.

What about anticipation? Work in experimental psychology and theories about expectation do exist, though not comparable in their elaboration to those on memory. But a neurophysiological understanding of what in the brain corresponds to long-term anticipation, as far as I can tell, does not exist.

The difficulties of identifying brain functions that correspond to our experience and idea of time are great. Although, as I have related, the potential number of different states of the human brain is immense, it is still difficult to imagine just what might constitute a record of "Kaaren and I will play apple pie-and-full-moon in broad daylight."

Over two decades ago the distinguished British neurologist Lord Brain raised the following question. If memory corresponds to certain brain states, expectation to some others, and the analysis of current sense impressions yet to others, and since all of these coexist, how can they be told apart from the point of view of neurology? This question has not yet been answered. The neurological sources of the great wealth of the mental present, from the point of view of scientific theory and experiment, are yet to be identified.

Language, the architect of time

All living things talk, each its own way. But none, save humans, possess the genius necessary for the formation of symbol and metaphor exemplified by language. It is human language that makes possible the articulation and the moving of imagined events, things, and feelings into the future and the past. Language thus serves as the architect of noetic time. This section begins with thoughts about our animal brethren, mere bricklayers in the architect-of-time figure of speech.

Microorganisms communicate by chemicals, the lower metazoan invertebrates by chemical and tactile means, mosquitoes by acoustical and chemical means, butterflies through visual and chemical channels, and birds by visual means and voice. The bill of the platypus detects electrical signals emitted by the muscles of its prey. Black-tailed deer have glands on their hindlegs that produce sex-attractants called pheromones. They rub their foreheads against their legs, then against twigs, leaving a message: "Come this way." Frog language includes chirps, chuckles, and thumps generated by pressing their pouches against the ground. It is a seismic communication, like the tapping of messages on the wall between two hotel rooms or jail cells.

Virgin females of a species of fireflies answer the triple-pulsed flashes of the male of their own species by adopting a courtship posture, then twist their lanterns toward the signaling male. After mating, they become predatory. They mimic the answers of the females of other species, attract the other males, and eat them. Figure 31 shows a more civilized behavior and more refined language in a species of grasshopper males. The illustration is the choreography and score of one cycle of a mating dance that may be repeated twenty or more times. At the end of the wooing song and dance, receptive females permit the Meistersinger males to mount.

Humpback whales sing their tunes as they travel; it is their way of keeping the guys and dolls of the group together. Their songs show family dialects and also change somewhat from year to year. At a time when there were still whales in the Mediterranean, ancient Greek mariners heard these songs reverberate in the hulls of their wooden ships, which made the songs sound as if coming from all directions. The mariners thought they heard sirens sing. Judging from the recorded humpback whale songs I have heard, I should have thought so myself.

Chimpanzee communication by gestural and vocal signals begins to

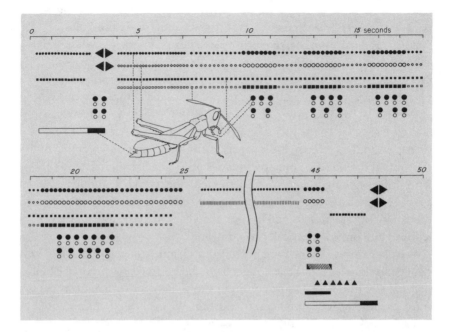

FIGURE 31 COURTSHIP SIGNALS IN A SPECIES OF GRASSHOPPER. The cycle between the heavy black triangles, from about 4 to 48 seconds, may be repeated twenty or more times. The filled black and open circles, as if on a musical staff, represent the motion of the palps (segmented sensory feelers for tasting food). The upper two rows of these circles are for the palps as shown, the lower ones, with a different program, for the palps on the other side of the head, not seen. The other circles of the score show the motions of the knees, the antennae, and the penis. Toward the end of the cycle, from about 45 seconds on, the crosshatch indicates sparring movements; the upward pointing triangles, body rocking; and the single solid horizontal bar, the sideling up to the female. Courtesy, Daniel Otte, Academy of Natural Sciences, Philadelphia.

resemble that used by humans, but when compared with man, chimpanzees fail on two accounts. First, their vocabulary is miniscule. Second, as do all animals, they treat messages as unviolable wholes; they cannot break them down into parts and reassemble them to form new messages. Although able to tailor their behavior to the observed and varying characters of friend or foe, they cannot delay their responses to stimuli. It is "Banana now!" but never "Play tomorrow," much less "What happened to the guy who tickled my nose last winter?" In experiments on ape language, it is the superior gift of the experimenter that pulls the animal toward the upper boundaries of its genetically limited skills. The situation is analogous to modern medicine, which has added to average longevity by permitting people to reach the genetic limits of their lifetimes. But no human will live for 200 years and no chimpanzee will learn to talk unless genetically reengineered.

People communicate via all the sensory channels they command. The courtship signals of the grasshopper have their refined cousins in the body language of humans. Tactile communication in hand-to-hand combat is murder, among acquaintances, friendship. Women can communicate by pheromones made at home, men by aftershave lotion made in factories. People send messages by throwing snowballs, fits, and intercontinental missiles. Gesticulation is a very old language. If done according to communally approved patterns, with chisel and hammer or pen in hand, it is called writing.

In humans, visual and auditory sensing is much more refined than chemical sensing, such as that by scent or by taste. The network of five million neurons that make sight possible and that process the visual signals is packaged in the retina and forms what is sometimes called the external brain. Those neurons are distributed over three-quarters of the inner surface of the human eye and their encoding is spatial. In contrast, the system that processes the audio signals is in the brain and its encoding is temporal. The vocal dexterity of human language—both in coding and decoding—is made possible by the advanced temporal organization whose refinement we already admired in connection with the decoding of musical sound.

Since speaking a human language demands the capacity for fine-structuring temporal behavior, it promotes a sophisticated physiology. The refined temporal control needed for speech is illustrated by the many different rates (events per second) at which a data acquisition system in speech research must monitor speech functions. Typically, in one second, the electric activity of the brain has to be sampled 200 times, sound 24,000 times, air flow 40 times, lip movement 1000 times, oral air pressure 100 times, jaw strains 60 times, larynx motion 600 times, and respiration 20 times. All the skills one must have before saying no more than "Get lost!"

These capacities are necessary but not sufficient for communicating by a human language. Using them in earlier, less developed forms, our ancestors had to learn how to utter words; words made stories possible and stories demanded and generated temporal frameworks. The expanding capacity for story telling brought with it story thinking, which helped push the horizons of time further into the future and the past. Through this process, language defined and refined the contrast between permanence and change.

A small child babbles, thinks out loud, and does not care whether anyone listens; he or she begins with monologues. As self-definition advances, the child begins using socialized speech. This changes speech from being all play to acting with a purpose. The child builds his cosmology: "The Universe According to Me." Out of the ever-changing chaos, permanent and therefore nameable features of the world separate out. The child is expanding the boundaries of his umwelt.

Artistic, literary, and scientific labor is a continuation of what the child began: the building up of the universe from elements judged permanent. Putting the same thought into different words, the humanities and the sciences create visible, audible, and tangible expressions to represent things, events, feelings, and relationships that are judged as unchanged through extended periods of time.[11]

But what are those features of the world that remain unchanged and hence nameable? Are they given by nature for us to discover? Confucius, writing twenty-six centuries ago, implied such a view, for he stressed that all things have their one and only correct name by which they should be called.

---

11   Recall the earlier statement that something which appears to be a slice of cheese for a split part of a second, the tone of a violin for the next, then a prairie dog, a painting, a toothache, then the smell of garlic, could not be given a name.

If names not be correct, language is not in accordance with the truth
of things. If language be not in accordance with the truth of things,
affairs cannot be carried to success . . . properties will not flourish
. . . punishments will not be properly awarded . . . people will not
know how to move hand or foot. . . .

Although there is general agreement that certain universal qualities of
human experience cut across all cultural boundaries—without them
communication among people would be impossible—the Confucian
stability is gone from our appreciation of reality. The ways in which
people cut the pie of their time experiences into their slices of permanence
and change are permitted, but not prescribed by nature; they are instead
matters of temperament and social conditioning.

In Sanskrit, the classic language of India, change itself is comprehended
as permanence. For instance, "everything flows," the famous phrase of
Heraclitus, becomes in Sanskrit, "everything is impermanent." The two
phrases describe the same experience. Or do they? The verb "flows"
conveys a sense of change; "is impermanent" speaks of eternal stability.
Where English perceives motion, Sanskrit perceives stasis. Although
Sanskrit is more distant to the daily life of modern India than classic Greek
is to American business, fundamental differences in attitudes toward time
and tempo between the heirs of the Greek and Sanskrit cultural traditions
remain.

For the ancient Hebrews, time was seen as a stable continuum with
several possible divisions; they had no word for "time" in the modern,
Western sense. Duration was always something concrete, something
expressible by numbers attached to natural or cultural cycles, such as to the
days, weeks, months, or years. Time as abstract flow did not exist.

The Iraqw people live in northern Tanzania. Recent work by the American anthropologist R. J. Thornton revealed that their native tradition
lacked a chronology along which events could be arranged in historical
order. Events are related to one another by spatial proximity rather than
by sequence. Since places are fixed with relation to each other, the Iraqw
do not posit a direction of temporal change. In Iraqw, the form of question
"when?" corresponds to "where?" There are no parallel forms to represent our distinction between "*Where* did she tell you about her love?" and
"*When* did she tell you about her love?"

The English word "escapement," the device that alternately stops and releases the motion of the driving wheel of a clock comes from the French *échappement,* from *échapper,* "to escape, become free." The German name for the same device is *die Hemmung,* "the inhibitor," "the restrainer." Watching an impersonal mechanical gadget, French and English speakers prefer to abstract from their sense impressions one kind of permanent feature: the striving for freedom. German speakers prefer to abstract from identical sense impressions a different kind of permanent feature: the necessity for restraint. Evidently, whatever is judged as permanent in that mechanical process is not an attribute of the process alone but a preferred view of the observer. It is not by chance that English people stand for office, Americans run for it.

How do languages perceive the reality of time itself?

First, there is the way "time" is identified or defined. This varies with temperament, profession, and custom, all of which are coded into the languages. This is the semantic level. Second, different languages employ time words and combine them with other words differently. This—the syntactical level—becomes immediately and glaringly evident to all who learn a non–Indo-European language.[12] Third, each statement about time includes the conscious experience of the speaker, making an already complicated situation more so. But the channels are well defined: through semantics, syntax, and socially conditioned ideas of selfhood, languages tell us how we are supposed to perceive time and everything else.

For instance, in Old Testament times, an oath had to be taken holding the genitals of the man before whom the oath was spoken. That custom seems to have gone out of vogue during the first millennium of our era, but we do learn from medieval English law that the words of women could not be admitted as proof "because of their frailty." They could not testify as witnesses. However, the deeper-lying reasons for this legal stance have nothing to do with frailty, as witnessed by our language. The English "testify" comes from the Latin *testis,* "witness," related to *terstis,* "a third one" present at an occasion. The diminutive of *testis* is *testiculus,* "little witness." In English, the word appears as "testicles." They are the

---

12   The English word "time" translates into Hungarian, a Finno-Ugric language of Asiatic roots, as *idö.* When queried quantitatively, *Mennyi az idö?* (What is the quantity of *idö?*), it means, "What time is it?" When queried qualitatively, *Milyen az idö?* (What is the quality of *idö?*), it means, "How is the weather?" The identification of time with natural change and history appears self-evident and therefore undeniable to the speakers of that language.

little witnesses to what one must assume to be the naked truth. (Please note that this writer did not make up the phrase "naked truth" just to be wittily employed on this page.)

Has the legal position of women in this and in other matters been a victim of language? No. It has been the victim of beliefs held firmly because they were self-evidently true. They were self-evidently true because words in their support, carrying their inner logic, came—and come—effortlessly to the speakers' lips, such as in: a person who's got no balls cannot be trusted. The continuity of the value judgment carried by this remark is stunning.

There is no easy way of saying "eotemporality is a kind of time that has no present and does not flow." Therefore, by the conservative momentum of language, that kind of time ought not exist. The difficulties in explaining the nature of the temporalities below the noetic, or the twin paradox of relativity theory, or the meaning of mental functions, do not guarantee that what is being said about time, relativity theory, or the sources of human behavior are so. But it does make the task of the truth teller (or liar) more difficult.

In everyday use, personal names carry a great deal of information on where people believe they belong and on their scales of value. Some time ago the public transit vehicles of Ontario displayed the following public interest advertisement.

> For Pete's sake, for Juanita's sake,
> For Horst's sake, for Liv's sake,
> For Dmitri's sake, for Maria's sake,
> For Nadia's sake, for Chi Mingh's sake,
> For Azis' sake, for Sol's sake,
> For everybody's sake, let's work together.[13]

As children grow, they are usually called by different names because they are judged to have become different persons. In all languages, names select permanent character traits desired for the infant or present in an adult. Paul means small, Claude lame, Jules downy. Black American

13   Quoted in Mary V. Seeman, "Name and Identity," *Canadian Journal of Psychiatry* 25 (1980): 129.

names often refer to the earth, as in Eartha, Clayton, or Ashton, because the earth was believed to be protective of the child. Position in birth order, sex, and social standing are coded into many names. Names, like all symbols, are bulwarks against change.

Names color feelings and prejudge associations. Each year 50,000 Americans petition the courts for change of names. The Armenian revolutionary Iosif Vissarinovich Dzugashvili preferred to be known, in his acquired (Russian) language, simply as a man of steel or *stal*. Therefore he called himself Stalin. Norma Jean Mortenson made her fame as Marilyn Monroe; Samuel Clemens as Mark Twain. As did St. Paul, each of these people began to rebuild his or her reality by picking a new name appropriate for his or her new self, that fictitious, all-important character.

What else is in a name? Personal identity maintained through time and approved by society. Dr. Zhivago in Boris Pasternak's novel by that title died of a heart attack in a tragic, human way. Lara, his life's light, had it worse: she was first changed into a nameless person and therefore died twice.

> One day Larissa Feodorovna went out and did not come back. She must have been arrested in the street at that time. She vanished without a trace and probably died somewhere, forgotten as a nameless number on a list that afterwards got mislaid, in one of the innumerable mixed or women's concentration camps in the north.[14]

Making people into nonpersons is the socially invented and enforced version of Alzheimer's disease. It deprives a person of that single symbol that he could otherwise have retained for life and held up against the change of everything else. It steals from him his sense of human time and regresses him into the biotemporal umwelt of living but mindless organisms.

What else is in a name? A great deal of emotion. Think, for instance, of a simple, almost journalistic, matter-of-fact report: "There is no God but God and Mohammed is the Prophet of God." In Arabic the sounds change into a hypnotic dance: *La illaha illa lah, waw Muhammadu rasul illahi.* The tape-recorded prayer heard through the public address system of the mosque says what it says more by the way it says it than by its reasoned

14   Pasternak, *Doctor Zhivago* (New York: Pantheon, 1958), p. 503.

argument. In some languages it is impossible to make certain statements without an appeal to collective passion. And collective judgments on how to handle time correlate with those passions, as all observant travelers know.

Is the nature of time victim to linguistic prejudice that is unwilling to admit the absolute, naked truth? No. There are no such final truths. But there are socially conditioned beliefs in the character of ultimate reality, including the nature of time. Those beliefs vary. Languages, through their self-consistent logics, convince their users that the nature of time implicit in the way they identify temporal experience is in harmony with the universe.

The time of dreams

Dreams speak a language made mostly of images. One of the striking aspects of dreams is that while we dream, the images appear as obvious forms of a complete reality. They recount stories and tell us that the temporal setting of those stories is the way the world is, and that's that. Only when dreams are examined in the waking state do their stories and their images seem primitive, the time and space of their universes odd.

To come to grips with the temporal umwelt of dreams, let us take a mini-course on Freudian depth psychology, as we had a mini-course on Darwinian evolution and will have one on Einstein's relativity theory.

It used to be believed that to understand human behavior, including our own actions, all that was needed was to monitor with honesty and intelligence one's own reasoning and behavior. It is nothing like that, said Sigmund Freud. The forces that control behavior and the reasons that make a person creative or destructive, happy, unhappy, or insane are normally hidden from his or her own awareness.

They may be discovered, however, by a careful descent into the deeper levels of the mind. One can find out how those deeper levels reason through logical deductions from whatever is available: thoughts, feelings, and patterns of behavior. By a simile, it is possible to determine the content of the earth's magma by examining the lava of spontaneously erupting volcanoes, but it is not enough just to take a look; it is necessary for a trained geologist to conduct a scientific analysis.

The depths of the mind that Freudian psychology explores are the archaic perceptions of reality that take form in the archaic levels of the brain (see fig. 26). Since we are not normally aware of these reality assessments, they are said to be done unconsciously. Contrary to popular belief, however, Freud's major discovery was not the existence of unconscious processes and perceptions—these were already known when he began his work—but rather the discovery of a method of exploration called free association. He realized that if a person freely associated feelings with feelings and expressed them as if no listener were present, his train of thought would show certain syntactical and contextual patterns. He would use metaphors in individual, self-consistent ways and practice recognizable patterns of repetition.

All of these, when taken together and carefully reviewed by a trained listener, would be seen to converge on the judgments and preoccupations of a person. There is nothing mystical to the method. It is a statement about the use of everyday language. The secret is in knowing what to listen to.

The royal road to learning about the policies of the archaic levels of a person's mind is free association with dreams. The scientific interpretation of dreams has shown that dream images reveal the biological, emotional, and intellectual needs of a person. To interpret them we do not need the language of superhuman spirits; it is enough if we learn to deal with human concerns about birth, sex, and death in the multitudes of their roles.

Listening to reports about a person's dreams is the first step. But those reports cannot be uncritically trusted because they are under unrelenting inner censorship. Free association can get around the censor, just as listening to people in the privacy of their apartments can reveal opinions that the political censors would not permit to appear in print.

The censor is a functional aspect of the superego. It is a portable cage whose bars are the do's and don'ts learned when the child's mind first began to expand its temporal horizons according to the socially conditioned views of reality, conveyed by parents and by language, when he or she first realized that different choices lead to different short- and long-term effects.

The mind's censors are as ruthless but more subtle than political censors. What they try to keep out of awareness is the presence in our minds of a not-so-friendly reptile and a number of mammals. Many people are hard put to believe this, imagining it to be some farfetched Freudian discovery. It is not. Christianity, one among many religions, has been teaching the same: there is an animal in you, don't let it loose.

But these ancient skullmates' assessments of reality remain with us and let themselves be heard whenever the higher brain functions are quiescent, such as in sleep. What social restraints do not permit us to say, or even feel, in the waking state appear in our dreams behind various masks. If you know how to remove the mask, you can find out what makes a person tick—and also how the archaic levels of the mind perceive the nature of time.

The temporal umwelts of the archaic brain functions are temporalities of the earlier forms of the brain. In no sense are those realities "timeless," as Freud himself and almost everyone else examining the unconscious has claimed them to be. They are only deficient in certain attributes of time when compared with nootemporality. To wit, they are mixtures of eotemporality (continuous but directionless time) and prototemporality (fragmented and directionless time).[15] Those little old ancestral brains with their umwelts close to the biotemporal try to represent through dream imagery their own concerns and also those concerns that are communicated to them from the more evolved layers of the brain. But events in those representations are not held together by what we would call cause-and-effect relationships, because early assessments of reality did not include the kind of connectedness we recognize as causation in our waking states. Instead, they connect through magic causation reminiscent of those in fairy tales. Sequences of events are often mixed because the arena available to the conscious mind—to wit, noetic time—does not exist in archaic reality.

Consistently, the predominant temporality of dreams is a sense of presentness. It is true that not everything happens at once, but the unfolding of events has an aura of only a vague temporal direction, if any. Although the content of dreams is profoundly influenced by long-term futures and pasts significant to the dreamer, the dreams themselves do not visually manifest those distant futures or pasts. Their images and emotions are always in a present. If you dream of your early childhood forty years ago, then in your dream you appear as a child or an observing adult *at that time*.

Freud has noticed that the long ago, if it appears in dreams, is transformed by the dream into space. Something that happened a long time ago

15   The concepts of eotemporality and prototemporality were introduced early in this chapter. Definitions can also be found in the Glossary.

is seldom manifest as a memory-in-the-dream. It is more likely to appear as something observed in the present, at a distance in space. This is consistent with the fact that space perception is older than time perception. Distances are an older category of reality than are periods of times, and hence more familiar to the ancestral brain. If a dream, played out in the archaic realities of the mind, wants to express feelings that in the waking state would refer to events in the distant past, it can only offer long distances in space to represent the temporal relation of those events to the dreamer.

Freud also observed that the times of day, such as clock readings, do appear in dreams but they seldom refer to time. They have the same role as do all other numerals one dreams about. They refer to emotionally loaded events. The thirty-seven dollars paid to a bartender for a Tom Collins stood for the age of the dreamer, divorced at age thirty-seven from an alcoholic mate whom he first met as Miss Collins.

In sum, the temporal world of dreams is not one of great sophistication, but rather one of primitiveness. While dreaming, our dreams are almost always taken as complete and real because they represent ancient understandings of the world that, to our older brain functions, are just as complete and real as the modern universe is for our modern brain functions. Dreams are complex mental processes, as distinct from person to person as are fingerprints, and like fingerprints, they also have certain common rules by which they are constituted. They may be disturbing because they can reveal through their atmospheres the world of the mammals and reptiles within us. In that world only the struggle of a present exists. Dreams can also be inspiring for many reasons, among which we must count the dawning in the dream world of a universe of conscious experience and, with it, the freedom of noetic time.

The experienced speed of time

It is known on good authority that "Time travels at diverse paces with diverse persons." It is known on equally good authority—such as that of the reader—that time travels at diverse paces with the same person in his or her lifetime and that it can even do so within a day or an hour.

To talk about time moving, traveling, or, if you wish, standing still while we move and change is a metaphor. But metaphors are not arbitrary but derive from the structural properties of the mind. They cannot just take any form. And once established, they can show a great deal of stability. That time passes or flows at different speeds is one of those stable metaphors. What guarantees the validity and the stability of the metaphor?

Answers to the question of why time seems to pass at varying speeds are numerous. They may be found in works on experimental and theoretical psychology, in biological approaches to time, in the study of language, and in the many branches of philosophy. But all the theories I know of share one fundamental, usually hidden, and quite unjustifiable assumption. They all mask an image of a moving cosmic present, a universal flow of time whose speed our minds can assess correctly or incorrectly. One of the manifestations of these judgments, so the arguments run, is the feeling that time passes too rapidly or too slowly, or perhaps not at all.

However, as I related in the introductory paragraphs of this chapter and as we shall soon see in detail, there is nothing in the physical world to which this image of a moving present or flowing time could correspond. There is no ultimate rate of passage in the cosmos to which some inner sense of passage could be compared. For the sources of the feeling of time passing, and for its abstract representations, we must look entirely within our heads.

Let me attempt to give an answer to the question of what is the basis of our experience of the changing speed of time, without an appeal to a flow of time in the physical world.

Let us recall that long-term planning and memory, conscious experience, selfhood, and concern with abstract and symbolic causes populate the umwelt of the neocortex. In contrast, the mammalian and reptilian brains attend to those needs that demand immediate satisfaction. The furnishings of their reality are those of concrete objects and goals. In their world, selfhood cannot be defined because their capacities for the formation of symbolic continuities are poor or nonexistent.

The two umwelts amount to two different ways of perceiving time, with the nootemporal subsuming the biotemporal. These two ways of assessing time may be associated with the ideas of time understood and time felt. We already encountered this distinction when discussing the underlying causes of St. Augustine's dilemma: he knew what time was as long as no one asked him to put his knowledge into words.

Time understood helps order and interpret sense impressions by our intellectual faculties, such as by the mind's skills for the symbolic transformation of experience, exemplified in language. Through the manipulation of symbols, the boundaries of time understood may thus be extended much beyond the biological time horizons of the body.

Time felt helps order and interpret sense impressions by our emotive evaluations of reality. Those evaluations are informed of, though not limited to, biological demands: hunger, sex, the possession of territory. The boundaries of time felt, as here interpreted, are limited to the reaches of organic expectations and memory.

In a simplified, schematic manner, let us next think of time felt and time understood as the manifestations of two very complex and distinct neuro-biological processes or two families of living clocks.

The reader will remember that every time measurement must involve at least two clocks as well as a theory, or belief, that joins their readings. For the two clocks of this three-cornered dynamics let us substitute the affective and intellectual dimensions of human reality. The "theorems" that connect their "readings" are made of the drives, fantasies, superego demands (moral and social convictions), and immediate needs of survival.

It will also be recalled that whenever we measure time by comparing two clocks, the one that is judged more trustworthy is usually the one whose functions are better understood. In the case of the affective and intellectual dimensions of human experience the decision as to which we should use to judge a situation often demands long and difficult deliberations, because all human actions involve both feelings and understandings. Sometimes, when decisions must be made rapidly, the reference clock that is selected without deliberation is the one invested with the greater mental and emotional energy, because its functions happened to be most relevant to meeting the challenges of the instant. But whether the decision is reached rapidly or slowly, the judgment selected may be thought of as deriving from the "clock" better understood as it were. It is in terms of this "clock" that other "readings" are then evaluated.

My speculative suggestion is that what emerges into consciousness is the "theory" that evaluates and connects the readings, the end product of an intricate process of weighted comparisons. We express these "theorems" in metaphors. That time may be said to flow is itself a broadly but not universally used metaphor. That time passes slowly, rapidly, or at the right speed are other "theorems." They report on the psychological significance of comparisons between archaic and modern assessments of experiences, disguised as quantitative statements.

Identifying the roots of our sense of passage, such as I attempted to do, is a necessary step toward an answer to "What is time?"

In that classic query "time" has always meant nootemporality with its clear distinctions between future and past, and with its future somehow changing into the past. It seldom if ever meant the biotemporality of the animal world with its limited temporal horizons, the presentless eotemporality of macroscopic physics, the prototemporality of the particulate world or the atemporality of the universe of light. Its usual meaning in Western intellectual history may be paraphrased as follows: "Why is it that my self-awareness includes an irreducible element that I can describe as one of passing and that this sense of passing is so very useful in my interpretation of the world?"

The reasoning I have offered suggests that the roots of our experience of time's passage should be sought in the complexity of the human brain and in its hierarchical assessment of reality. It further suggests that the reason why the idea of passing time is so useful in the struggle for survival is because it is a much selected for organizing principle, appropriate in its nature to the human umwelt. The reader familiar with the history of ideas will recognize in this stance a combined Kantian and Darwinian view, placed into the frame of the hierarchical theory of time.

The thoughts sketched must yet be worked out in critical detail. Until that is accomplished, the delightful best we can do is to conclude as we began with a quote from the great knower of time and man. Here, through the characters of Orlando and Rosalind in *As You Like It*, Shakespeare holds forth on the experienced speed of time.

> *Ros.* . . . Time travels in diverse paces with diverse persons. I'll tell you who Time ambles withal, who Time trots withal, who Time gallops withal, and who he stands still withal.
> *Orl.* I prithee, who doth he trot withal?
> *Ros.* He trots hard with a young maid, between the contract of her marriage and the day it is solemnized; if the interim be but a se'nnight, Time's pace is so hard that it seems the length of seven years.
> *Orl.* Who ambles Time withal?
> *Ros.* With a priest that lacks Latin, and a rich man that hath not the gout; for the one sleeps easily because he cannot study, and the other lives merrily because he feels no pain. . . .
> *Orl.* Who doth he gallop withal?

Ros. With a thief to the gallows; for though he go as softly as foot can fall he thinks himself too soon there.

Orl. Who stays it still withal?

Ros. With lawyers in the vacation, for they sleep between term and term, and then they perceive not how Time moves.

Let Rosalind's comments close this section. Instead of a topical summary, let us have a bird's-eye-view of this section on "Nootemporality, the Brain's Way of Minding the Body."

The emergence of the mind, and with it, the human gift of perceiving the world in terms of noetic time, was identified with the appearance of our species. In very general lines, the growth of each infant mirrors the history of the steps to mind.

Compared with the young of other mammals, human infants are born too soon. They have their brains, but those brains are unformatted. The formatting is done through socialization: the nine months' gestation period continues in the womb of family and society.

The brain of the newborn has far more nerve cells and nerve cell connections than does that of the adult. Very early impressions not only promote the internal organization of the brain but also help whittle down the neuronal supply to those that are found useful in the new environment. The early days, the mother's care, the routine of the environment, the manner of doing things, and the mother tongue become permanently marked in the infant's brain, and therefore in the way he perceives reality and time.

As the seasons pass, the child begins his lifelong search for personal identity through a complex mental process involving expectations and memories, integrated with sense impressions in the mental present. In due course he will beget other human beings resembling himself and at least one other person. Ten or twelve scores of seasonal changes later his mind will cease to maintain its mental present; the person will have lost his identity. After another while the body will cease to maintain its organic present, and from that moment on, it will begin disintegrating into the chemical elements that make up the earth.

We say that the person was born, learned, loved, struggled, aged, and died.

Throughout this drama of becoming and remaining human, the person has been distinguished from other persons by his selfhood or identity. In a circular, feedback fashion that characterizes all organic processes, the framework of noetic time that was necessary to create and maintain self-hood also demanded the existence of communities. Without the development of collective human groups, selfhood and noetic time could not have come about in the first place. It is the community that can secure the survival of peoples' biological and brain children beyond the boundaries of individual lives. Through language, art, and artifact, it is society that makes possible the peculiarly human form of struggle against the inevitability of passing that we know as the building of civilizations.

Sociotemporality: The Socialization and Collective Evaluation of Time

John Donne was a poet, priest, father of twelve children, and master of words with passionate feelings. He penned the following passage sometime in 1624 in his *Devotions:*

> No man is an island, entire upon itself; every man is a piece of the continent, a part of the main . . . any man's death diminishes me because I am involved in mankind; and therefore never send to know for whom the bell tolls; it tolls for thee.

Every bell tolls for everyone and every facet of our lives shows aspects of the civilization and society in which we live.

This section is about sociotemporality, the temporality unique to people living and working in societies. Its study encompasses two time-related functions of society.

The first is the timing of collective actions through synchronization and scheduling. This process may be called the *socialization of time.*

The second one is the creation and maintenance of value systems that guide the conduct of the members of a society. Such systems derive their authority from the history and the plans of the group, as seen by its members. This process may be called the *collective evaluation of time.*

Whereas the socialization of time is quite unmistakably the evolutionary continuation of the highly developed rhythms of animal societies, the collective evaluation of time is built on the symbol-making skills of the mind and has no more than its seeds in our animal ancestry.

Twenty-four centuries ago Aristotle began his famous book on ethics by remarking that, "Every art and every inquiry, and similarly every action and pursuit, is thought to aim at some good. . . ." The good, for Aristotle, was whatever in fact was being aimed at. The good for man was whatever man would naturally seek, such as the fulfillment of his or her capacities. My interest in this quote is not in the question of what is good, but in stressing that human actions have aims that may be subjected to moral judgments. Whether the aims are good or evil, right or wrong, whether they are informed of justice or burdened by injustice, the human sense of time in human actions is always presumed.

Ethical rules pertain to the most sophisticated behavioral issues in which the human sense of time may be manifest, for they apply to the workings of the highest known integrative level of nature, that of society. Responsibility and moral judgments are intrinsically social concepts. Ludwig Feuerbach, a Bavarian natural philosopher, put it this way in his *Principles of the Philosophy of the Future,* written some 130 years ago:

> the single man for himself possesses the essence of man neither in himself as a moral being nor in himself as a thinking being. The essence of man is contained only in the community and unity of man with man; it is a unity, however, which rests only on the distinction between I and thou.[16]

That "essence of man" contained in the community may be retained only in a viable society, and societies can retain their viabilities only if and only as long as (1) they can maintain their social present and (2) they are able to direct the behavior of their members toward collectively beneficial goals.

The themes selected for this section include issues of both the socialization and the evaluation of time. The section begins by explaining the nature of the social present, the third member of the hierarchy that already includes the organic and the mental presents. It then turns to problems of collective identity, which is the social version of personal identity: it is a name we call ourselves, a fictitious character on the stage of collective imagination, a symbol that helps overcome the finiteness of individual life. These

---

16 Feuerbach, *Principles of the Philosophy of the Future,* trans. M. H. Vogel (Indianapolis: Bobbs-Merrill, 1966), p. 71.

thoughts lead to the consideration of four of those issues of conduct that have always been under collective control: to wit, the perpetuation of the self (food), perpetuation of the species (sex), the education of children, and the conduct of cruel conflict (war). Individual behavior in these matters is controlled by the community through economic pressure, violence, and education. Its methods are said to be authorized by the goals of the community, themselves derived from one or another evaluation of the past and the future.

The chapter closes by considering the nature of social goals, perceived as the most advanced symbol making of which our species is capable and the most sophisticated means available for humans to overcome the individual's passage.

The social present

Human societies are coeval with the species, and there have been many and different ones around, even in historical times. To become and remain a tribe, society, or civilization, it was necessary for each to create and continuously maintain its *social present*. A carefully kept sequence of social presents is called a *schedule*.

Animals and plants that share the same ecological niche must coordinate their biological rhythms: there must be a chasing time, an eating time, and drinking, mating, and building times (fig. 32). It is the same for humans, except that times for working, thinking, music making, and other civilized activities must be added to the list of events that must be scheduled.

Social cycles are rhythmic schedules, cyclic variations in the amplitude or nature of one or another of the variables, such as work, of the social present. In complete analogy with biological cycles, some social cycles correspond to certain external rhythms of the astronomical or living worlds, but some have no such correspondences. For instance, while the daily and yearly cycles are tied to the sun and the months derive from lunations, market days held every Tuesday are not tied to any rhythms external to society.

The modern system of synchronization and scheduling that helps keep the collective integrity of the industrial state is a foremost example of social cycles: it does not originate with but only allows for organic cycles. Its historical roots may be found in *The Rule of Saint Benedict,* a directory designed for the spiritual well-being of the brothers. We previously encountered it in the context of calendars. It specified the time of collective

FIGURE 32 COLLECTIVE PRESENT OF MEMBERS OF SPECIES SHARING
LIVING QUARTERS. Cover-drawing by Richter; © 1973 The New Yorker
Magazine, Inc. Courtesy *The New Yorker* and the artist.

bathing, headwashing, bloodletting, the refilling of mattresses, and the daily routines arranged by the canonical (that is, procedural) hours.

There were seven canonical hours to coordinate the monks' lives. The day began with the Matins and Lauds (at midnight), continued with the Prime (first hour) at 6 A.M., Terce (third hour) at 9 A.M., and Sext (sixth hour) at 12 noon. None (the ninth hour) followed at 2 or 3 P.M., Vespers (evening worship) at 4 P.M., and the Compline (for "completed") at 7 P.M. After that, the brothers went to bed. Distributed across the day were appointed times for reciting psalms, reading versicles, saying Mass, and reading privately.

By and by, the canonical hours came to rule not only the Benedictine but all monasteries, and through them, the large households. Through those households, they gave a collective rhythm to the life of the land. The wealthy of late medieval times paid for making of illuminated manuscripts called Books of Hours to remind them of the passing of time and the permanence of the Church—and to have an object of art to admire and be proud of. Some of the many beautiful ones, such as *The Hours of Catherine of Cleves, The Visconti Hours,* and the *Très Riches Heures* of Jean, Duke of Berry (fig. 33), have been reprinted recently. The canonical hours still help us maintain the social present: we have lunch at Nones, though we call it noon and have shifted it to midday.

The ringing of bells synchronized the labor of the countryside. The large bells of the steeples pealed, smaller ones jangled, very small ones tinkled. Taken together, they helped extend the schedule–conscious values of the monks to one and all. The bells were regulated by sundials, by sandglasses, by water clocks, and later by mechanical clocks. The scheduling created a collective beat, promoted the rationalization of life, and reflected a belief in a universe well ordered in time. The significance of scheduling and timekeeping to the well-being of all was infused into the Christian way of life; it came to complete fruition with the focus of Protestant pragmatism upon the importance of number and time in the conduct of daily life.

The organic present, as the reader may recall, maintains life by securing that biochemical events that must take place simultaneously do so and those that should not, do not. The social present has a similar task: it assures concerted action or its absence.

For instance, while mating animals must be at the same place at the same time, the individual prey does well to absent itself when the predator's inner clock readies it to stalk. This second kind of synchronization is known as temporal segregation or isolation. Nocturnal and diurnal species, for instance, fill segregated temporal niches of the 24-hour day. There are many examples of seasonal and even yearly isolation that involve whole populations. The distribution of cicada broods in Illinois serves as an example. In any given location, cicadas will emerge every 13 or 17 years. Both 13 and 17 are prime numbers, divisible only by one and by themselves. No predator with a shorter life cycle can adapt to make its appearance cyclically coincide with that of either group of cicadas.

Scheduling by temporal segregation is the routine work of appointment secretaries. The opposite of temporal isolation is the uninterrupted availability of one person to another. It is a sign of great devotion if done freely, of slavery if done involuntarily.

Among religions, the interweaving of feasts constitutes another example of temporal segregation. So does the traditional separation of holy days from civil days or, in the words of Mircea Eliade, of sacred time from profane time, regarded as qualitatively different times. Shakespeare, the great poet of time, beautifully described in *Hamlet* the Christian distinction between the two times.

> Some say that ever 'gainst that season comes
> Wherein our Saviour's birth is celebrated
> The bird of dawning singeth all night long;
> And then, they say, no spirit can walk abroad;
> The nights are wholesome; then no planets strike,
> No fairy takes, nor witch hath power to charm,
> So hallow'd and so gracious is the time.

Christmas carols still speak of a holy night, meaning exactly what they say: time is holy on that night and not profane as it is on other nights. The holy times are perceived as cyclically recurring events and are celebrated in rituals of the eternal return. Christ is born again each Christmas, is crucified on each Good Friday, and resurrects on each Easter Sunday.

Secularization made no difference in the celebrations of eternal return. Countries have their national holidays; men and women have their birthdays and wedding anniversaries. In that presumably most secularized state, the Soviet Union, there is a body of centrally designed rituals

FIGURE 33   A BOOK OF HOURS: THE RHYTHM OF ETERNITY. Books of Hours were illuminated private devotional works popular in the Middle Ages. They contained prayers or offices appointed to be said at the canonical hours. The most famous of them, and probably the most widely known illustrated manuscript, is the *Très Riches Heures* of Jean, Duke of Berry. It was made for him by the Limbourg brothers—well-known artists of their time—and an unknown scribe during the period 1410 to 1416, though some of the pages, such as the one reproduced here, were not finished until sixty years later by other artists.

This particular Book of Hours begins with calendar pages for the twelve months, each with an illustration of an activity characteristic of that month. Then follow illustrated tracts from the Gospels, Psalms, and the life of Christ, and even a map of Rome.

This page, for the month of September, shows a grape harvest in a picturesque scene set before the Château de Saumur. Aproned young men and women fill their baskets with grapes, load them into hampers on mules, and then into vats on wagons.

The inner semicircle on the top shows the thirty days that hath September; the outer, the change of the signs of the zodiac from Virgo to Libra.

Original at Musée Condé, Chantilly. Courtesy, Giraudon/Art Resource, New York.

intended to replace the ecclesiastical and non-Russian rituals of tradition. The ceremonial for the Registration of the Newborn resembles baptism, with "invited parents" replacing godparents. The infant receives a name-star while the official recites, "Let this star light the path of your son (or daughter) as the star of October lights the path for the whole world." It is a feast of perennial return, celebrating the event that put the Bolsheviks in power.[17]

The organic and mental presents both involve simultaneities of necessity rather than chance, as does the social present. To create and maintain it, it is necessary to have communication among group members. Cells communicate by chemical and electrical signals; members of a species or several species among themselves, by anything they can use for signaling. Male dogs come from miles to be at the right place at the right time after the message "I'm a bitch in heat" has reached their nostrils. It is not difficult to get the message of grizzly bears, because they have halitosis; when humans get such a message they do well to absent themselves, in an example of the social present secured by synchronized separation.

A group of people is able to form a tribe, a nation, or a civilization only if they can cooperate in some ways and share certain values. These conditions can only be met if there exists a social present maintained through communication.

The width of the social present is determined by the time necessary to make people take concerted action. In its turn, that period depends on the distances involved and the speed with which messages may be carried. I would imagine that the earliest messages were grunted and shouted, later ones drummed or carried in memory, and only much later written. For a given region, a faster horse could narrow the width of the social present. When mounted messengers were replaced by modulated radio waves, the social present of all distances on earth narrowed, potentially, to a small part of a second.

★

17   The day of the Russian Revolution in the Julian calendar, still in use in the Russia of 1917, was October 25. In the Gregorian calendar, the date was November 17. For the non-Russian world of 1917 and for modern Russia (with its Gregorian calendar), the October Revolution took place in November. But it is still celebrated in October, because the scaffolding of the collective past is the calendar, with its recurring names, and not an abstract, independent flow of time.

Using the domain of social time, let me illustrate in some detail the idea of the width of the present, in anticipation of the later use of that concept in special relativity theory.

In July 1775, Benjamin Franklin was appointed to head the American Postal Service at a salary of $1,000 a year. He increased the number of post offices, introduced stage coaches, and started a packet service to England.

Imagine a Norwegian immigrant living in Philadelphia at the end of the eighteenth century, exchanging letters with his family in Bergen, Norway. It takes 40 days for a letter to go in either direction. Each letter he receives can tell him what happened to his family up to 40 days earlier. He will live his life from instant to instant with a 40-day period of ignorance of his family. His "now" in Philadelphia corresponds to the "now" in Norway 40 days earlier. Likewise, if his opinion is needed in Bergen, each letter he sends off is followed by a 40-day period of impotence because for that period of time he is powerless to influence anything over there. His "now" in this case corresponds to the "now" in Norway 40 days after his present in Philadelphia.

The mental present of the lonely Norwegian is the time needed to take conscious action in response to a stimulus; let us say, 0.2 second. This is his "now" in Philadelphia. At the distance of Bergen, the 0.2 second opened up to a total of 80 days because it took that long for him to respond to a stimulus originating in Bergen. His present-at-a-distance comprised 40 days of postal signal, 0.2 second of nerve signal, and another 40 days of postal signal travel. Our friend was continuously surrounded by an ignorance–impotence bubble of 80 days.

Instead of the Philadelphia, U.S.A.–Bergen, Norway, distance, we may think of the well-documented stretch of London to Bath, England. Through the middle of the eighteenth century, the London–Bath social present was about 80-hours wide at its ideal best, because it took 38 hours to get a letter from one town to the other town by horseback mail. In 1784 the first mail coach began to run along improved roads, cutting the ignorance–impotence bubble at either end to an optimal total of 20 hours: 10 hours for a letter to go, 10 hours for it to come (fig. 34). For the 100-mile distance, the speed averaged 10 miles per hour. For the American pony express eighty years later, running between St. Joseph, Missouri, and Sacramento, California, a distance of 1,800 miles across rough terrain and high mountains, the remarkable goal of an average speed of 7.5 miles

FIGURE 34   PARING OFF THE SOCIAL PRESENT: THE ORIGINAL
LONDON–BATH MAIL COACH OF 1784. Its speed narrowed the temporal width of the social present between the towns of London and Bath from about 80 to about 20 hours. The coach was built by a certain Mr. John Palmer, Jr., for a number of purposes, among them the need to get a quick exchange of actors between the Theatre Royal at Drury Lane, Covent Garden, London, and the Theatre Royal in Bath, managed by Mr. John Palmer, Sr. Every mail coach carried a guard who rode shotgun with a blunderbuss and two pistols, wearing the royal livery of scarlet and a beaver hat. He also carried a horn to warn traffic on the road ahead and to signal 250 yards before each turnpike gate, so that it may be opened for the coach to pass through without delay. The horns of the Swiss mail buses still play the traditional Swiss mail coach tune.

Courtesy the British Post Office, with appreciation to P. G. Howe, Photographic Librarian. Crown Copyright.

per hour was achieved by the legendary Buffalo Bill. The St. Joseph–Sacramento social present was 480 hours (20 days).

The clutch of eggs of the South American ostrich tinamou hatch at the same time, though not laid at the same time; the chicks call to each other so as to synchronize their first look at the world. It is in that kind of a setting that the newborn tinamou begins its life and learns the practical significance of the collective present. Human infants about to be born do not call to each other, nor are they all delivered at the same time. But whenever a child is born there are, or at least ought to be, other people around, falling over each other in the process of establishing their social present.

The clutch of nations that make up our globe did not become nations at the same time, but they have been calling to each other. Through the increasingly tighter synchronization of human affairs made possible and forced upon us by the myriad communication links that crisscross the earth, they are signaling their simultaneous emergence into a single postindustrial world. Once synchronicity has been established, the hierarchy of presents—the organic, mental, and social—will have a new sibling, the global present.

A note on simultaneity

Common sense suggests that the "now here" corresponds to and is simultaneous with a "now there" and a "now everywhere." I wish to examine this idea of simultaneity, because we will need that concept later in connection with sending light signals across the universe. It is easier to grasp the nature of that idea in the context of social present because social time is a part of our experience, whereas beaming light signals to distant galaxies is not.

We have come a long way since Buffalo Bill as far as message speeds go: telephone and radio can cut the ignorance–impotence bubble between any two points of the globe to less than a second. But whether a present-at-a-distance is 80 days, 20 days, or 20 hours, a rational view holds that somewhere within that period, perhaps halfway through it, there is an instant-at-that-distance which is simultaneous with the present instant here. The same rational view suggests that somewhere within the second necessary for the radio signals to go and come, the wink of a present millisecond here is simultaneous with the wink of a certain millisecond there.

The existence of such simultaneities seems to be an obvious fact of nature not subject to the availability or unavailability of communication gadgetry.

All this is intuitively evident. Except that, as I remarked earlier, an understanding of time based entirely on intuition is inadequate for the appreciation of its nature as revealed by contemporary science. The idea of simultaneity between two events at a distance from each other is a relevant example for it has proved to have no meaning, even in principle, unless we take into account the mode of the available communication, as we shall see in our survey of special relativity theory in Chapter 4.

In the history of ideas, it often happened that the scientists and the artists of an age became curious about and sensitive to identical aspects of the world. It is thus probably not by chance that the novelist William Styron became fascinated by how little actual meaning may be attached to assumed simultaneities in the social context. He expressed that fascination in a concrete and powerful manner through the words of Sophie in his novel *Sophie's Choice*. The narrator of the story recalls one of Sophie's remarks.

> "On the day I arrived in Auschwitz," I heard her say behind me, "it was beautiful. The forsythia was in bloom."
> I was eating bananas in Raleigh, North Carolina, I thought, thinking this not the first time since I had known Sophie, yet perhaps the first time in my life aware of the meaning of the Absurd, and its conclusive, unrevocable horror.[18]

The length of the ignorance–impotence bubble of the social present for global distances may be decreased by inventing faster means of communication until the global present becomes shorter than the mental present. In fact, we have already passed that point. But in cosmic dimensions, the temporal length of the ignorance–impotence bubble, as we shall see, cannot be decreased below certain lower limits whose value depends on distance. This will make mincemeat of what our common sense told us about the meaning of "at the same time."

18   Styron, *Sophie's Choice* (New York: Bantam Books, 1980), p. 567.

Social identity: Temporal horizons beyond death

We have found that the self, the "I," is a symbol standing for a living object that shares the external world with other, similar objects: stray dogs, fat cats, other men, other women. But unlike those objects, the self has an inner component that holds executive powers over the body. It can order conduct detrimental to parts of the body and even deadly for it as a whole. The same symbolic object can also be imagined as surviving the body.

Social or collective identity is also a symbol. It stands for an object—let us say, the United States of America—made up of many people and things. This imaginary object shares the world external to itself with other, similar objects: "Last year the U.S.A. played an important role on the stage of international politics, together with Iceland, Burma, Mexico, and Italy." But unlike those external objects, the collective self we call by the name America—short for the United States of America—also has an inner component. This component holds the executive powers of social action and is known by many and different names such as the law of the land, the collective purpose, national honor, and the will of the people. Group identity can demand behavior of the body social that is detrimental to its individual members—as in a terrorist suicide attack—and even deadly for society as a whole—such as Nazism was for Germany. The same symbolic object can survive the demise of the society itself, as in "The glory that was Rome."

The collective self is represented by a name or a symbol, as is the individual self. Americans pledge "allegiance to the flag of the United States of America and to the Republic for which it stands." For Englishmen it is, "Gentlemen, the King!" for Christians, the Cross, for Communists, the hammer and sickle. It is necessary to use signs and symbols because it is difficult, if not impossible, to give a rigorous definition of what makes people members of this or that community.

The future and past of the collective self acquires meaning only with respect to a viable social present. The citizens of ancient Greece cannot work together any more to maintain their social present because they are all dead. For that reason, ancient Greece has neither a future nor a past; we alone have a past of which the past and future of ancient Greece, such as before and after Socrates, is a part. To determine the significant features of Greek culture is not simply a matter of looking at old artifacts and writings. Their contents must be referred to our social present through our interpretation of their meaning for the Greeks. This is a very tricky task.

Historians must construct the story of the past from primitive and not-so-primitive material and assess its details in terms of ever-changing value judgments. Ancient documents that supposedly recorded events are of many kinds: wall painting in caves, artifacts, written scrolls, and books. But none were made without collective guidance and personal judgment as to what was important. Many documents were written for the explicit reason to cheat a bit or a lot. As seen from the present, all documents are garbled, because it is impossible to know all the circumstances that influenced their making and that should be known for a correct interpretation of their significance.

History is also a series of uncertain dates. For centuries, the birth of Christ was off by four years and until recently, the age of the world was off by twenty billion years. The collective past is as ambiguous as is the story of "me, the name I call myself." What do people consider to be the true facts of their first, young love, as they look back at it from the ages of 25, 35, 55, and 85 years? Was World War I started by the assassination of Archduke Ferdinand in Sarajevo, or was that murder only a part of an ongoing struggle, already the war itself, whose "real" causes lay elsewhere? In Christian tradition, the history of man is the Fall from grace; some people label the past of our species the ascent of man. Which one is it?

With all these uncertainties, it is not surprising that the past of the collective self and, with it, the present character of that self have often been redrawn. The changes of view so created may be, but need not be, political machinations in support of an ideology, profession, class, or other concern. Usually they only demonstrate the steady but unpredictable changes of social reality. Whenever the assessment of man's position in the world changes, so does history and the character of the name we call ourselves. History is not something fixed to be discovered; it must be created and maintained in the social present.

Likewise, social identity is not something that can be discovered the same way as the source of a river may be, and then fixed forever on a map. Instead, it must be created and maintained, and remains subject to the changing evaluations of subsequent generations. Sometimes those changes take place very slowly, sometimes very rapidly. But steady beneath the development of all societies has been the desire of people to secure the survival of their children and their brain children, and thus defy the finality of their personal demise.

Social control of conduct: A collective evaluation of time

The kaleidoscope of ethical standards explicit in the international and national events of our days shows a wildly divergent pattern. It surely does not suggest that all people on earth naturally tend toward the same norms of conduct. I do not believe that was ever the case, and certainly it is not the case today. From group to group, moral codes are as varied as the groups' respective evaluations of time. For, as this subsection will try to show, it is the socially held evaluation of history and of the nature of time that gives rise to ethical systems in the first place.

Before embarking on a discussion of time, society, and ethics, it is necessary to recollect some ideas about the hierarchy of freedom of action.

Stones—for which read "nonliving objects"—have no freedom of action. They do not actually do anything; their fate is something that happens to them. Therefore we cannot speak about stone behavior as we do about animal behavior, and not at all about stone conduct. Because of the absence of freedom, the world of the stone is deterministic and predictable. Its time has no direction.

Animals have a certain degree of freedom of action, depending on the species. To the extent that they do, there exists for them a distinction between future and past. For that reason we can talk about animal behavior, but still not about animal conduct.

Only when it comes to humans can we talk about conduct, classify it by value judgments such as being right or wrong, describe its goals as good or evil, and exert social pressure over individual behavior by demanding adherence to established moral standards. Implicit in all this is the belief that responsibility makes sense and that the decision made by opening or not opening a door includes an element of freedom. In other words, that the future is to an extent unpredictable and hence qualitatively different from the past. This distinction is one of the hallmarks of nootemporality and sociotemporality.

The need for collective guidance in matters of conduct arises from the creativity of imagination. The mind continuously invents recommendations about what to do next: Sweep the garage; seduce your pretty cousin; build a bridge; suggest to your senator that all social security numbers be multiplied by 3.14159; kill everyone who disagrees with you on abortion; go back to sleep.

Animals must also make choices, but only a time-knowing creature can comprehend that a person killed in a bar brawl will be dead not only for a few minutes but forever. Animals also copulate and the males and females of some species stay together for life, but only humans can perceive a "From this day forward," recognizing that a child begotten is a life-time affair.

In the 1952 movie *High Noon,* Gary Cooper, his quiet, male self, was faced with the kind of dilemma that has concerned moral thinkers, writers, clergy, and poets through the ages: social obligation versus personal desire. The young Quaker woman involved was Grace Kelly, making the conflict understandable.

> Oh to be torn twixt love and duty,
> What if I'd lose my fair-haired beauty,
> Look at the big hand creep along,
> Nearin' high noon. . . .

Overlooking the world of mice and men, a cat resting on the shelf could not be bothered by long-term consequences of anything, such as how it is going to live with the memory of a decision it has made. Not so for men and women, the knowers of noetic and social time.

This subsection reasons that codes of conduct, offered as selection guides from among imaginable actions, gain their authority from collectively endorsed evaluations of time. It is not possible to offer here a treatise to support this claim, but it is useful and interesting to sketch a few examples in order to make the link appear plausible. The constellation of conducts at which we will take a closer look are: (1) the taking of food, (2) sexual behavior, (3) the raising of children, and (4) the conduct of war.

Deciding what *food* to eat and how, provided there is a choice, consists of much more than filling a hungry stomach or stuffing a fat face. I am referring to the ethical and social values attached to taking food, not gourmet cooking.

The Roman conquerors of Britain imported for their table small squirrels they called by the Latin name, *glis*. It is our dormouse. Each was kept in an earthenware called glirarium, which was a place for keeping a dormouse to be fattened until the pot had to be broken to get it out. After the dormice were cooked, they were served with honey.[19] A similar custom has been reported in the China of the early twentieth century, where the feast consisted of eating newborn mice in honey.

Is it right or wrong to eat dormice, beef, lobster, beans, or pork? The busy men and women of the industrialized West tend to hold that what to eat, and how, ought to be determined on a pragmatic basis—health and food value, market conditions, and personal taste—and not on the basis of abstract ethical stances. Such an opinion is itself a value judgment, however, the hallmark of a civilization that evaluates history not as a dialectic of doctrines and ideas but of biological processes.

The 470 million Hindus, 560 million Muslims, 17 million Jews, and millions of other people who follow dietary laws insist that taking food involves a great deal of moral judgment. Cows eat only grass, they might say, but people are omnivorous. The mind suggests hot dogs or hush puppies—a decision must be made. The principles guiding those decisions derive from different readings of the nature of reality and the significance of history and different evaluations of the destiny of man. A good illustration of this point is the prohibition of the Koran against the eating of animals that have been slaughtered as offerings to other than the Muslin god: a clearly political rule. Although many dietary laws have important hygienic significance, it is not hygiene but abstract images, certain symbols relating to the social self, that regulate the taking of food.

Ethical doctrines on the taking of food are enshrined in language, art, artifact, and commerce and tend to appear as unquestionable elements of correct conduct. If you do decide to have dunken dormice instead of dunken doughnuts, you may not find the paraphernalia in the local store. And there are many places where such a diet would be judged unappetizing, which also goes for eating other people.

Whatever the specific customs and preferences may be, only if it is allowed that a person, acting in the present, can make his future not copy

19   For an illustration of the custom, the reader may wish to take a second look at Tenniel's famous drawing of the Mad Hatter's Tea Party in *Alice in Wonderland*. The glirarium turned into a teapot, into which the Mad Hatter and the March Hare were trying to insert the Dormouse head first.

his past, can the taking of food be handled as an issue of ethics. A knowledge of human time is a prerequisite for dietary laws.

From rules on how to perpetuate the self, let me turn to rules that govern the perpetuation of the species. In our days of short words, all related issues go under the heading of *sex*.

At the turn of the sixth century, the Anglo-Saxon king Ethelbert of Kent issued the first code of Anglo-Saxon laws. One part of it is known as the Catering Wages Act. Among other matters, it regulated the charges for sexual companionship.

> If a man lies with a maiden belonging to the king, he shall pay
> 50 shillings of compensation. If she is a grinding slave he shall pay 15
> shillings; if she is of the third class, 12 shillings. . . . If a man lies with
> a serving maid of a ceorl [a free farmer] he shall pay 6 shillings.[20]

The ravisher of a "free woman with long hair" shall pay her guardian 30 shillings as compensation because his protection has been flaunted. The Code witnesses a scale of values that holds maidens for goods, measures behavior in monetary units, and thus grants a numerical scale to sexual conduct. In contemporary Western ethics the notion of maidens-as-goods in unpopular, but measuring human behavior by numbers of monetary units is encouraged. The practice is at the very foundation of modern economic theory, making the moral opposition to sex for money rather hypocritical. Drawing a line between what kind of human action may and what may not be measurable in numerical terms has been the responsibility of judges and juries adjudicating issues in medical ethics, life insurance, and family laws. The decision, whether in our age or in that of Ethelbert of Kent, depends on the ways that those who decide evaluate the course of human destiny.

Through language, art, artifact, law, and custom, moral codes have been built into all cultures. Figure 35 shows a thirteenth-century warning against the unauthorized use of the flesh. The devil is taking a woman to Hell on Judgment Day because she did not heed the official guidelines. Five centuries later, the German poet Goethe had different guidelines to offer in "Answers at a Parlor Game":

---

20   D. M. Stenton, *The Englishwoman in History* (New York: Schocken, 1977), p. 7.

FIGURE 35   PRESENT ACTS HAVE LONG-TERM CONSEQUENCES:THE DAME
AND THE DEVIL. Only humans can appreciate the long-term effects of present
behavior. This grants our species great advantages in its struggle against other
species, but often makes it difficult to select a path of action from among the many
possible and impossible ones imagined by the mind. This task of selection has no
equivalent in the animal kingdom. To help make a decision, societies have formulated general guiding principles. This thirteenth-century statue can be found along
the central bay of the south portal of Chartres Cathedral in France. The statue
shows the horrors of being taken to Hell by the Devil on Judgment Day for having
misbehaved on earth. To this writer, the couple looks more like two people preparing to have their private tryst, which may be exactly what the unknown sculptor
had in mind. Illustration, No. 324 from a series, reproduced courtesy of Editions
Houvet.

> Show to women gentle care
> It's the way their hearts to master.
> And who has the pluck to dare
> May get on a good deal faster.
> But who does not seem concerned
> Stays remote and quite unbending,
> He seduces by offending.

Sara Teasdale, in one of her sensitive love songs called "Four Winds," written in 1911, gave a woman's version.

> "Tell me then, what shall I do
> That my lover will be true."
>
> .   .   .   .
>
> "In the tempest thrust him forth,
> When thou art more cruel than he
> Then will Love be kind to thee."[21]

The moral lessons of the Dame and the Devil derive directly from salvation history. The conduct prescribed by the two poets have no apparent connection to cosmic purpose. They represent individualized and secularized evaluations of sexual behavior, paralleling the larger, social trend that secularized salvation history into the idea of redemption through scientific and technological progress.

Unlike the philosophical and religious traditions of the West, with their linear, directed time, the Japanese, Chinese, and Indian heritages embrace the idea of cyclic time and construct their moral tenets on love and sexual conduct accordingly. Consider, for instance, an example from the genre of Japanese stories known as the legends of the struggle for a mate. It tells of a woman called Cherry Blossom Girl, who was unable to bear the fate of being loved by two men and killed herself.

Cherry blossoms have a particular significance in Japanese literature. Just as the full face of the moon denotes the beginning of its waning, a full blossom signifies decline and death, the eternal cycle of nature. Cherry Blossom Girl is the Japanese female version of Dylan Thomas' western,

---

21   Teasdale, *The Collected Poems* (New York: Macmillan, 1942), p. 24.

Faustian male: "Wild men who caught and sang the sun in flight, / And learn, too late, they grieved it on its way."

The story presents a perception of existence often associated with the cyclic view of time. It does not see death as cutting short the forward progress of life, but rather as becoming realized, having kept it company all along. Living and dying are two sides of the same process, with either side available as a standard of conduct. Thus, perishability, impermanence, and fragility are hallmarks of Japanese esthetics. They, together with the praise of living objects such as flowers, grant Japanese art its air of ruthless beauty. The same conviction in the unity and complimentarity of life and death informs the traditional Japanese rules of conduct, making it difficult to reconcile them with their Western counterparts. Those Western codes have grown out of and together with beliefs that judge time as a linear and cumulative progress, and give life a position of absolute primacy over death.

Whatever the specific customs and preferences may be, only if they admitted that a person can create his future more or less freely can sexual behavior be classed as an ethical issue. The notion of sexual responsibility presumes a belief in the reality of noetic and social time, continuously opening a partly undetermined future.

If food has been regularly taken and sexual acts performed, with or without collective approval, *children* are often born. How are they to be raised?

The care of the young has its roots in biology: most species care for their offsprings. They have no choice but to do so or else to abandon them, depending on what their genes say. A mother cat knows the sick and the runt, separates them from the rest, and stops feeding them, in a programmed postpartum abortion.

In humans, the care of children is guided by biological instincts as well as by moral values. Compared cross-culturally, moral judgments show great differences. The justification given by the Ayatollah Khomeini for the driving of Iranian children, chained together, onto Iraqi mine fields is that of a holy war. It speaks of an evaluation of the past and future quite different from the ones that prompted the passing of child labor laws in civilized lands. But it is insufficient to say that murdering children is wrong: there are those who believe that it is right, and the view is not limited to some residents of the Middle East. Untold numbers of children were slaughtered in the West before, during, and after the Children's Crusades.

Quite obviously, attitudes toward children have economic, cultural, and personal components. Across all these variables, however, we find an ever-present element—the social evaluation of time—reflected through the opinions and behavior of the parents. Let me illustrate this point by comparing the idea of childhood in the Middle Ages and the Renaissance.

In medieval society, the idea of childhood as we understand it did not exist. This does not mean that affection and care did not exist, but rather that caring for small children was an unprofitable investment of time and effort because two out of every three children died before the age of one. Frequent pregnancies made up for the loss, but were also the cause of the very high mortality rate among young wives.

Parental tenderness, the readiness to sacrifice, and a great deal of curiosity about these little humans were also present, yet children were left to their own devices to fend for themselves in the squalor of country or castle until they were five or six, a practice that is suspected to have been responsible in good part for the emotional neutrality of medieval man to death. Medieval art was quite indifferent to the kind of joy we tend to associate with healthy childhood. All this was consistent with the view of a closed and finite universe and a natural world that had hardly changed since Creation and would hardly change until Judgment Day.

Attitudes toward children began to change rapidly during the Renaissance. Judgment Day was pushed into a distant, hence irrelevant, future, people became interested in earthly goods and wanted to know how to pass them on to their offspring. Children came to be valued as guarantees of better things to come. The future opened up; Renaissance art is full of happy little humans. All this was consistent with an understanding of the universe as vast and perhaps even infinite, and with a natural world in which Renaissance scientists and humanists came to appreciate the element of change.

The socially endorsed evaluations of future and past, so this discussion suggests, have been guiding parental attitudes toward children. The relationship is, of course, not a mechanical and precisely traceable one, but more like a leitmotiv with an infinite number of variations.

Food, sex, children, and everything else are involved in that epitome of passion and folly we know as *war*. Because of the awesome human dimensions of violent and cruel conflicts, people's opinions about war can serve as litmus tests of their ethical beliefs. In terms of what has been said, it would seem that opinions about socially approved violence should also be found to correlate with people's evaluation of the nature of time and man. I believe that such a correlation does exist and shall illustrate my point by examining an exchange of letters between Albert Einstein and Sigmund Freud on the causes and possible remedies to violence and war.[22]

In 1932, Einstein wrote to Freud asking whether there was "any way of delivering mankind from the menace of war?" This, he contended, had often been attempted but never succeeded. "As for me, the normal objective of my thought [the behavior of nonliving matter expressed through mathematics] affords no insight into the dark places of human will and feeling." Being "immune from national bias," he was puzzled by the fact that a small group "can bend the will of the majority, who stand to lose and suffer by the state of war . . .". Had Freud made any "recent discoveries" to "blaze the trail for new and fruitful modes of action?"

"It is a general principle," answered Freud, "that conflicts of interest among men are settled by the use of violence." After tracing the story of the use of force from muscles to tools and weapons, he noted that the final purpose remained the same: each side wanted to gain its objective. And the safest victory occurred when the opponent has been killed. In the course of evolution, the community took over the guarding of the peace by directing its violence against those persons who challenged its power. The violence of the members of a society could be directed against other members of the same community; the violence of the community at large, against other communities.

The reason why a minority can sway a majority to favor war may be found in the nature of human instincts. Popular understanding, Freud continued, sees love and hatred as polar opposites, yet neither is less important than the other. Human behavior is characterized by "the concurrent or mutually opposing actions of both." The instinct of love also calls upon a desire to master a person and, with it, upon an element of hatred. It follows that "there is no use in trying to get rid of man's aggressive inclinations."

---

22   The exchange may be found in Sigmund Freud, *The Standard Edition of the Complete Psychological Works*, ed. J. Strachey (London: Hogarth Press, 1964), vol. 22, pp. 199–215.

[Civilization consists] in the progressive displacement of instinctual aims and the restriction of instinctual impulses. Sensations which were pleasurable to our ancestors have become indifferent or even untenable to ourselves; there are organic grounds for the changes in our ethical and esthetic ideas. . . .

[In conclusion] I trust you will forgive me if what I have said has disappointed you, and remain, with kindest regards, Sincerely yours, [Vienna, September, 1932] *Sig. Freud*

It will be useful now to give a thumbnail sketch of the two men's evaluations of the nature of time. Their differences are not along party but along professional lines.

Einstein's theory of the physical world revealed that there was nothing in that world to which our idea of time's passage could correspond. He generalized this finding into the belief that time was unreal. His view of the world was Platonic: all things and events were governed by geometrical laws, and as in geometry, the correct answer to any question worth asking must be unambiguous.

What kind of answer did he expect? He wanted to learn of laws of human behavior revealed through "recent discoveries." As we would say it today, he was looking for a breakthrough. The intellectual focus of his life were the laws of the physical world. Those laws can be precisely stated, are inviolable, and can be used to make reliable predictions. He assumed that the laws of human behavior, if properly understood, are of the same kind. "Is it possible to control man's mental evolution," he asked in the same letter, "so as to make him proof against the psychoses of hate and mental destructiveness?"

In Freud's theory of man, time was taken as central to all endeavor, something of a universal quality of the world, and most definitely real. He saw human actions as ruled by a cauldron of bubbling and boiling feelings whose course was unpredictable even in principle, except perhaps in very broad lines. He also held, as implied above, that moral judgments were guided by those chaotic feelings and therefore showed great differences.

Freud spent his life dealing with and thinking about people. As far as I know, he never so much as attempted to discover any law of the mind that would mimic the laws of physics. Instead, he formulated and tested general principles, such as the background to his claim that there was "no use in trying to get rid of man's aggressive inclinations." Beneath this assertion lay Freud's belief that the forces that drive people to be destructive are the same that fire them to be creative.

Any proposal concerning remedies to war must assume that people may be held responsible for what they do. But responsibility makes sense only if the future is taken to be at least partly open; in other words, only if time may be validly described by the metaphor of flow or passing. If the time of the physical world is mistakenly identified with the temporalities of man and society, proposed remedies to war will be valid only for wars of robots.

The dangers implicit in such an approach was the theme of Anthony Burgess' *Clockwork Orange,* made into a successful motion picture in 1971 by Stanley Kubrick. Its hero is a despicable young man: a rapist, murderer, gangleader, and thief. He is apprehended and then brainwashed at the State Institute for the Reclamation of Criminal Types so as to make him a useful citizen. After the treatment is finished, he undergoes a public test to demonstrate that he now instinctively rejects criminal acts of violence and sex. The story is told in the words of the hero.

> "Choice," rumbled a rich deep goloss. I viddied it belonged to the prison charlie. "He has no real choice, has he? Self-interest, fear of physical pain, drove him to that grotesque act of self-abasement. . . . He ceases to be a wrongdoer. He ceases also to be a creature capable of moral choice."
>
> "These are subtleties," like smiled Dr. Brodsky. "We are not concerned with motive, with higher ethics. We are concerned only with cutting down crime. . . ."
>
> There was a lot of govoreeting and arguing then and I just stood there, brothers, like completely ignored by all these ignorant bratchnies, so I creeched out:
>
> "Me, me, me. How about me? Where do I come into all this? Am I like just some animal or dog?" And that started them govoreeting real loud and throwing slovos at me. So I creeched out louder, still creeching: "Am I just like a clockwork orange?"[23]

23   The hero spoke Nadsat, a disturbingly ugly Russianized English slang. Goloss means voice; viddied, saw; charlie, chaplain; govoreeting, talking; bratchnies, bastards; creech, to scream; slovos, words; clockwork orange, automaton. Burgess, *Clockwork Orange* (New York: Norton, 1963). The quote is from p. 126.

Is it possible to control man's evolution," we heard Einstein ask, "so as to make him proof against the psychoses of hate and mental destructiveness?" The question is ill-conceived, for it mistakes the temporal umwelt of man for that of a clockwork orange. Therefore, the answer is no, unless people are reduced to live in a biotemporal umwelt. Then they could kill without hatred and build without love.

In our epoch, science has been taking over the role of religion, and scientists, accomplished or otherwise, have been thrust into the position of being the high priests and arbiters of all things: right and wrong, good and evil, peace and war. But not unlike butchers, bakers, and candlestick makers, scientists are faithful members of their fraternities. They tend to approach, analyze, and propose solutions to the great problems of mankind in terms of the assumptions, methods, and models that have been found useful in their specialized trades.

But approaching, analyzing, and acting upon problems of individual conduct using, for instance, the specialized field of biology known as sociobiology is bound to remain an incomplete and incompletable enterprise. For it cannot address behavior rooted in the mental manipulation of future and past through imagination, and the consequent creativity and destructiveness of individuals.

Likewise, dealing with the collective conduct of societies by means and models useful in the study of statistical aggregates in physics, as is increasingly done through the employment of numerical analysis—for instance, in game theory—is also bound to remain an incomplete and incompletable enterprise. For, it cannot address the dynamics of intentionality peculiar to collective responses to the call of symbolic causes, which is a hallmark of sociotemporal reality.

A conclusion to this subsection emerges. Problems in the social control of conduct must take into account individual and social evaluations of time, because it was those evaluations that gave rise to ethical teachings in the first place.

A different drummer, or the evolutionary office of moral codes

What is the evolutionary office of moral codes, of all those rules of conduct that were born from individual and collective evaluations of time?

It has been broadly believed that enforcing behavior judged correct by a society has helped promote the keeping of social stability. As far back as it can be traced, people have certainly demonstrated the readiness to formulate rules of conduct, to follow them, and to break them deliberately. The cruelty that members of our species have been ready to mete out for breaking those rules and the suffering they have been ready to endure for keeping or for breaking them are nothing short of incredible. But the readiness to legislate and play the moral animal has produced little stability. On the contrary. I believe that ethical rules have been responsible for those two states of instability that uniquely characterize human conduct: creative madness and destructive craze.

The destructive frenzy of men and women has aspects of nightmares, their creative ecstasy aspects of wish-fulfilling dreams. The insecurity that comes from the knowledge of passing manifests itself in the generation of dreams and nightmares, beautiful Helenas and frumious Bandersnatches, maintained in a ceaselessly annoying fashion. This is the brain's way of minding the body and of keeping itself in training. I believe that the role of morality is not at all to create stability and balance but rather to maintain the mind as the viable designer of acts good and evil.

The beliefs and moral codes so designed are embodied in every language and in every piece of art, machine, scientific theory, and rule of conduct. Taken together, they assure the dynamic viability of society, while they also make all groups conservative: every society wants to do things certain ways and not in any other way. Those signs and signals of language and the environment that represent approved conduct after a while blend into the background, there to become unnoticeable, obvious, and obviously correct instructions. Even if a society is bent on change, such as the people of America, the change itself must take place according to approved rules. Noting the staying power of calendars in the chapter on time measurement, I called this phenomenon cultural inertia.

"We hold these truths to be self-evident, that all men are created equal, that they are endowed by their Creator with certain unalienable Rights. . . ." These are self-evident propositions in the Western tradition of freedom and individualism. They are not facts of nature in spite of the reference in the Declaration of Independence to "the Laws of Nature and of Nature's God." They are symbolic causes to fight for, beliefs that the West has worked out from a number of abstract concepts, mainly from those upon which Christianity is based. Those tenets—for instance, that we are all created equal in the image of God—are not self-evidently true either. They are also symbolic causes, and as such, their powers are awesome.

Why would a person dedicate his or her life to the memory of someone who has been dead for millennia? Why build houses for symbols such as "the people"? Why would anyone be ready to die for Mao Zedung or the king whom he could not hope ever to see, even from a distance? Why the frenzy to kill all Jews, by people who have never seen a Jew, unless "Jew" was a symbol that stood for something more than people of that faith? Why to murder or maim another man to revenge a sister's lost virginity, which in itself is an insignificant anatomical change? Why arm to conquer America by people who have never seen an American? Why build a space-craft to take people to places more unfriendly than anywhere on earth?

The answer is that humans live by and for symbolic causes. Symbolic causes help secure the continuity of society through conserving patterns of behavior and belief, and for that precise reason, they are also the most efficient means by which social motion may be redirected. Doing so—changing prevailing social norms—is never easy and, in some form or other, it usually demands sacrifice. Socrates had to drink hemlock to make a point for the freedom of speech, Christ drove himself to cruci-fixion to make his point on what was right and wrong in human affairs, the Chinese needed two revolutions to make their country leave feudalism for a postcapitalistic state. Not all sacrifices need to be tragic, however. The endorsement of an unpopular phrase or artistic expression may be a form of sacrifice that helps change the direction of cultural motion.

All political, religious, and philosophical systems have formulated ultimate truths and moral principles, holding them up as permanent, unchanging tenets of right and wrong. The contradictory and changing character of such presumably permanent and universal standards is so great that their claims to absoluteness and finality are impossible to believe. The best anyone can do is to regard them as valid for a time and place.

The conclusion jumps to mind.

Human moral tenets are created by people as the need for them arises, and for that reason, they cannot be discovered as can planetary orbits or the rules of animal behavior. Moral laws are ideas that suggest permanence in the chaos of ceaseless change, but by their very nature, they do not secure stability for great lengths of time. Instead, they accelerate change.

The ethical journey of our species is not the story of walking along a
laid-out road but rather of blazing a trail into the non-existent and thus the
creation of new reality. The Greek novelist Nikos Kazantzakis wanted to
know how it was all done.

> How can you reach the womb of the Abyss to make it fruitful? This
> cannot be expressed, cannot be narrowed into words, cannot be sub-
> jected to laws; every man is completely free and has his special libera-
> tion. No form of instructions exist, no Savior exists to open the
> road.[24]

The drawings in figure 36 can serve to tell the story of all descendants of
my hairy ancestor, the one who first conjured up in his mind the images of
possible and impossible long-term futures and began to arrange his
present so as to control that future.

The drawing on the left, "The Carmagnole," is by Käthe Kollwitz,
made around 1900. The carmagnole is a folk dance, sometimes introduced
by a caller, such as the drummer boy on the right, who joined his contemporaries in the celebration. It is danced around a real or symbolic
victim. In this drawing the dance is around the guillotine, a device
invented by Dr. Guillotine, who subsequently almost lost his head on
account of it. The drawing is believed to have been inspired by the words
of Dickens in *A Tale of Two Cities*.

> There was no other music than their own singing. They danced to
> a popular revolutionary song, keeping a ferocious time that was like
> the gnashing of teeth in unison.

In the drawing on the right, the drummer boy is the son of a different
place and age. He still steps in unison with his contemporaries, but unlike
them, he sports a mustache and a beard and has decided to wear an ascot
and also sneakers instead of boots. More important than the sartorial
details, however, is the sound of his drum which may be heard because the
sound of the others has become an unnoticeable backdrop.

24   Kazantzakis, *The Saviors of God*, trans. Kimon Friar (New York: Simon and
Schuster, 1960), p. 129.

FIGURE 36 THE CALL OF THE DRUMMER. On the left, Käthe Kollwitz, "The Carmagnole," circa 1900. Courtesy, Galerie St. Etienne, New York. On the right, courtesy 3M Corporation, St. Paul, Minnesota.

The two drummers are alter egos of the selfsame drummer in all of us, the marcher for symbolic causes, presumed conquerors of time's passage. Shakespeare spoke for all such causes, each claiming the future and each based on memory. This is the talk of King Henry V to his troops on the eve of battle; in Shakespeare's world as in ours, humans are always at war because of their unresolvable conflicts.

> This day is called the feast of Crispian:
> He that outlives this day, and comes safe home,
> Will stand a-tiptoe when this day is nam'd,
> And rouse him at the name of Crispian.
> He that shall live this day and see old age,
> Will yearly on the vigil feast his neighbours,
> And say, 'Tomorrow is Saint Crispian.'

And so the open-ended process of creating ideas about what is right and wrong goes on. Some principles are found more lasting than others, but none can remain stable compared to the laws of biology or physics. When ethical precepts need repair, as they always do, the direction of cultural development must be changed through new symbolic causes and hence through sacrifice. This was done not only by Socrates and St. Thomas More but also by the people listed on the Plaque of Martyrs in the Cathedral of Canterbury; by Che Guevara, Malcolm X, Cardinal Mindszenty, Dietrich Bonhoeffer, the seven Challenger astronauts, by people not listed anywhere, by each and every individual worth his or her salt.

The heavily evolutionary outlook of this chapter and of this book should not be misconstrued to constitute a statement that, because of their evolutionary origins, life or mind are somehow less precious than they have been held to be in the humanistic tradition. Quite the contrary. It demonstrates the uniqueness of both and makes it clear that there is no storage room anywhere from which life, people, and thoughts, once killed, could be replenished by writing a purchase order. There is only the travail of the immense journey: the mini-clockshop that defined now-ness in a now-less world; the gene that learned to sing a program and pass it on with variations; the human brain that expanded the limited horizons of biotemporality to the unlimited horizons of noetic time; and the society that experiments with ethical codes to help it overcome the finiteness in time of the individual and fulfill his and her aspirations in the lifetimes of others.

The thirteenth-century Persian poet Jalāl al-Dīn al-Rumī was the spiritual founder of the dancing dervishes and the author of 30,000 quatrains (rubayat). In poetic metaphor he recognized what in prosaic language we might describe as nature's evolutionary increase in complexity and the increasing wealth of reality so created.

> I died from mineral and plant became,
> Died from the plant and took a sentient frame.
> Died from the beast and donned a human dress,
> When by my dying did I e'er grow less?

In al-Rumī's images, this chapter dealt with plant, with the sentient frame (animal), and with the beast that donned human dress (the mental functions of the brain). I take exception to al-Rumī only in that I view the earlier forms of evolutionary passage as surving and changing, rather than dying. In the following chapter we will turn to the issue of time and the "mineral."

FOUR

Ask the stone to say

TIME IN THE WORLD OF MATTER

The art of conversing with stones is called physics. The question-and-answer periods of the conversations are called experiments. The usual talk is about sizes, temperatures, densities, motion, causes and effects, and the nature of physical space and time. The language spoken is mathematics.

This chapter sketches what physicists have learned about time in the physical world. They found no "supertime," by which one might mean a kind of temporality in which grand and good things impossible for us living and thinking creatures would become possible, such as eternal life or time travel. They found, instead, that time in the physical world is so primitive that it cannot accommodate the ideas of a present, with respect to which one could speak of a future and past. Our experience of time's passage—from past to present to future—is a notion we must bring to physical science as living beings: it cannot be extracted from what is known about time in the physical world.

Current advances toward a unified field theory suggest that the initial universe or singularity, as it is sometimes called, was governed by a single principle of unified forces. Further, all particles of that universe shared with the particles of light the features of masslessness, and they all traveled at the speed of light. I assume that the state of energy corresponding to those primeval conditions was the first and remained the fundamental integrative level of nature.

The creation of the universe is identified in physical cosmology with the beginning of the expansion of the universe, known as the big bang. After the big bang, some of the available energy began to jell into stable elementary particles: they make up the next organizational level of the physical world, the corpuscular level. Even later, particles began to form solid bodies. Together, these massive bodies constitute the third organizational level of the physical world, that of the astronomical universe, with its ten billion galaxies, each with trillions of stars.

We begin our survey of time in the physical world with a sketch of the nature of time in the world of speed-of-light particles, as revealed by Einstein's special theory of relativity. We then go on to learn about time in the world of particles, using quantum theory as our guide. Next we focus on the time of the universe at large—its beginning, history, and ending—as suggested by the teachings of general relativity theory. We then turn to thermodynamics—originally the study of heat, now the science of the transformation of energy in all its forms. We will find that the existential tensions of life and mind that give rise to our concern with time have a very distant, unevolved ancestry in the coexistence of orderability and disorderability in the physical world. The closing section reflects upon the power and limitations of physics as far as statements about the nature of time are concerned.

The Atemporal Kingdom of Light: Special Relativity Theory, the Physics of Fastest Signals

Gamma rays, X-rays, visible light, microwaves, and radio waves are all electromagnetic radiation differing only in frequency. Following general practice, I will use the name "light" to stand for all of them. In the universe of light none of our conventional ideas of time hold. Nothing in it corresponds to our idea of past–present–future (the flow of time) or earlier–later. Our guide to an understanding of this alien, atemporal reality will be Einstein's special theory of relativity, abbreviated as SRT.

Boy chasing light

Albert Einstein is familiar as the white-haired, kindly looking scientist with frightening mental powers. If he were to play Trivial Pursuit, one may be tempted to say, he would win it on the supergenius level with no effort. But nothing could have been further from the real Einstein than the spending of his life on trivial pursuits.

He was a boy of transparent honesty who could neither conceal his contempt for his school's regimented mentality nor stop asking questions his teachers found difficult to answer. Before the final exam of his last gymnasium year he dropped out. With a friend he hiked through the Apennine Mountains to Genoa, where he settled with relatives. In the words of his biographer and colleague, Banesh Hoffmann:

> Museums, art treasures, churches, concerts, books and more books, family, friends, the warm Italian sun, the free, warm-hearted people—all merged into a heady adventure of escape and wonderful self-discovery.[1]

In 1895 he took the entrance exam of the Department of Engineering of the Swiss Federal Institute of Technology in Zürich—and failed. Later he passed and finished his schooling, but he could never bring himself to study what did not interest him. "Like most men of genius," wrote the distinguished British cosmologist and historian of mathematics, G. J. Whitrow, "he was largely self-taught." In Zürich he married Mileva Maric, a young Serbian classmate. Later, when family life began to demand too much of his time, he separated from Mileva, and in 1919 they were divorced. He awarded her any money that should come to him from a Nobel Prize three years before he received it. When it did come, the funds were duly sent, but much of it was lost in inept currency exchanges.

There are two major ways for humans to overcome passing: by nurturing their children and by nurturing their brain children. Einstein opted for the latter way.

At the age of sixteen, he happened upon a puzzling feature concerning the propagation of light. It had been known that light was the oscillation of an electromagnetic field and that it propagated at the rate of 3.10^{10} cm/sec (300,000 km/sec, or 186,000 mi/sec). If one could run after a light beam and catch up with it, thought young Albert, one should observe a stationary wave spread out in space.

Throughout the universe light waves are everywhere, running in every which direction, coming from innumerable moving and stationary sources: candles, stars, fireflies, the sun. Someone somewhere should have observed a stationary wave or at least a slow one. But no one ever did. Nor did prior theory permit the existence of light beams that would propagate

---

1   Hoffmann, *Albert Einstein* (New York: Viking Press, 1972), p. 26.

at velocities less than c, the symbol used to represent the speed of light. The boy of sixteen, chasing the light beam, could not catch it.

He did so, figuratively speaking, after a ten-year mental incubation period. Then in five weeks, he wrote his epochal paper that bore the prosaic title, "On the Electrodynamics of Moving Bodies" (fig. 37).

Great steps in science, such as the one taken by Einstein, represent sudden, discontinuous leaps of imagination. They are always unpredictable from what was known before. A new scientific system usually employs a number of interlocking, mutually reinforcing ideas that make its arguments look circular: the complete theory cannot be understood until its details are appreciated, but those details cannot be appreciated until the whole theory is understood. The mind is capable of making such leaps. The circularity I described is the case for SRT, just as it is for quantum theory and just as it was for the heliocentric ideas of Copernicus in his age.

Revolutionary new insights also tend to upset people because they attack beliefs that appear to them as obviously and necessarily true. Copernicus did not dare print his argument in favor of a sun-centered universe for fear of ridicule; Johannes Kepler needed very deep convictions to be able to insist that the obviously circular planetary orbits were elliptical and that the planets did not travel at constant speeds. Darwin's idea of organic evolution by natural selection is still unaccepted and even resented by many people, as are Freud's notions that we are usually unaware of the mainsprings of our actions.

Einstein never upset too many people outside of the physics fraternity because at first only they could understand what he was saying. Throughout his life he remained a thoughtful intellect, a fiddler on the roof of the brain, an artist by temperament, whose greatest praise for a physical theory was not that it was accurate but that it was beautiful. In human relations he remained naive, as we have already seen demonstrated in his exchange with Freud. When he was offered the first presidency of Israel, he turned it down, saying, "I have neither the natural ability nor the experience necessary to deal with human beings."

From absolute rest to absolute motion

Imagine lying on your back on a grassy meadow, watching the clouds drift by. Suddenly, they stop, as the earth, with you on it, begins to move while the clouds remain fixed. Then again the clouds move and again the

X-Achse unter der Wirkung einer elektrostatischen Kraft *X*, so ist klar, daß die dem elektrostatischen Felde entzogene Energie den Wert $\int \varepsilon X d x$ hat. Da das Elektron langsam beschleunigt sein soll und infolgedessen keine Energie in Form von Strahlung abgeben möge, so muß die dem elektrostatischen Felde entzogene Energie gleich der Bewegungsenergie *W* des Elektrons gesetzt werden. Man erhält daher, indem man beachtet, daß während des ganzen betrachteten Bewegungsvorganges die erste der Gleichungen (A) gilt:

$$W = \int \varepsilon X d x = \int\limits_0^v \beta^3 v \, d v = \mu V^2 \left\{ \frac{1}{\sqrt{1 - \left(\dfrac{v}{V}\right)^2}} - 1 \right\}.$$

W wird also für $v = V$ unendlich groß. Überlichtgeschwindigkeiten haben — wie bei unseren früheren Resultaten — keine Existenzmöglichkeit.

Auch dieser Ausdruck für die kinetische Energie muß dem oben angeführten Argument zufolge ebenso für ponderable Massen gelten.

Wir wollen nun die aus dem Gleichungssystem (A) resultierenden, dem Experimente zugänglichen Eigenschaften der Bewegung des Elektrons aufzählen.

1. Aus der zweiten Gleichung des Systems (A) folgt, daß eine elektrische Kraft *Y* und eine magnetische Kraft *N* dann gleich stark ablenkend wirken auf ein mit der Geschwindigkeit *v* bewegtes Elektron, wenn $Y = N \cdot v/V$. Man ersieht also, daß die Ermittelung der Geschwindigkeit des Elektrons aus dem Verhältnis der magnetischen Ablenkbarkeit A_m und der elektrischen Ablenkbarkeit A_e nach unserer Theorie für beliebige Geschwindigkeiten möglich ist durch Anwendung des Gesetzes:

$$\frac{A_m}{A_e} = \frac{v}{V}.$$

Diese Beziehung ist der Prüfung durch das Experiment zugänglich, da die Geschwindigkeit des Elektrons auch direkt, z. B. mittels rasch oszillierender elektrischer und magnetischer Felder, gemessen werden kann.

2. Aus der Ableitung für die kinetische Energie des Elektrons folgt, daß zwischen der durchlaufenen Potential-

earth. This is a well-known play of the senses. What does "really" move, the earth or the clouds?

Motion once meant motion with respect to the absolutely resting earth. The sun, the moon, the planets, the stars, ships, horses, and slimy snails were all said to move because they changed their positions with respect to the earth. The Copernican revolution altered this firmly held view by selecting the sun as the new reference at absolute rest. After Copernicus, it was the sun that rested; from then on ships, horses, snails, and the planets were all said to move because they changed their positions with respect to the nonmoving sun.

By letting the earth loose among the planets, Copernicus created a scientific perception of the astronomical world that showed a higher-order of unity than was possible to attain earlier. Astronomical theory broadened, the description of planetary orbits became simplified. Should someone have given evidence that the sun itself moved with respect to the Milky Way, then the Milky Way might have been declared the object that remained at absolute rest.

In Newton's physics—the next step after the Copernican understanding of planetary motion—the role of absolute rest ceased to be attributed to any specific body. Instead, it came to be an attribute of absolute space.

> Absolute space, in its own nature, without relation to anything external, remains always similar and immovable. . . . Absolute motion is the translation of a body from one absolute space into another.[2]

In a remote region of the fixed stars, wrote Newton, there might well be an object that is at absolute rest. But even if there were none, we still may and even must abstract from our senses the idea of absolute space. After Newton, motion came to mean the changing of position with respect to absolute space. Mail coaches, buffalos, the earth, and the planets were all said to move because they changed their positions with respect to the absolutely resting space. They also moved with respect to each other, which made it unnecessary to refer continuously to absolute space.

---

2   This and the following two citations are from *Sir Isaac Newton's Mathematical Principles of Natural Philosophy and His System of the World,* trans. Andrew Motte, rev. Florian Cajori (Berkeley: University of California Press, 1966).

An appropriate twin to absolute space was

absolute, true and mathematical time [which] of itself, and from its own nature flows equably without regard to anything external, and by another name is called duration. . . .

Since events also happen one with respect to another, it is unnecessary to refer continuously to absolute time. It is sufficient to have

some sensible and external (whether accurate or unequable) measure of duration by means of motion, which is commonly used instead of true time; such as an hour, a day, a month, a year.

The physicist's *t,* the symbol that stands for time in the equations of physics, has been thought to represent the concept of time defined above: a uniform flow unperturbed by what happens in time. In Newtonian thought, my time is your time; it is the bedbug's time, the moon's time, everybody's and everything's time.

Newton's absolute space and time, held together by the relationship $v=d/t$ (velocity=distance/time), allowed for a scientific perception of motion that provided a higher-order of unity in our understanding of nature than was possible with the sun-centered universe of Copernicus. After Newton, there were no privileged heavenly bodies; the physical laws of motion could be extended to every piece of moving matter across the universe. There was no need to refer to the sun, much less to the earth. The broadened view permitted Newton to formulate the first truly general scientific principle, the universal law of gravitation.

Newton died in 1727. By the nineteenth century, the idea of absolute space became reified into the idea of an all-pervasive ether, a notion that probably originated with Descartes. Light could be imagined as the wave motion of ether, a substance at rest through which all objects moved without disturbing it. The earth, comets, joggers, and raindrops could now all be said to move because they changed their positions with respect to the absolutely resting ether. In 1887, A. A. Michelson and E. W. Morley, two physicists working at the Case School of Applied Science in Cleveland, Ohio, decided to measure the velocity of the earth with respect to the absolutely resting ether, represented by light, which was assumed to be—as I just mentioned—its wave motion. They set out, therefore, to measure the velocity of the earth with respect to light beams. Their experiment gave a null result: as far as they could tell, the earth did not move with respect to the postulated framework of absolute rest. Let us leave M & M and the physics community of their age puzzled, and take a new tack.

Einstein's 1905 paper was not written to explain the negative outcome of the Michelson–Morley experiment—it was much broader in its intent—but among many other matters, it did resolve that puzzle.

Einstein asserted, though not in these words, that there was nothing in nature that corresponded to our idea of absolute rest, but there was something that corresponded to the concept of absolute motion. It was the motion of light. He then proceeded to rewrite physics with absolute motion rather than absolute rest as the reference frame for the laws of motion.

What was to be understood by absolute motion?

It is easy to understand what absolute rest could mean. It would mean that an object existed or could be constructed so that all objects in the world could be said to move if they changed their positions with respect to it, and that the laws of motion written on this assumption were empirically correct. But as I have noted, no such object could be identified. And if one was constructed in the form of absolute space, the laws of motion so written were empirically incorrect, as demonstrated by the Michelson–Morley experiment.

However, there was something we could identify as absolute motion. It is a motion—a physical process—such that everyone, everywhere, at any time will describe it as having the same speed, regardless of his or her motion with respect to it or to anything else. And the laws of motion written on this assumption—those of SRT—were empirically correct.

The idea of any motion having a speed independent of the motion of an observer goes against common sense derived from common experience. If on a straight road, a bicycle moves away from me at 10 miles/hour and I walk in the opposite direction at 3 miles/hour, our relative velocity will be 13 miles/hour. If the bicycle and I move in the same direction, our relative velocity will be 7 miles/hour. How am I to imagine our relative velocity remaining 10 miles/hour, regardless of my motion?

Yet this is exactly what the equations of SRT tell us about a pulse of light. No matter how rapidly and in what direction anyone moves, the measured speed of light will remain the same, $c = 3.10^{10}$ cm/sec.

Michelson and Morley measured the speed of the earth's motion with respect to light beams. But nature is so constituted, says SRT, that the speed of any motion with respect to light beams is always the same, be it the motion of the earth, that of a rushing galaxy, or a crawling snail. It follows that it is not possible to find a speed other than c with respect to light.

Stop searching for a star at absolute rest or even imagining that an absolutely resting space or ether exists, states SRT. It does not. Fix your eyes on absolute motion. Ask not what your velocity v is with respect to something or other; ask, instead, what fragment of the speed of light that velocity is. Ask what is your v/c ratio?

If you are a good long-distance runner (10 km/hour) your v/c while running is 9×10^{-10}. When the v/c ratio is as small a number as this, relativistic physics is indistinguishable from Newtonian physics.[3] The difference between the two physics becomes significant, however, as v approaches c, that is, as v/c approaches one. For instance, elementary particles called muons can be observed traveling at 99.999,999 percent of the speed of light, making their v/c ratio with respect to the observer 0.999,999,9. The various special relativistic changes, discussed below, then become substantial.

Special relativity theory is the science of light-referred physics. The consequences of this rewriting of physics are far-reaching.

Curiouser and curiouser: The atemporal level of the physical universe

Sam Walter Foss, a New England poet of the end of the nineteenth century, must have had in mind a home with curtains, a bedroom, a kitchen, a fireplace, and kindness when he wrote,

> Let me live in a house by the side of the road
>   Where the race of men go by—
> The men who are good and the men who are bad
>   As good and bad as I.

---

3   If you run at the speed of a car (100 km/hr), your v/c ratio is 0.000,000,009; at the speed of a rifle bullet (1 km/sec), your v/c with respect to the rifle is 0.000,003. As a fellow traveler of our earth, moving at 30 km/sec with respect to the average motion of all other bodies in the galaxy, your v/c with respect to that average motion is 0.000,1.

That was then. Today we live in a laboratory by the side of a space-road where the space-race goes by. Let us imagine a spacecraft of the light-chaser class, carrying a laboratory equipped with future state-of-the-art computers, communications gear, and scientists. Let me then present the approved, science-fiction version of special relativistic travel. It is science because it is based on the well-established principles of SRT. It is fiction because its gadgetry is nowhere in sight.

We are going to watch a series of light pulses rushing by, pursued by a light-chaser. We could not actually see the vehicle because it would be going too fast for our eyes, but we could remain in radio contact with it. By telemetry, let us measure times, distances, and masses on the light-chaser. For instance, we could measure the mass of a standard lollipop on board, the length of a standard broom, and the time it takes for an egg to acquire the consistency of a standard three-minute egg.

First we begin monitoring the speed of the light-chaser, making a measurement every hour. We express our findings in terms of the v/c ratio. They are: 0.999; 0.999,9; 0.999,99; 0.999,999; 0.999,999,3; 0.999,999,4; 0.999,999,45; 0.999,999,48. . . . After the last figure, the light-chaser stops gaining speed. The reason is soon found: the standard 1-ounce lollipop weighs 62 pounds, the weight of the spacecraft increased a thousandfold; its rockets are not powerful enough to further accelerate the spacecraft at any appreciable rate.

But the craft does travel rapidly: at 99.999,948 percent of c, to be specific, at 299,999.82 km/sec. At that speed, the light pulses whose velocity we determined to be 300,000 km/sec should be drifting by the light-chaser at a leisurely 180 m/sec, according to classical kinematics. We ask the travelers to confirm that speed. What they find, however, is 300,000 km/sec.

The reasons for this curious state of affairs are, however, soon found. Their standard broom is now much shorter than ours, and it takes much longer than three minutes to prepare a standard three-minute egg. In other words, their clocks tick slower than ours do. The method they have used to measure the speed of light was the same as ours. They laid down their standard broom so many times to establish a standard distance, then measured the time taken by a light pulse to cross that distance. They obtained for the value of c the same number we did because their broom and their egg (clock) were in cahoots and conspired to make the magnitude of c remain the same.

The Michelson–Morley experiment involved two domains of physics that were regarded, until Einstein, as separate. One was Newtonian mechanics; it dealt with such problems as the motion of the earth in the astronomical universe. The other was James Clerk Maxwell's electrodynamics, the science of electricity, magnetism, and the propagation of electromagnetic fields such as light. SRT combined the two and through the combination explained the absence of slow light waves. There are no light waves that travel at speeds less than c, because the value of c is a constant of nature.

A number of constants of nature exist whose value remains the same whether the person making the measurement is young or old, male or female, rushing back and forth, or standing on his head. They include the electric charge on an electron, Planck's constant, the constant of gravitation, and, since Einstein, the speed of light.

For the astronauts on board, light-chaser life is as usual: Weight is appropriate to height and build, pulse registers 72 per minute, length of gestation for an Arabian camel remains 406 days. It is only in comparing time measurements between systems in relative motion that special relativistic effects enter. And those effects are mutual. They are symmetrical. The people on board the light-chaser will conclude that our standard lollipop weighed 62 pounds and that our clocks ticked much slower than theirs did.

But within the confines of each of the two systems—laboratory and light-chaser—where relative velocities compared with c are small, the world of time, space, and matter behave identically. SRT tells us interesting things about time measurement under some unusual conditions, but does not say what time is. It assumes that we already know what it is.

The light-chaser experiment is technologically too farfetched, but we do not really need to perform it. The validity of SRT has been tested through innumerable experiments and found valid. The tests normally involve very rapidly moving small objects, because except for galaxies (and they make for a different kettle of heavenly fish), only particles can approach the speed of light. An electron, for instance, weighs 10^{-27} grams. Even if its mass increases ten-thousandfold, it will still weigh only 10^{-23} grams; the energy it needs for further acceleration is not yet prohibitive. But even electrons cannot reach c, because their masses increase without limits as the upper boundary of possible relative velocities is approached.

The only objects that can travel at velocity c are photons, the particles of light, because they have zero rest mass. For the same reason (zero rest mass), they cannot travel at velocities less than that of light. They are always on the move and always at the same speed.

Although no object of however small mass can reach the relative velocity c, we may imagine my friend Georgie the Ghost riding an accelerating muon. We may watch him hop over to ride a photon like Billy the Kid would jump from the back of his swift horse onto the roof of a coach, rolling along with the even-swifter train. If ghosts won't do, the scientific interpretation of the equations that govern the world of particles and photons will, because we are able and entitled to think ourselves into their umwelts.

No clock in itself can measure time or be judged slow, fast, or just right. Imagine, therefore, a traveling and a stay-home particle. Since all particles tick (oscillate), we have two clocks. As the speed of the traveling particle increases with respect to the stay home one, the rate of its ticking with respect to the stay home one will decrease. That slowing is not the running down of an unwound clock. The rate of time's "flow" in the particle's umwelt has slowed. As the particle's velocity approaches that of light—so the equations tell us and experiments confirm—the particle clock, as seen in its own framework, will tick slower and slower until, upon changing into a photon, it stops ticking.

For the photon, the passage of time has no reality. In its world nothing corresponds to future–past–present or before–after. Its umwelt is atemporal.[4] In the life of the photon—using the world "life" for lack of a better word—all events happen at once. If a photon is emitted at one instant and absorbed a year later (as perceived by us people, using the hibernation clocks of bears to measure out a year) the two events, as far as the photon goes, happen simultaneously. Since no two events may be closer in time than being simultaneous, light propagation must and does constitute the fastest possible signal. Furthermore, since in the framework of the

---

4   Before going on, the reader may want to review the discussion of level-specific temporalities that occurs early in Chapter 3 and recall the visual metaphor of the arrow that lost its head and tail, then broke to pieces, and finally vanished. The "finally vanished" form of time is the atemporality of electromagnetic radiation.

photon, all distances shrink to zero (remember the shrinking standard broom?) the umwelt of the photon is not only atemporal but is also aspatial. There is nothing in the photon's reality to which our ideas of time or space could correspond.

May a person gain eternal life by becoming a part of the atemporal world of light? No. Because to get there, his body must first change into a cloud of particles, then each of those particles must change into electromagnetic radiation. The surest way to get on with such a project would be to be dropped into the center of an atomic explosion. Unfortunately, long before the desired state of timelessness is reached, those coordinated functions of matter we know as life will have ceased. So much for entering the atemporal world of light to secure eternal life.

All this madness of special relativity theory about atemporal processes is beautifully self-consistent, agrees with experiment, and is completely sane.

Missing arrowheads and absent presents: The eotemporal level of the physical universe

Let us leave Georgie the Ghost riding on a photon in the atemporal integrative level of the physical universe, ignore for the moment the integrative level of elementary particles, and enter—in our minds—the umwelt of massive matter. That is the world of all bodies made up of trillions of tightly packed particles, such as cherry pits, the rock of Gibraltar, the moon, the planets, the sun, and all the stars.

The two-way, presentless time of this organizational level, as the reader will recall, may be represented by the shaft of a head- and tail-less arrow and is called eotemporality for Eos, the goddess of dawn. The prefix *eo-* has been used by scientists to stand for the earliest, oldest of developing forms. Eohippus is the earliest ancestor of the horse, eolithic the oldest period of the Stone Age. Eotemporality is the oldest continuous kind of time. It came about in the course of inorganic evolution when, after the big bang, the wildly rushing particles of hot gases began to jell into solid matter.

How is one to understand eotemporality, a kind of time that has no preferred direction?

Think of a clock on the wall. It does not tell us what we are to mean by "now": on the contrary, we must tell it. The clock only gives a name to our "now," calling it "3:47" or "7:19." Neither does it say anything about future or past. Its hands could be made to rotate counterclockwise if one put in an extra gear. Someone else may prefer to redesign the dial so that, going clockwise, noon would be followed by 11, 10, and 9 o'clock. Nothing in the clock could tell us whether an instant was later or earlier than another one. People looking at a backward-rotating clock would exclaim, "See that crazy thing?" They would not think they were getting younger.

The equations of physics dealing with the macroscopic world report about this presentless, directionless quality of the temporal umwelt. The physicist's t, as actually used, does not have a preferred direction. Here is an equation we already know, $d=vt$: distance traveled = velocity times time. Let us call the present instant 0 hour and use the equation. A plus sign will signify the future, a minus sign the past.

Multiply 60 miles/hour by +2 hours, and get +120 miles. Multiply 60 miles/hour by −2 hours, get −120 miles. How do I know that "+120 miles" means that two hours from now I will be 120 miles away? How do I know that "−120 miles" means that two hours ago I was 120 miles away? By knowing what is meant by future and past with respect to the present. The formula did not tell me. Even the plus and minus signs are arbitrary. Depending on the direction along which I decide to measure distance, it might be that in two hours I will be −120 miles away and that two hours ago I was +120 miles away. Future, past, and present, and the flow of time as a phenomenon of the world were smuggled into the formula through the time experience of a living being.

In all equations of physics t enters with the same ambiguity: it represents time flowing in either or both directions. Physicists say that the variable t does not respond to the direction of time. It informs us of the nature of time in the macroscopic physical world.

The human notion and experience of time, as we have learned in the earlier chapters, relates to the experience of passing. Therefore, when philosophically minded physicists speak of time in physics, they habitually identify it with the time of life and mind. But as I just explained, the peculiar way time enters physics does not allow for time's flow. If time's passage is not represented in physics, so many philosophers of science hold, then it cannot be real. The conclusion often stated in no uncertain terms is this: time's passage is an illusion.

Upon the death of his close friend, Michelangelo Besso, Einstein wrote to Besso's son and sister. Michael, said Einstein,

> left this strange world just before me. This is of no importance. For us, devout physicists [*gläubige Physiker*], the distinction between past, present and future signifies only an obstinate illusion.[5]

The obstinate illusion is biotemporal and noetic time. Einstein must have felt supported in this view by the common interpretation of the laws of physics as reporting on a timeless world. But the world of massive matter is not timeless but eotemporal.

Let me illustrate the relationship between the directionless, presentless time of the physical world on the one hand and biotemporal and noetic time on the other by the famous clock problem—also known as the twin paradox—of relativity theory.

In 1911, writing about the implications of relativity theory, Einstein made the following remark.

> If we placed a living organism in a box . . . one could arrange that the organism, after any arbitrary lengthy flight, could be returned to its original spot in a scarcely altered condition, while corresponding organisms which had remained in their original positions had already long since given way to new generations. For the moving organism the lengthy time of the journey was a mere instant, provided the motion took place approximately with the speed of light.[6]

How may it be possible that under certain conditions you and I can come up with different time measurements for the amount of time that has passed between the same two shared events, and that the difference is not an error but is, so to say, really real? Why is it that my traveling twin,

5   *Albert Einstein–Michael Besso Correspondance* (Paris: Hermann, 1972), p. 638. My translation.

6   Einstein, *Vierteljahrschrift der Naturforschende Gesellschaft, Zürich*, vol. 56, pp. 1–14. Translation in A. Kopff, *The Mathematical Theory of Relativity*, trans. H. Levy (London: Methuen, 1923), p. 52.

having returned from a trip taken at very great speeds, aged, say, only ten years while I aged thirty-four years?

The time dilation, as the differential passage of time is known and as experimentally confirmed beyond any reasonable doubt, demonstrates the validity of the Lorentz transformations of SRT. These are equations (instructions) for the calculation of time for a clocklike process—or even empty space, such as a room—that is in motion with respect to the measurer.

The best way to elucidate the clock problem is to examine the equations in their mathematical form or, as is done in Appendix 1, through their geometrical representation. The Lorentz transformations do not elucidate the passage of time; on the contrary, they assume that we already know what time is before we turn to them.[7] Let me attempt, therefore, to give a description of the clock problem in a manner that shows how we bring to it our experience of time's passage.

Imagine a clock and a rod (refined versions of an egg and a broomstick) sent on a round trip to a faraway star. Before they leave, the traveling and stay-home rods are placed side by side; they are the same length. The two clocks, also side by side, tick at the same rate. While on the journey, the rod will undergo motional contraction—it shrinks, for real—while the clock will undergo time dilation, also for real. After their return the rod and the clock are again placed next to their stay-home twins. The rods are again the same length, the clocks again tick at the same rate. By comparing the rods' lengths and the clocks' rates, nothing could be known of their past adventures.

Next, let us attach a cumulative counter to each clock. After the return of the adventurous one, the two will beat time once again at the same rate but they will sing different songs. The stay-home clock may recite: 2,032,260,686 . . . 2,032,260,687 . . . 2,032,260,688 . . . 2,032,260,689 . . . while the traveling one says 4,593,474 . . . 4,593,475 . . . 4,593,476

---

7   To repeat, the Lorentz transformations say nothing about what we are to mean by time. They give us only the rules for transforming time readings between coordinate systems in a uniform, relative translation. An understanding of what we are to mean by the present, the future, the past, and the flow of time cannot be extracted from them. The situation is analogous to knowing the exchange rates between any two currencies on earth, but knowing nothing whatsoever about the purchasing power of either of them anywhere or, for that matter, knowing nothing of what is meant by currency.

. . . 4,593,477. . . . It is now possible to tell which clock was away: the one that generates smaller numbers.

The meter rod that traveled is no worse off for wear. Since it stands for distance (without any reference to time), it cannot have memory. Neither can the clock: its reality is eotemporal. It has no way of recognizing whether a tick is before or after another one, whether it is to be added to or subtracted from a count. To render a cumulative count, there must be a summation along a temporal direction, going from the past to the future, with respect to a present, such as the organic present. For intellectual work, the mental present is also necessary, but not for the counting of ticks, such as occurs in aging.

For instance, the gestation period of an Indian elephant is 645 days, that of a house mouse 19 days. Let each mate with its respective partner at the same time, as determined by an experimenting relativist. Then let the mouse go on a round trip and return home to deliver its pink offspring, just as the elephant produces its long-nosed one. In this experiment it was the animals' organic clocks that did the counting, not their minds.

This science-fiction experiment permits us to learn of a unique relativistic time concept: that of *proper time*. It is the time indicated by a clock-like process that is at rest with respect to the object for which the passage of time is to be measured. The clock is usually imagined as being next to the object or even as a feature of the object itself. For instance, the 645-day and 19-day gestation periods were measured in the respective proper times of the Indian elephant and the house mouse.[8]

In Orwell's *Animal Farm,* all animals were equal but some were more equal. All proper times are proper as defined, but some are privileged. Specifically, the proper time between two events in the life of a clock that is at rest with respect to the average motion of matter in the galaxy will always show the longest possible elapsed time between those two events. The proper time of all other clocks having the same two events occur in their lives, but moving with respect to the first clock, will show briefer time intervals. If the traveling clock is a photon, the time interval decreases to zero: the photon's umwelt is atemporal. We will return to this asymmetry in a subsequent section entitled "Cosmic Time: Mainstreamers and Deviationists."

---

8   The original German term for proper time, *Eigenzeit,* is self-explanatory. *Eigen* means "belonging to the self"; *Zeit* means "time."

Since the elephant's proper time is privileged in the sense sketched, is it somehow the one, and the only true elapsed time between the mating and delivery days? No, it is not.

No clock in itself can measure time. One needs at least two clocklike processes and a theory to connect the two readings in a meaningful, stable way. In this case, the expected readings could be calculated ahead of time from those equations of relativity theory that connect the readings.

If the numbers do not agree with what the theory predicted, then one must suspect an error in the clock readings, in the theory, or in both. In any case, a valid time measurement has not been made. But if the calculations and the readings of the clocks agree, be they atomic, mouse, or elephant clocks, then a valid time measurement has been made. In the example given, the Indian elephant's 645 days and the house mouse's 19 days are both the true lapses of time for the conditions given.

As the distinguished relativist J. D. North wrote, in a paper about time and relativity theory, "The moral of all this is that we should not speak of relative clock rates at all without specifying the events with respect to which they are measured." Let me add that for a complete understanding, we must keep in mind the role of the organic present in introducing the directional flow of time, with respect to which cumulative counts of the ticks of physical clocks become possible in the first place.

The issue of simultaneity, or "I wonder who's kissing her now"

This melancholy tune of the 1940s goes on, "I wonder who's teaching her how. . . ." The kissing and teaching could not be going on next to the wonderer, for then he would not need ask; it must be happening instead, in Shakespeare's sober words, "from me far off, to others all too near." The question of the 1940 song pertains to "Who?" That is easy to answer: it's that creep. The difficult question to answer is this: what is to be meant by "now" somewhere "from me far off"? In other words, what do we mean by an event somewhere else being simultaneous with an event right here?

I have already spoken about the question of simultaneity in the context of the social present. In this section we are going to extend what we have learned there, to include cosmic distances.

It has been assumed that now is now, across the world. One may be ignorant of what is happening at a faraway place right now, but common sense said that was only because nobody could run fast enough or shout loud enough to get the message across 20, 200, or 2,000 miles. Light, so scientists used to hold, was an exception: it could propagate instantly. In the seventeenth century, the physicist Christian Huygens (whom we already met in connection with the improvements he made on the Galilean pendulum) demonstrated that it could not. The sudden brilliance of a star seen at night is not an event occurring "now" but one that happened in the past.

But there were no theoretical reasons why a physical process other than light, one that propagated instantly, could not some day be discovered. The situation changed with Einstein's discovery that *c* constituted a natural upper boundary to the speed at which energy (and hence messages) may travel. For reasons I shall try to explain with the help of an example, this discovery introduced an ambiguity into what was to be meant by a distant event being simultaneous with a here-and-now.[9]

Let us recall the plight of the man in eighteenth-century Philadelphia separated from his family in Norway. Every instant of his life was preceded by 40 days of ignorance and followed by 40 days of impotence as far as events in Bergen were concerned. He lived in the center of an 80-day ignorance–impotence-bubble. In his experience of time (in Philadelphia), future and past were separated by the brief period necessary to make a conscious response to a stimulus, say, by 0.2 sec. That 0.2 sec, when projected to the distance of Bergen, opened up to 80 days because that was the lapse of time which, at that distance, separated stimulus from response. With the invention of the telephone, the 80-day bubble could be reduced to a small part of a second that may be neglected for nonscientific use.

However, as distances increase, the limiting velocity of light ceases to be negligible. The mental present of 0.2 sec, at the distance of the moon opens up to 2.5 sec, at the distance of the sun to 1,000 sec, at a star one light year away to two years, at the edge of our galaxy to 160,000 years. And there is no possibility of reducing these periods through technological improvements. The boundaries are set by nature.

It has been a rational belief, as we learned earlier that somewhere within the duration of an opened-up present-at-a-distance, perhaps half way

9   In Einstein's paper the sequence of reasoning goes from (1) questioning the meaning of simultaneity-at-a-distance to (2) the privileged position of the speed of light. These two issues—the need to examine what was to be meant by now-at-a-distance and *c* as a limiting velocity—are two sides of the same coin: the replacement of absolute rest by absolute motion as the reference state for the laws of motion.

through it, there was an instant-at-a-distance that was simultaneous with my present instant here. But SRT has shown that such an instant at the edge of the galaxy, to use an example just given, may be anywhere within a 160,000 year span. Nature does not define an instant that, within that span, must necessarily be simultaneous with my "now" here. Furthermore, two events at a distance—e.g., one to my right and one to my left—that I judge to have been simultaneous because I saw them at the same time, will not be judged as simultaneous by someone moving with respect to me. And vice versa.

This family of not-common-sense affairs is known as the relativity of simultaneity. It refers the phenomenon of the "now" back to the living organism whose internal coordination defines the organic present; to the healthy brain whose synchronized functions create and maintain the mental present; and to society that creates its social present.

It has already been said but let me say it once again: there is nothing in inanimate nature to which the idea of "now" could be applied. Self-consistently, there is no future or past and no arrow or flow of time.

By way of review, I wish to note that as do all great scientific theories, SRT also has aspects of the art of fugue: it is made of a number of intricately interwoven, repeating patterns.

It revealed that the umwelt of light is atemporal. It replaced absolute rest by the absolute motion of light as the reference for the laws of motion. It demonstrated the limiting character of light velocity and the motional changes of rods, clocks, and masses. It took over from classical physics the absence of a present in the physical world and maintained the related absence of temporal direction. It discovered the temporal boundaries of actions-at-a-distance (anchored to the organic present, smuggled in from biology). All this is a far cry from the reassuring security of absolute time and space, held as obviously and necessarily true facts of nature by earlier, more stable epochs.

The Prototemporal World of Particles: Quantum Theory, the Physics of Ordered Randomness

From the world of SRT, we next turn to the universe of particles.

Shortly after the big bang, some of the energy began to jell into particles that moved at speeds less than that of light. Even later, these particles began to form solid, macroscopic matter. The subject of this section is prototemporality, the time of the integrative level between the atemporal universe of speed-of-light particles and the eotemporal universe of massive matter. The prefix *proto-* has been used by scientists to signify "first formed" or "parent substance," as in protoplanet, protoplasm, or protozoa.

The parrot and the salad: Duality and indeterminacy

Like special relativity theory, quantum theory comprises a number of mutually supporting principles. Each of them was new when first proposed, and each was unpredictable from what had been known earlier. This subsection discusses two of those interlocking principles and what they reveal about time in the universe of particles.

Quantum theory teaches that elementary forms of matter may be given two very different but equally valid descriptions. One is that of waves. As with all waves, these are specified by their frequencies, wavelengths, and amplitudes. The other representation is that of small, solid, material objects. As with all solid bodies, these are specified by their positions in space, their masses, and their momenta. The simultaneous validity of these two different descriptions is known as the principle of *duality* or *complementarity*.

Speaking about ambiguity in art, Picasso once remarked that, "A green parrot is also a green salad and a greet parrot. He who makes it only a parrot, diminishes its reality." An object in quantum theory is a particle as well as a wave. He who makes it only a particle or a wave diminishes its reality and will not be able to construct a theory of the atomic world that can be experimentally verified.

It is difficult to talk about the elementary forms of matter. Our language has no word for something that is a solid body like a watermelon and also a wave like those of the sound of music. I will call them elementary objects.

Anthropologists have similar difficulties when they want to describe cultures different from their own. They must first learn to understand the world as seen by the people of those cultures in their own terms. Only then, through the use of new concepts, can anthropologists make the beliefs and views of the other culture intelligible to their own. In an analogous situation, the umwelt of a small child is so different from that of an adult that children sometimes seem to belong to a different species. If instead of children, we try to understand the world of the ape, the snail, or the virus, it is necessary to construct new concepts and speak languages that will be alien to the uninitiated.

Leaving the virus umwelt for that of elementary objects, even our common modes of visualizing reality become useless. We can describe the nature of that universe only through mathematics, the most abstract of all languages. The work of quantum theorists, therefore, is by necessity entirely mathematical. It consists of identifying symbols whose mathematical behavior corresponds to the behavior of instrumental observations. Quantum theory makes statements about those symbols and not about the objects whose behavior they are believed to represent. This is the only way we know how to handle that world. What follows is an attempt to translate those symbolic relationships into other and more familiar symbols, to wit, garden-variety words.

Elementary objects are both particles *and* waves at the same time. Which of these features manifests itself depends entirely on how an experimenter puts the question. If the test asks, "Are you a wave?" the answer will be, "Yes, I am." If it asks, "Are you a particle?" the answer will also be "Yes, I am." It is impossible to design an experiment that could provide the answer that a single elementary object is both a wave and a particle, but that is not really a problem, because elementary objects of the same species are completely and indistinguishably identical. There is no mark by which one electron could be distinguished from another. If I used an electron to tell me that it was a wave, I can use another one to tell me that it is a corpuscle (a "little body").

Imagine a watermelon seen as taking up space and the rhythm of flamenco heard in time. Then imagine, if you can, something that is both a watermelon and a flamenco rhythm at the same time. *Not* a dancing watermelon, but a thing that is also a melody.

The strange state of affairs of the prototemporal world, represented by the watermelon–flamenco, does not make it somehow superior to ours. On the contrary. It is a level of nature so primitive that in it processes and structures—time and space—are not yet completely distinct.

There is a haziness to the prototemporal world, recognized in quantum theory by the principle of *indeterminacy* or *uncertainty*. That principle specifies the reciprocal manner in which certain conjugate (paired) variables behave: as the experimental precision with which one of the variables is determined increases, the precision with which the related variable may be determined decreases.

One pair of variables is energy and time, another is momentum and space (position). As the experimental accuracy in measuring the energy of a particle is increased (experimental uncertainty decreased), the accuracy of determining the time (instant) when the particle had that energy decreases (experimental uncertainty increases). And vice versa: As the experimental accuracy in specifying the time (instant) when the energy of a particle is to be known is increased, the accuracy with which the value of that energy may be determined decreases (the uncertainty in its value increases). The same kind of reciprocal relation holds between the momentum and the position of a particle.

To discover the exact behavior of a particle in a manner that such determinations are made for macroscopic bodies, it would be necessary to know precisely where it was and when, and what energy and momentum it had at that instant, at that spot. But the conspiracy of the uncertainty principle places a limit upon the accuracy to which members of the pairs of variables I have mentioned may be jointly known; the numbers obtained may only be ranges of values.

According to the Copenhagen interpretation of quantum theory—so named after the great Danish physicist Niels Bohr and generally embraced as correct by mainstream physics—these limitations to precision in our measurements are not due to imperfections of our instruments or to the insufficiency of our understanding. Instead, they reflect certain limitations to our garden-variety concepts of space and time.

The conclusion I wish to draw is that in the netherworld of particle-waves, the nature of time and space themselves are in a sense hazy and ill-defined.

The prototemporal world with its particle-waves is all around us. It is made of 75 percent hydrogen and 24 percent helium, with the rest of the elements taking up the balance. When particles combine to form massive bodies, duality and indeterminism do not cease to be true, they only become insignificant; the laws of quantum mechanics connect with those of macroscopic physics in a continuous, unbroken fashion. The quantum mechanics of large bodies is equivalent to classical mechanics and electromagnetism. In the world of large bodies, the hazy, fragmented time of the prototemporal world becomes the continuous (still undirected) time of the eotemporal universe represented by the physicist's t.

Probabilistic laws: Perfect reporters about an unevolved world

If we replaced a proton by another one, no one could tell and we could not prove it because, as I remarked earlier, particles of the same species are indistinguishable. They are absolutely identical. We may talk of one or another electron if there are two, but we cannot talk of "Electron 1" and "Electron 2" as if each carried different labels. If they did, they would not be identical. The indistinguishable identity of particles cuts down the number of ways systems of two or more particles can behave. What would mankind's options be if all humans were indistinguishably identical? Very limited.

Speaking with the wisdom of hindsight, it is evident why all particles of the same species must be exactly alike. They are the firstborns of chaos, the first steps in creating order out of complete disorder. From these elementary objects, the larger, distinguishable bodies of the eotemporal world came to be built. The chances of there being two macroscopic bodies in the universe that are exactly alike is negligibly small. Among the living, it is nil. Even viruses of the same species differ among themselves, as witnessed by the fact that they are subject to natural selection.

The collective behavior of objects that are exactly alike can only be governed by statistical laws, because there is no way to single out any of them for individual attention. Again, it is evident why the earliest laws of nature had to be statistical. They were the first steps along the road toward the deterministic (nonstatistical) laws of macroscopic physics, on the way to the governing principles of life that allow for the freedom of organic creativity.

The behavior of aggregates of very many identical objects is governed by the law of large numbers: what chance is in the behavior of a few, is a non-negotiable demand by the many. There is no way of telling which atom will do next whatever a statistical law requires, even though statistical laws are deterministic for the behavior of the aggregate of many atoms. For large numbers—and even a cubic centimeter of water contains 10^{22} molecules—the statistical character of nature jells into the deterministic world of macroscopic physics. The probabilistic, unpredictable behavior of individual particles cannot be used to justify the unpredictable behavior of certain massive bodies, such as is manifest in the creativity of humans.

That the atomic world demands the use of statistical laws for the description of its behavior tells us that their spatial and temporal umwelt is a more primitive one than that of massive objects. Specifically, the following argument holds for all aggregates that are made up of identical objects. The probabilistic distribution in the behavior of many identical systems made of such objects, and spread out in space, is the same as the probabilistic distribution in the behavior of any one single system in time.[10] A spatial pattern in design, as it were, is interchangeable with a temporal pattern of a program. Space and time are not yet distinct.

James Bond always examined the prior runs of the roulette wheel even though he knew that each time the croupier flicked the ivory ball, he began a game independent of the games before it. Independent but not lawless: the series of roulette games spread out in time remained controlled by an iron-fisted, deterministic law that happened to have a probabilistic form. Had Bond assumed that all roulette wheels in the Casino were identical, he would not have had to spend time on finding out the history of a particular wheel. He could have photographed from the ceiling the positions of the ivory balls on a large number of (stopped) wheels at a single instant. The temporal program of one wheel-and-ball (one system) would have been interchangeable with the spatial pattern of many wheels-and-balls (many systems) at an instant.

Einstein never felt at ease with the probabilistic character of quantum theory. He preferred to understand the physical world in terms of well-defined causes, positions, and instants, allowing for precisely predictable effects, and remarked once that God does not play dice. But the statistical laws of the prototemporal world do not hide God's inscrutable plans behind a screen made impenetrable by the probabilistic character of quantum principle. Quantum theory does not report about a universe of superior evolutionary order. On the contrary. It is statistical because its principles govern the most primitive organizational level of the physical world above chaos. The babbling of an infant is not an attempt to hide his thoughts and thus mislead his listeners. He does not have well-defined thoughts to be hidden.

I said earlier that the prototemporal world of particles is all around us.

---

10   The connoisseur will recognize this as a statement of the ergodic theorem: for dynamic systems of distinct but indistinguishable members, ensemble averages and time averages yield the same number.

Statistical laws are also all around and beneath us, as it were, and the atomic world is not the only domain to which they apply. They are also appropriate to all aggregates of objects that are indistinguishable, in one or another way, and apply to them with respect to that feature of identity. Examples are all taxpayers, all unwed fathers, all postal patrons, and all drivers. The statistical predictions of the American Automobile Association on the number of fatal accidents on a particular holiday weekend are usually close to what happens, even though no one can tell who, when, and where will be involved. In a different domain of social life, it is the practical indistinguishability of dollar bills of the same species that makes cash economy possible.

Think-tanks often analyze conflicts of warfare, commerce, or politics in terms of game theory, using statistical laws. Assumptions are made about the goals of the opposing camps of individuals. The situation is then phrased in terms of probabilistic equations and a computer program is prepared. Scientists, public opinion pollsters, and marketing analysts then sit back and wait to see who will win the war, get elected in South Dakota, or sell the most burglar alarms. Such predictions are often reliable because mass behavior—consisting of certain patterns in sociotemporality—has its prototemporal components.

But humans as individuals have dimensions of freedom that cannot be subsumed under the rubric of statistical unpredictability, but rather under the name of behavioral and mental unpredictability. Manfred Eigen and Ruthild Winkler in their book, *Laws of the Game* (New York: Harper and Row, 1983), complain about the present limitations of game theory because of the human psyche. "The most unpredictable factor in the practical application of game theory," they write, "is the actual behavior of the players" (p. 16). Indeed so. People are able to beat in a creatively unpredictable manner many probabilistically predictable rules. That is why statistical generalizations about people as individuals do not work. The umwelt of statistics is the prototemporal world, that of the human mind is the nootemporal one.

Probabilistic events, that is, events of controlled randomness, have fascinated people as far back as such a concept can be traced. The Old Testament has many examples of casting lot as an aid in deciding among a number of possible choices. Through dice, for instance, Fate could send messages to us people. Dice was a way of shifting responsibilities to the Almighty, such as when the soldiers cast lot for the garments of Christ or when a member of a firing squad uses a dummy bullet. Figure 38 is a happier example.

Once upon a time on a Scottish heath, two men dared three witches to tell them the future by examining a set of identical seeds. The metaphor is powerful and flexible.

> *Banquo.* If you can look into the seeds of time
>   And say which grain will grow and which will not,
>   Speak then to me . . .
> *First witch.* Hail!
> *Second witch.* Hail!
> *Third witch.* Hail!
> *Macbeth.* Stay. you imperfect speakers, tell me more. . . .

Macbeth assumed that the witches knew more than what they said. Perhaps they did, perhaps they did not; the play is magistrally ambiguous about that issue. The prototemporal world definitely does not know more than what the laws of quantum theory say. Those laws are perfect speakers reporting on the behavior of an unevolved world.

The boundaries of time

Boundaries of time are not surfaces in space, but limitations to freedoms of action. Thus, when a society collapses, the freedom of its collective action is lost. When a person loses his mind, his freedom of human action is lost. When a living organism dies, the freedom of doing the myriad things that life can but nonliving matter cannot do is lost. When a solid body changes into the gas of its molecules, the freedom of assuming an infinite variety of well-defined shapes that solid matter has but gases have not is lost. In the sequence I described, several boundaries of time were crossed: that between the socio- and the nootemporal integrative levels, that between the nootemporal and biotemporal organizational levels, and those between the adjacent levels of the bio-, eo-, and prototemporal worlds.

Let me attend here to the interface between the proto- and atemporal integrative levels of nature, that is, between the universe of particle waves with mass, and that of particle waves without mass. Other boundaries of time will be discussed later.

As our friendly ghost hops from a very fast muon to a photon, it will lose its freedom to be in a world of space, time, and controlled random-

FIGURE 38 PROBABILITY IN QUANTUM THEORY: THE PERFECT REPORT
FROM AN UNEVOLVED WORLD. People have always tended to regard the laws of
chance as imperfect expressions of a complete but hidden plan. This engraving is
on the inside of a bronze mirror made in Greece between 350 and 300 B.C. It shows
Aphrodite and Pan playing five-stones. Aphrodite was the goddess of love. "Her
enchantment came from this: allurement of the eyes, hunger of longing and the
touch of lips that steals all wisdom from the coolest of men." Pan was a woodland
god of pastures and flocks who lived in the romantic paradise of Arcadia. The
winged youth is Eros, the goose is an attribute of Aphrodite (as the serpent in figure
I is an attribute of Shiva). Five-stones was a game of chance, originally a magical
means to divine the future. Its throws are controlled by statistical laws. In contemporary physics, the behavior of particles is also governed by statistical laws.

From the Collection of Bronzes, the British Museum. Reproduced by courtesy of
the Trustees of the British Museum.

ness, because the world of light is atemporal, aspatial, and completely chaotic. But these specifications do not mean that it does not exist. It does, and the boundaries between it and the prototemporal world is a region bustling with particle-waves of many kinds.

From among them, six types of particles are called stable because they do not spontaneously decay to other particles and radiation. They are the graviton, photon, electron-neutrino, muon-neutrino, electron, and proton. To these we may add the antiparticles of the last four (gravitons and photons being their own antiparticles).

All other species are unstable. From among them some are called long-lived because they exist for periods longer than 10^{-23} sec, the length of time that used to be regarded as the shortest meaningful period. The rest are short-lived; they decay in periods shorter than 10^{-23} sec. The dance is continuously going on along the boundaries between the world of radiation and particles, even as you turn your flashlight on a raccoon raiding the garbage pail.

Early modern physics thought of electrons, protons, and neutrons—of particles in general—as tiny solid balls. This model was replaced by the particle-wave duality recounted above. Early modern physics also thought of the world of particles as being held together by gravitational and electrical forces that could reach over great distances. When dealing with matter on the macroscopic scale, physicists still speak of these forces and employ them in their equations. But quantum theory perceives the interaction among particles not by such forces but by yet other elementary objects called gauge particles. A less technical descriptive name for them would be "messenger particle-waves." Their prototype is the photon, the quantum of light that carries the forces of electromagnetism. Gravitons, the quanta of gravitation—thus far only conjectured but not actually observed—carry the forces of gravitation. Other messenger particle-waves carry the so-called weak and strong forces within the atomic nucleus. The range of photons and gravitons is infinite, those of the nuclear gauge particles is limited to nuclear dimensions.

As the reader will have surmised, the happy world of a few, well-defined particles—the electron, proton, and neutron—is gone.

At present, the only particles that appear elementary, in that they seem to contain no internal subcomponents, are the leptons (electron, muon, neutrino, massive tau lepton, and their associated antiparticles), and the quarks. Quarks are geometrical objects identified with quantized combinations of physical properties, not unlike the Platonic or regular solids were identified with the Greek elements (see fig. 4). The six known, or suspected quarks are distinguished by their flavors, called *u* (for up), *d* (down), *c* (charm), *s* (strangeness), *t* (truth), and *b* (beauty). Each flavor comes in the three primary colors of red, yellow, and blue. The use of qualities to describe numerical parameters is also rather ancient: it reminds one of the ten heavenly stems of the Chinese (see chap. 2 n. 6).

Quarks combine by twos or threes into hadrons, which is the collective name for all particles that are not leptons. In quark description a proton, for instance, is a *uud* triplet. Each quark also has one of three strong inter-action charges, also labeled by the three primary colors of red, yellow, and blue. Quarks have their respective antiparticles, called antiquarks.

It is assumed that various combinations of quarks make up all members of the particle world, a veritable zoo with more inhabitants, bearing stranger names, than one may comfortably imagine. Many members of that zoo are called resonances, because they are too short-lived to leave a trace on a bubble chamber photograph and may only be identified as resonance peaks of various curves.

It has been suggested that there may be no such things as elementary particles, because every particle is made up of every other.

The physical properties of those "whatevers" that make up the proto-temporal world are so far from human experience that they may only be handled by abstract means such as numbers, symmetries, and geometrical configurations. The incongruous descriptive names given to the different variables remind me of the strange and many-colored clothing people in some of our large cities wear because they feel that, were it not for their regalia, they would lose their identities, and become nothing but numbers.

We leave the cauldron of boiling quarks with a quote from Edward R. Harrison's splendid book, *The Masks of the Universe*.

We have reached the point of postulating fundamental particles that, in principle, cannot be observed directly as isolated entities existing in their own right. They are beyond the reach of direct verification. This is something new in science, tantamount to postulating the mythical gods of long ago.[11]

11 Harrison, *The Masks of the Universe* (New York: Macmillan, 1985), p. 134.

The computer data base called HEP (for High Energy Physics) of the Lawrence Berkeley Laboratory of the University of California contains 150,000 particle physics papers; that number is increasing at the rate of 15,000 papers per year.

Along the atemporal–prototemporal interface, short-lived particles continuously hop out of and back into the atemporal world sometimes described as a vacuum foam. Not that these objects have been waiting their turn, ready-made: they were only potentially present. For that matter, anything that the universe above the level of the atemporal can permit to exist could hop out of the foam, such as a Thanksgiving celebration with turkey, cranberries, and people. But the probability of particles spontaneously combining to create that feast is negligibly small. Nothing of true interest to human life ever emerges directly from the atemporal world of light.

To produce a configuration of living and thinking matter, twenty billion years of evolutionary labor was needed, including, near its tail end, the travail of human history and the order of Governor Bradford of Massachusetts to celebrate the harvest of 1621. To get from the atemporal chaos to the Thanksgiving dinner, several boundaries of time had to be crossed.

Motion, rest, and the flight of Zeno's arrow

I promised in the section in Chapter 1 titled "Ideas about Time" to return to Zeno's paradox of the flying arrow in the context of quantum theory. We are now ready to do so by considering what has been mistakenly called "atoms of time."

It used to be believed that no physical interaction could take place in periods of less than 10^{-23} second, a lapse of time that used to be called a chronon. It is the time needed by the fastest signal (light) to cross the shortest meaningful distance (the radius of an electron as a particle). Later theory suggested that the shortest meaningful lapse of time may be 10^{-43} second. This is a figure that comes from cosmology (to be discussed later) and is known as the Planck time. Since no temporal organization may be given meaning within this period, the Planck time—measured by us temporal creatures—must be regarded as atemporal. One may, of course, think of shorter periods, such as 10^{-1492} second as a mathematical exercise, and a number with no physical significance.

Whatever its actual value, the shortest physically meaningful period is often called an atom of time. This is a very bad name for it, because atoms have always meant elementary constituents of some kind out of which more complex objects may be constructed. But it is not possible to construct biological and human time from elements of atemporality. The only way to proceed is through the evolutionary creation of consecutive integrative levels, each with its peculiar temporality.

The primordial state of the universe was one of ceaseless motion and change, a state of affairs that still describes the world of light. Only with the jelling of energy into particles did motion at speeds less than *c* become possible. And only after the particles began to form massive macroscopic bodies did physical structures come about to which our concepts of rest, permanence, space, and time could be applied.

The fact that the physical world has evolved from motion to rest—rather than from rest to motion—helps to explain why Zeno's paradox of the flying arrow has continued to annoy thinkers and why, after each proposed solution, people have come up with alternative explanations, each to be knocked down. The persistence of the paradox does not reflect the cumulative, dialectical enlargement of knowledge, but rather indicates that something is fundamentally wrong about the question itself.

Zeno's point was that since motion cannot be put together from elements of no motion, and since the arrow is at rest during each instant (or certainly was some time ago), motion itself must be an illusion. The paradox resided in the contradiction between the sense experience that told us that the arrow moved and the convincing logical conclusion that, for the premises stated, it could not possibly move.

Solutions offered since Greek antiquity fall into two families of views: (1) there ought to be a way of constructing motion from elements of rest, therefore we must keep on looking, and (2) there is no way that motion can be constructed from rest, therefore there is no use in trying. Very importantly, the two groups of solutions have one basic assumption in common: they both hold that rest is a more fundamental state of nature than motion, and for that reason, it is reasonable to ask how nature puts motion together from elements of rest. But it is not.

As demonstrated both by special relativity and by quantum theory, the fundamental state of the world is motion. Along the foundations of the physical universe nothing ever stops; the particle-waves keep on moving even after they have jelled into solid matter. A question that does correspond to the reality of the universe is this: How is rest constructed from elements of motion?

One can solve this question analytically by adding two equal and opposite velocities. But this is only a formal, abstract method. Our notion of rest is not constructed by vector analysis, but is created through the conceptualization of perceptual impressions, guided by our psychology.

The integrative nature of our sensory apparatus prevents us from recognizing the ceaseless motion at the foundations of the physical world. For instance, the sensitivity of the human ear cuts off frequencies just above that of the Brownian noise of the fluid in the cochlea. Frequencies characteristic of the workings of the nervous system, such as carrier frequencies of signals and discharge rates of neurons, remain below the threshold of awareness. Keeping the noise of chaos out of the behavioral controls of man and beast is a necessary condition for organisms to be able to form impressions of continuities and thus survive.

Psychologically, the idea of rest is in the same category as that of selfhood. It is created by the mind in its need for unchanging continuities, for permanent identities. Without permanent identities—that is a cat, this is a window, my name is Aram—civilizations could not have been built.

In sum, the idea of permanence is a mental construct, permitted by the necessary coarseness of our perceptual faculties, and refined and guided by the ever-present desire to lessen the burden of passage.

We will recall that Zeno's purpose in constructing the paradox of the flying arrow was to support an argument advanced by his teacher Parmenides. Zeno wanted to show that ultimate reality is permanence, that motion and change are virtual. Heraclitus of Ephesus, a contemporary of Parmenides whose thoughts we also encountered, held that on the contrary, ultimate reality is ceaseless change, permanence is virtual.

The discoveries of special relativity and quantum theory suggest that motion is the fundamental state of the physical world and that rest acquired meaning only after massive bodies were formed with respect to which other massive bodies could be at rest. Permanence, which is the ideational twin of rest, is even a later evolutionary form of reality. It acquired meaning only with life and the mind of man.

The Parmenidian-Heracletian dialectic on the nature of time has been a conspicuous and pervasive feature of Western social and intellectual history. The same dialogue may be identified in various forms in various other civilizations. The religions and social teachings of certain epochs tend to stress changelessness as the significant aspect of time, whereas others—such as our own epoch—side with change; yet others view permanence and change as forming a unity of balanced opposites.

The findings of special relativity and quantum theory support a different view from all of the above. They suggest that permanence is an emergent quality of the world, having arisen in the course of evolution from the fundamental restlessness of the universe, and that permanence and change are hierarchically related, in that rest and permanence may be constructed from elements of motion and change, but not the other way around.

Cosmic Time: General Relativity Theory, the Physics of the Universe

White dwarfs, black holes, and red giants; a cast of 10^{21} (sextillion) stars and 10^{81} (quattrodecillion) particles; massive matter gathered into 10^{10} (ten billion) galaxies scattered wide and far between but fairly uniformly, floating in a sea of light. This is the world of physical cosmology. The universe, which is the largest possible object, reaches another boundary of time because it is also the oldest possible fossil. Its structure and behavior are governed by the laws of general relativity theory first described by Einstein in 1916. This section employs that theory to help explore the nature of cosmic time.

How to make the universe both limited and unlimited: A short course in astral geometry

In one of his famous letters to the classical scholar Richard Bentley, Sir Isaac Newton remarked that if space were a finite sphere, then, under the forces of gravity, all matter across the world would come together

> into the middle of the whole space, and there compose one great spherical mass. But if matter was evenly disposed throughout an infinite space [then it would] make an infinite number of great masses, scattered at great distances from one to another throughout all that infinite space.[12]

12   Newton to Bentley, 10 December 1692, in *Correspondence of Isaac Newton*, ed. H. W. Turnbull (Cambridge: Cambridge University Press, 1961), vol. 3, p. 234.

Newton's law of gravitation was not negotiable, and a huge blob of matter was not seen anywhere. There were good reasons to believe, therefore, that the universe was infinite.

But during the nineteenth century, attempts to reconcile an infinite, homogeneous distribution of stars with Newtonian gravitation ran up against apparently intractable theoretical difficulties. Weighing the various issues, it appeared to Einstein that the universe as a whole could neither be infinite nor have finite boundaries (edges). It would have to be an object that is both limited (that is, not infinite) and unlimited (that is, not bounded).

A geometrical object that has such apparently contradictory properties is the surface of a sphere. Its area is finite, yet it is also unbounded: one may move on it without ever coming to an edge. But the surface of a sphere is two-dimensional, whereas our experiential space is three-dimensional. For our 3-D space to have similar properties, it would have to curve back upon itself. It would have to become an object describable as a sphere with its center everywhere, its boundary (surface, edge) nowhere.[13]

Einstein succeeded in constructing a set of equations that described the 3-D space of the universe as finite and unbounded. He also showed how the geometry of the universe was determined by its matter and energy content. Special relativity theory could not be used for the writing of a cosmic physics because that theory could not accommodate matter, gravitation, and the joint action of the two, which is accelerated motion. Thus Einstein felt it necessary to invent a set of new principles, which he called the generalized or general theory of relativity (GRT).

In that extended theory, light retained its privilege as the reference of absolute motion while it acquired an additional role: The path of the light beam became the basic element, the straight line of the geometry appropriate for a description of the universe.

What is meant by the straight line being the basic element of a geometry —of all geometries? Ask the Egyptians, who invented geometry for the

13   These requisites remind one of the thoughts of Nicolas of Cusa, the fifteenth-century German cardinal and mathematician who concluded from metaphysical rather than what we would call today scientific reasons that the world must be both finite and infinite. As I noted in the earlier discussion (Chapter 1, the section called "Ideas about Time"), he did not foretell modern physical cosmology. We are witnessing here an example of variations to a pattern of human thought—the handling of ideas of finity and infinity—whose ultimate communality to all people resides somewhere in the neurological organization of the brain and in social conditioning.

purpose of measuring land. As early as 2300 B.C. they used *harpedonap-tai* ("rope-stretchers") to help them draw the straight lines necessary for land survey and astronomical work. A stretched rope was then, as now, a good way to trace the shortest distance between two points. And by definition, the shortest distance between two points is called a straight line. Throughout its history, geometry depended for its rules on what was to be meant by a straight line, that is, by the shortest distance connecting two points. Let me offer some exercises that will help illustrate some of the geometrical concepts involved.

> 1a Draw three points on a sheet of paper, connect each with the other two by using the shortest distances. These lines will be the straight lines of plane geometry. The angles of a triangle in plane geometry add up to 180°. The area of any such triangle is $\frac{1}{2}b \times h$, where b is the length of the base and h is the height of the third point above the base.
>
> 1b As another exercise, draw a straight line on the same sheet of paper and erect two perpendicular lines to it. The two perpendiculars are said to be parallel.
>
> 2a Draw three points on a sphere. Connect each with the other two, using the shortest distances that may be drawn on the sphere: they will be arcs of great circles. Since they are the shortest distances on the surface of a sphere, they must be recognized as the straight lines of spherical geometry. The angles of spherical triangles add up to more than 180°; in the limiting case, when a spherical triangle becomes a hemisphere, its angles add up to 540°. The area of a spherical triangle depends on the radius of the sphere and the number of degrees the sum of its angles have over 180°.
>
> 2b As another exercise, draw a straight line on the sphere. Erect two perpendiculars to it at two points. These two straight lines must be called parallel because they are both perpendicular to a third one. Yet they will meet somewhere on the sphere, not even once but twice.

These two geometries—plane and spherical—are identical in their rules yet different in important ways, such as how to calculate the areas of triangles and whether their parallels do or do not meet. Both geometries are correct for the world to which they apply. Which one is right for the earth, for instance? It is easy to test: hire rope-stretchers, let them draw long straight lines, erect perpendiculars to them, see whether they meet. If they do, our earth must be curved. Next, let them draw a large triangle on the earth and measure its angles. If the angles and the area relate by the rules of spherical geometry, then our earth is a sphere. From the number of degrees by which the triangle's angles are over 180° and from its area, one may calculate the radius of the earth.

From the Greek geometer Euclid around 300 B.C. to the nineteenth century, geometry was judged to be entirely an art of logic. The mathematician Carl Friedrich Gauss signaled a change in this belief when he wrote in 1817 that geometry should not be classed "with arithmetic which is entirely a priori [purely mental] but with mechanics." Gauss worked out the foundations of a new geometry that we call non-Euclidean. He first called it anti-Euclidean, later astral geometry. The name implied that the differences between Euclidean and the new geometry would become significant only at astronomical distances. His hunch proved correct. The geometry that turned out to be appropriate for the cosmic physics of Einstein was a non-Euclidean, astral geometry.

Non-Euclidean geometry is only one of two hallmarks of GRT. The other is the use of 4-D space for representing the history of moving points. How is that done? Mathematicians have enlarged the meaning of "space" to stand for any continuum in which processes may be recorded. It is possible to construct spaces in the mind with 4, 5, 15, 69, 2,001, or 5 trillion dimensions. It is not necessary to visualize them. It is enough to know the rules for erecting structures in them. The freedom thus afforded by being able to use simultaneously that many independent variables—one for each axis—allows for many-dimensional bookkeeping.

The rural railroad stations of Greece utilize a simple geometrical bookkeeping of the motion of trains. Whereas our timetables are vertical columns of train positions and times, theirs are graphs in which the horizontal axis of the graph plots the train's distance from Athens; the vertical axis, the time of day. From a curve, the train's progress may be read as a function of time.

Figure 39 shows one of the grandparents of the Greek timetable. Graphs of this kind are also currently used in Japan for the day-to-day operation of the *Shinkansen,* the bullet train.

Instead of Greek, French, or Japanese graph timetables in two-dimensions—one for distance, one for time—think of a topographic map

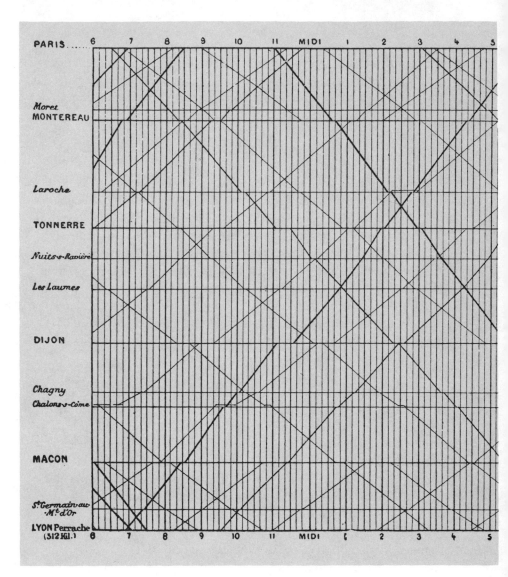

FIGURE 39   GEOMETRICAL BOOKKEEPING IN A 2-D PLANE OF MOTION THAT TAKES PLACE ALONG A 1-D LINE. This graph shows the movements of trains between Paris and Lyon with eleven intermediate stations where some trains stop and others do not. The time of day from 6 A.M. on one day to 6 A.M. the next day is plotted along the horizontal axis. The positions of the stations along the 512-km stretch are laid out along the vertical axis.

The express train that leaves Lyon just before 7 A.M. reaches Paris at 6 P.M. the

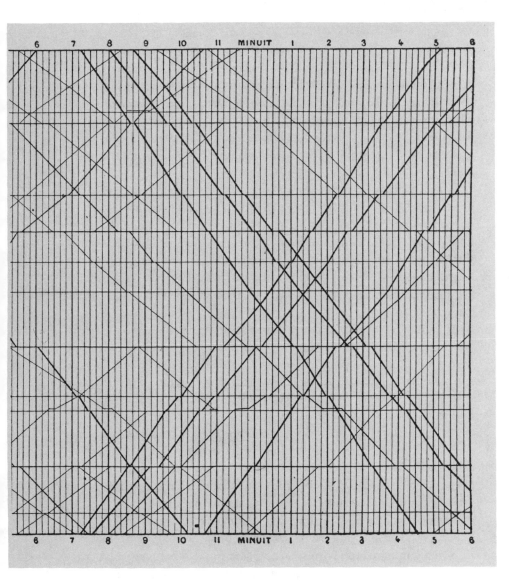

same day, with intermediate stops at stations where the train's time–distance line is parallel to the time axis. The steeper the line, the faster the train. The time–distance line of a science-fiction train that travels at infinite velocity would be parallel with the distance axis.

From Etienne Jules Marey, *La Méthode Graphique dans les Sciences Expérimentales et Principalement en Physiologie et en Médecine* (Paris: G. Masson, 1885), figure 7.

of the U.S. Geological Survey with two axes (dimensions), and a third dimension perpendicular to it for recording the time of day. Let the map show in heavy black overprint a railroad track running through the area. The history of the motion of a train may now be displayed by a curve in 3-D: each of its points above the railroad track will represent both a 2-D position on the map and a 1-D instant of time measured along the vertical axis.

Can an analogous scheme be devised for the 3-D space, 1-D time of our experience? Yes. Use a 4-D space. Think of it as having four mutually perpendicular axes. It is not necessary to imagine it, it is enough to know how to use it: it is a mental construct like the Jubjub bird. Three of its axes fix the coordinate positions of a point in space where a moving body is, the fourth one gives the time when the body is there. The space of this clever arrangement is the 4-D space of relativity theory, spoken of simply as a 4-space.

A point in such a 4-space represents an *event,* that is, a point in 3-space at an instant of time. One 4-point (as points in 4-space are called) may stand for "7 A.M. at home," another one for "9 A.M. at work." The 4-D line (or simply 4-line) connecting these points represents motion: "Traveled from 7 A.M. at home, via the Golden Gate Bridge, to executive offices of Bandersnatch Inc., where I arrived at 9 A.M."

As do all spaces, a 4-space also has its shortest distances between any two of its points; these, by definition, are its straight lines. A shortest distance in this case represents the history of the fastest possible travel between two events. We know from SRT that the fastest possible motion is that of light; light beams, therefore, will trace out the shortest distances—straight lines—of the 4-space of GRT's astral geometry.

Sometimes such a shortest distance may be shown to be a line that is curved if compared to a corresponding 4-line of another 4-space. A space whose shortest distances (straight lines) are curved is a curved space, and its geometry has to be non-Euclidean. The examples I gave above of plane and spherical geometries are both Euclidean, but they give a hint of what is meant by a curved 4-space.[14]

Einstein assumed that the way the straight lines (shortest distances) in

14   Those who do not care for curved straight lines might still enjoy Mother Goose: "There was a crooked man, and he walked a crooked mile. He found a crooked sixpence, against a crooked stile. . . ."

the actual 4-space of the universe were curved depended on the distribution of matter and energy in the neighborhood of motion. He constructed a master equation that stood for a family of equations; solutions to those equations could then be used to describe the motion of objects, including photons.

What does the master equation that could perform all this magic look like? Here it is: $-\kappa T_{\mu\nu} = R_{\mu\nu} - \frac{1}{2}Rg_{\mu\nu}$. One has to learn what the letters mean, but it is still only an equation, as was $v=d/t$. The left-hand side speaks of the physics of 4-space, the quantities on the right speak of its geometry. The equation tells us how the two relate. It is an instruction for learned bookkeepers of motion.

Important to our concern is that the equations may be used to explore the past of the universe and also to limit the possibilities of its future fate. It is the only known scientific statement that can tell us about the nature of cosmic time, using observational and experimental data.

The reader is now familiar with the 4-space of GRT, knows that light retained its privileged position as that of absolute motion, that it became the straight line (shortest distance) in the 4-space of GRT, and that in the presence of matter and energy, the 4-lines that represent motion become curved (in comparison with the lines of another 4-space without matter and energy). How is a curved 4-space used in the construction of a model of a universe that is finite and unbounded?

As I already said, to make a 3-space analogous to the surface of a sphere, it must be made to curve back on itself. In such a curved space, a straight line may reenter itself. Physical cosmologists are still doing some comparison shopping to decide which of several curved 4-geometries are best suited to describe the world as it is. The leading candidate, when used to describe a static universe, allows us to hop on a light beam, rush along its straight trajectory and find ourselves back where we started.

Cosmic time: Mainstreamers and deviationists

"Starting from fish-shaped Paumanok where I was born, / Well-begotten and raised by a perfect mother," wrote Walt Whitman, "I strike up for a New World." Starting from pear-shaped earth, where we were all begotten and raised by earthly mothers, we strike up to travel across the universe to find out what it is that physical cosmologists must account for. Then we can talk about "it" and time.

Riding a photon, we arrive at the sun in 499 seconds (as measured by an observer on earth) to visit with 99.86 percent of all the mass of the solar system. It will then taken five years (again, as measured from the earth) at the speed of light to leave behind Oort's cloud, a spherical shell of comets that thins out as the gravitational pull of the sun drops off. The sun with all its hangers-on travels an almost circular orbit around the center of the galaxy, making one revolution every 220 million years. The solar system is now at about the same spot in the galaxy where it was when dinosaurs first began to roam the earth.

The Milky Way looks like a flying saucer: a disk with a bulge at its center. Its diameter is 10^5 years by light, the bulge at its thickest is 1.3×10^4 light years across. There may be a black hole at its center. The interstellar matter of our galaxy makes up no more than 10 percent of its mass; 90 percent of it is in the relatively small bodies we know as stars. The galaxy is surrounded by a halo of dense globular clusters of very old stars, moving on elliptical orbits about the galactic center; Kepler's laws (fig. 5) apply to worlds he never dreamt of.

It is estimated that there are about 10^{10} galaxies in the universe. Intergalactic distances, on the average, are ten to a hundred times the sizes of galaxies. This makes the universe a mostly empty object, with ten billion dots distributed across it more or less uniformly. Of intergalactic material very little is known, but it is believed that there are a hundred million times as many photons as all the chemical elements taken together. The island universes (galaxies) float in a sea of atemporal light.

What can be said about the physical universe as a whole with any degree of certainty? That the atemporal and prototemporal worlds across the universe behave according to the principles of relativity and quantum theories; that the chemical elements behave everywhere as we know them on earth. That the distribution of the galaxies is isotropic: in whatever direction we look, they are, on the average, spread out uniformly through space. And that all galaxies are running away from us at velocities proportional to their distances from us. It follows from the observed expansion that in the past the galaxies had to be closer, the universe smaller. Since every galaxy is at the center of the finite–unbounded cosmos, all astronomers on all galaxies should come to the same conclusion.

If the present rate of expansion is extrapolated backward and is fortified by the assumption that the rate itself has been constant, then some 15 to 20 billion years ago the world had to be very small and dense. That smaller, denser universe was not located somewhere inside the present one. The universe remained the one and only universe, always including all of space.

Two items ought to be mentioned here because they are very rarely discussed in popular and even semipopular books on physical cosmology.

First, that the expansion of the universe is that of its substratum, the intergalactic sea of light. Intragalactic distances, such as between the earth and the sun, or the length of your pencil, are not free to expand with the universe and, hence, they do not change. Second, the recession velocity of galaxies is not limited by the speed of light because galaxies are not moving *through* space but are wafting apart *with* space.

To be able to appreciate the meaning of cosmic time for this curious universe, we start by remembering the special relativistic phenomenon of time dilation. Stated in a schematic manner, SRT revealed that moving clocks ran slow. To this motional slowing of clocks, GRT added a gravitational slowing. The rate of a clock now also depended on the strength of the gravitational field in which it was: clocks near a large mass tick slower than those near a smaller mass. An atom on earth oscillates faster than does the same kind of atom in the sun. As the strength of the gravitational field increases because there is more matter around, clocks tick slower and slower.

Directing our gaze backward in time, we find an increasingly denser universe whose clocks tick slower and slower. No macroscopic clock could survive the gravitational pull of the early days, but atoms could. However, as they reach the big bang, even they must change into speed-of-light particles that make up the initial universe; their ticks, in their own frames, slow to a halt. At the end (in the beginning) they enter (they were a part of) an atemporal world. [15]

Let me again return to SRT and recall that its time dilation effects became significant only for speeds close to c. How fast do the various stars travel? The billions of stars of our galaxy do travel every which way, but compared to the average motion of all the stars, none moves rapidly. The velocity of the solar system, for instance, referred to the galaxy at large, is negligibly small: its v/c ratio is around 0.000,06. It follows that our clocks measure galactic time.

Identical arguments hold for all other star and planetary systems in any

---

15   For those familiar with the main issues involved in the physics of the big bang, Appendix 2 gives the reasons why I regard the state of the initial universe as atemporal.

galaxy if they are at rest or move very slowly with respect to the average motion of stars in their galaxy. All such galactic clocks are mainstreamers, they all measure cosmic time. Very rapidly traveling macroscopic clocks are rare exceptions; they are the deviationists. (In my earlier example of the stay-home elephant and traveling mouse, the elephant was the mainstreamer, keeping galactic time, the mouse was the deviationist.) A public opinion survey of clocks on a representative number of randomly chosen galaxies (say, on a hundred million, which is 1 percent of them), taken across the universe, would reveal worldwide agreement on cosmic time.

In sum, the universe possesses a scale of cosmological–common time.

The beginning and ending of time

Since light does not travel instantly but takes time to propagate, the further a galaxy that is being observed, the further we are looking into the past of the universe. What astronomers photograph is not the way galaxies are now, but the way they were hundreds of thousands, millions, and billions of years ago. And the further the galaxy was when it emitted the light we now see, the faster it will be seen receding from our Milky Way. If astronomers could look into a distance of 15 or 20 billion light years, to the primeval boundaries of time, they would see everything running away from us at speeds close to that of light.

Approaching the big bang, the galaxies would be seen to be replaced by hot gases whose atomic clocks tick very slowly compared to those of identical atoms on earth, because of the strong gravitational fields. Close to the singularity, the universe would become opaque; a photograph of creation would be black, its darkness penetrable only by mental images, suggested by mathematical formulas that are based, in their turn, on accessible observations and theory.

The popular image of the big bang is that of nuclear firestorms across an immense world. But that does not describe the universe in its early stages: the big bang happened to a very tiny world whose space was, nevertheless, finite but unbounded.

A feature of the big bang models of the universe is an initial singularity of an extentionless, mathematical point. This amounts to saying that at the "absolute zero of time," to use the words of the relativist Charles W. Misner, the universe had infinite density. But physicists do not like to work with infinite physical quantities because it is difficult to give them physical meaning; they describe them as "pathological." The problem here is that classical GRT breaks down at atomic distances and times, long before the dimensions of a pointlike universe are reached.

An understanding of the physics of very small objects, as we have learned, is the task of quantum theory. It follows that before the physics of the singularity could be understood, it became necessary to quantize general relativity theory. This has been a difficult task because GRT is a theory of the eotemporal world and for that reason it is deterministic, whereas quantum theory—comprising the rules of the prototemporal world—is probabilistic. But work on quantum gravity has been progressing, being created by some of the most brilliant minds of an already select group of scientists. What we can say about the primeval cosmos has the hallmarks of quantum mechanical statements: they specify only boundaries and probabilities. The universe of the big bang at a period known as the Planck time (10^{-43} seconds after the theoretical zero) had to be less than 10^{-33} cm across (known as the Planck length) and had to have a density of at least 10^{94} grams/cc or 10^{88} tons/cubic inch. Examples of what this means do not help; numbers may serve only as guides. Figure 40 shows the actual size of the universe at the instant of creation and a very brief period thereafter.

Assuming that a beginning of time, the concept of $t=0$, makes sense, what was the timetable of creation?

In physical cosmogony creation is identified with the beginning of the expansion of the primeval atom. Its initial temperature was high, being somewhere between 10^{11}°C and 10^{12}°C. As the universe began to expand it also began to cool, for the same reason that the freon gas in every refrigerator cools when it is injected into a volume much larger than the one it had occupied before. The volume of the universe began to expand very rapidly; it was continuously injected into larger and larger volumes of its own self. All along it remained finite and unbounded; a world that was not finite could not have had this history.

The pre–big bang universe is often called the singularity. (For the meaning of "pre-" in this context, see below). In that primeval state, so I reasoned earlier, all objects had zero rest mass, traveled at the speed of light, and hence made for an atemporal and aspatial universe. This cosmic state of affairs came to an end at the 10^{-43} second limit. ("End" implies a flow of time and hence cannot apply to that early universe; the word is

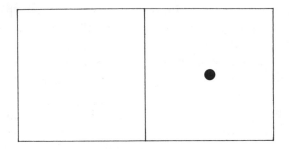

FIGURE 40   ACTUAL SIZE OF THE UNIVERSE AT AND SOON AFTER THE
INSTANT OF CREATION. The size of the universe at the instant of the big bang
(left) had to be 10^{-32} cm across or less, too small to be seen except through the
mind's eyes. Soon after the big bang, for a very short period, the universe had the
actual diameter shown on the right. The drawing is tricky only in this respect: that
smaller universe was not somewhere within our present world. It was the same
finite but unbounded world, but at an earlier stage of its evolution. The illustration
represents a mental image, as does the Jabberwocky but with a greater degree of
believability. The invisible point universe on the left was atemporal. The dot
universe on the right was prototemporal as well as atemporal.

used *faute de mieux*.) What happened between 10^{-43} second and about one
second is being coldly debated by hot-headed physicists, but there is general agreement that the state of the universe corresponded to a mixture of
what I have been describing as the atemporal and prototemporal states of
matter.

As the temperature continued to drop, different forces of nature that
were earlier undifferentiated "froze out" (in the language of physics); they
changed from potentiality to actuality (in the language of philosophy).
Ten seconds after $t=0$, the temperature dropped to 10 billion°C. Around
1,000 years after the singularity, the universe cooled to 500,000°C; by
10,000 years, to 5,000°C. Since iron vaporizes at 2,800°C, the place was
still warm. But by then, the universe was sufficiently large to become
transparent. The average photons the first intergalactic astronauts will
meet will have been around since that time. A hundred million years after
$t=0$, galaxies began to form, followed by the formation of stars. This was
the epoch when eotemporality was born.

Except for the remarks on temporalities, what I described is the standard physical model of creation. Using that model, what are we to make
of the beginning of time? One way to get an answer is to imagine taking

a journey back in time, imagining that our idea of a present, with its future and past, makes sense at each instant of the journey.

Let us begin with our births. No one has much difficulty imagining people who lived before him or her: we may easily fly over decades, centuries, millennia, and even cultures of the past. All through this time human nature remains the same; the nootemporal umwelt of the caveman was hardly different from ours. But somewhere in the hazy past, nootemporality vanishes as biotemporality becomes the highest, the most advanced form of reality. No creatures then dreamt of millennia. They thought only—if that is what they did—about the immediate future, past, and present.

At three and a half billion years, all living forms vanish. There is nothing on earth that could define the organic present and, with respect to it, give meaning to future and past. The highest form of reality is eotemporal. Approaching the big bang, stars and galaxies boil into hot gases, eotemporality vanishes, and the highest form of reality is prototemporal. Time and space are not distinct, all laws of nature are probabilistic, and there is nothing to correspond to our concepts of well-defined instants and locations. Taking one step further back, the prototemporal world vanishes as the universe becomes a very small and rather hot atemporal chaos.

Each of the crossings between adjacent temporalities may be described in the language of physics as a phase change in the state of matter and energy. An ordinary version of such a phase change is the freezing (or melting) of water. As we gaze back along that cosmic history of phase changes, the universe is seen to gain "symmetry," in the sense that word is used in physics. Physical phenomena are said to be symmetrical if the equations describing it remain invariant under all changes of the system that are of interest to the physicist. The most perfect symmetry is that of the atemporal chaos, because it is homogenously lawless. Whenever along the path of evolutionary development a new integrative level arose, some of the earlier symmetries were broken. The most evolved forms and functions of nature—those of life, mind, and society—are also the most asymmetrical forms of nature in the mathematical connotation of symmetry. The literature of time in physics often speaks about the breaking of earlier symmetries.

The question has often been asked by theologians, philosophers, scientists, and everyone else who ever thought of it: How did time come about from no time? "Time" invariably meant nootemporality, "no time" invariably meant non-existence. But the atemporal world of the big bang should not be mistaken for the metaphysical notion of nonexistence. As I already stressed, it designates a real world, but one to which our ideas of time and space do not apply. From the atemporal world arose the proto-temporal one with its incompletely differentiated space and time; this was the first symmetry breaking in the genesis and evolution of time. Later (judged in terms of human reality) as the astronomical world of massive matter was formed, space and time completely separated and became continuous, breaking thereby the prototemporal symmetry. In its turn, the eotemporal symmetry was broken with biogenesis, when life came to define a narrow-horizoned but clearly distinct future and past, separated by the organic present. Even later, the organic present evolved into the mental present, possessing broad temporal horizons, thereby breaking the biotemporal symmetry. Nootemporality and atemporality never had a common interface, hence one cannot ask how the former came from the latter.

Earlier I spoke of pre–big bang, the pre–$t=0$ conditions. In terms of the ordinary concept of time, the instant of $t=0$ was "followed" by the coming about of the prototemporal world. But the nature of the proto-temporal world does not permit events that relate as before-and-after— nothing could follow or precede anything—therefore, all we can say in that the instants near the singularity were contiguous with it. Continuity of time (the eotemporal) was not born until solid matter began to precipitate; the arrow of time, not for another twelve billion years. It is still useful to speak of "right after the big bang" because that is the way our mind likes to organize the world. Post–big bang then means matter organized by laws of structure and function, pre–big bang means structureless chaos subject to no laws.

We have encountered several different physical boundaries between the prototemporal and atemporal worlds. One was the motional boundary of SRT; the other, the gravitational boundary of GRT; the big bang was a cosmic boundary. All these boundaries are theoretically related to each other and to another kind called black holes.

Black holes are regions of space with matter and energy densities sufficiently high to make the escape velocity for their regions greater than c. It follows that not even light can escape from them and for that reason they cannot be seen. Hence their name. They are not empty of matter and energy the way cheese holes are empty of cheese; on the contrary, they are stuffed, yet remain hungry. They swallow everything that gets close to them: meteors, planets, solar systems, national banks.

The radius (R) and the mass (M) of a black hole relate through the simple formula $R=GM/c^2$, where G is the Newtonian constant of gravitation. The radius of a black hole that has the mass of the earth is about 1 cm, that of one solar mass is 3 km. There are reasons to suspect the existence of black holes with 10^6 to 10^7 solar masses with appropriate radii of 1 to 10 million km. There may also be microscopic black holes of the size of a nucleon (a proton or neutron), with a corresponding mass of 10^{15} grams or a billion tons. The temperature of a black hole is calculated to be only 10^{-8} degree above absolute zero temperature, whereas miniature black holes are very hot, approaching the temperature of the big bang. Mini-holes are imagined to comprise matter trapped under its own gravity at very early ages, making them surviving samples of the primordial fire. The radius of a mini-hole divided by the speed of light gives the approximate age of the universe when the black hole was formed. An atom-sized mini-hole in the present day universe would have been formed when the universe was 10^{-18} seconds old.

The proper time of a black-hole traveler will remain proper time, of course; measured in that time, the gestation period of an Indian elephant astronaut will remain 645 days, that of a house mouse 19 days. An outside observer, such as one on earth, will find that the biological clocks of these animals tick slower and slower as they approach their destination; it will take an infinitely long time in the observer's reality before the animals totally disappear from sight. The elephant and the mouse will be seen hanging suspended just outside the surface of the black hole, ghosts of frozen time.

In the reality of the astronaut animals, the falling-in happens rapidly but smoothly, with nothing unusual except that they will be compressed to atomic sizes and broken to their elementary particles. If the particles of the former elephant and mouse make it to the center of the black hole, then the animals will have been changed to speed-of-light particles whose proper times are zero and whose umwelts atemporal.

As a means for achieving eternal life, the conditions that black holes offer are no more comfortable than those available through special relativity theory. Mistaking the atemporal integrative level of the world for our heavenly home would be a fatal mistake.

Can or will the world have an end?

Creation has an extensive literature in physics, papers on world endings are few. This relatively meager interest cannot be explained by saying that there is not enough data because it did not yet happen. Useful data is always the result of intentional search and prior, theoretical exploration, and is not an accidental contribution by nature. If physicists were as much interested in eschatology as in cosmogony, then someone would already have matched the best-selling *The First Three Minutes* by *The Last Three Minutes*.

"Cosmogonies educate, entertain, account for the world and society as we find them to be, appeal to the intellect by explanation. Eschatologies restrict, frighten, moralize. . . . The asymmetry is a projection of our mind's unequal concerns with the beginnings and endings of our lives," so I remarked earlier. Birth is interesting, death is foreboding. This has not stopped artists from being concerned with eschatology and passing. But it seems to have stopped most physical cosmologists, intellectuals who by temperament prefer to deal with deathless, quantitative relations rather than with passing and death.

Recently, however, a number of physical cosmologists have offered what they regard as scientific support for the idea that the world was created so as to make the coming about of humanity possible. This belief is termed the anthropic principle. Having thus resurrected teleology (goal-directedness in evolution) under an acceptable name, they have provided the kind of aura that is necessary to make eschatology a scientifically reputable subject.

Whatever their work may eventually reveal, the instant ending of the universe is prohibited by the finite velocity of light: even a self-destruct signal needs time to propagate. The amount of time needed for a command of universal cataclysm to reach our galactic shores, should such be given by nature or God, equals the age of the world. The longer the cosmic past, the longer the cosmic future. The universe of our epoch has a guaranteed future of 15 to 20 billion years, growing by a day each day. But the rate of expansion seems to be slowing down; it may eventually stop and the world remain maximally expanded. Or the world may begin to collapse. This second scenario has a greater scientific believability. John A. Wheeler called the end process of such a collapse the "big crunch."

Since no branch of physics outside cosmology displays an arrow of time to serve as a possible source of our experience of passing (see the following section), it has been suggested that the roots of our sense of time be sought in the expansion of the universe. But none who maintain this view has said, thus far, how our bodies learn of that cosmic process. How does the biology of an aging turtle know of the expansion of the universe? How did the mind of the caveman, already concerned with passing and death, take note of the growing volume of the universe, even if not in these terms? And what is supposed to happen to the human sense of time during the period of collapse?

It is interesting to pursue this physics–fiction game, however, because it shows what happens if one tries to understand the nature of time from a single discipline alone. The scenario is this: the distant galaxies begin to rush toward us, the world gets denser and hotter, the galaxies disintegrate into structureless clouds of particles. "The last scene of all, That ends this strange and eventful history," is a black hole of destruction, no more than 10^{-32} cm across. Depicting the events in this manner assumes that the immense inhalation, following the cosmic exhalation, may be described in the terms of noetic time. But if so, then noetic time could not depend on the expansion of the universe.

Alternately, let us assume that our sense of time would change and what used to be experienced as past–present–future is now experienced as future–present–past. I do not know what this could possibly mean. But if I elect to take the arrow–of–time metaphor seriously and imagine that the opposing directions of time relate the way two opposing arrows drawn on a piece of paper do, then a simple situation emerges. A universe expanding in forward-flowing time becomes indistinguishable from one collapsing in backward-flowing time. But then again, noetic time could not depend on the expansion of the world. The whole thing is awfully muddled, and I would not even dare ask what was to be meant by your time flowing backward while mine flowed forward.

Instead of insisting that passing has its roots in the cosmic expansion, it is simpler to maintain that the physicist's *t* remains self-consistent across all of physical science, including cosmology, and that the physical universe is eotemporal. Cosmic time has no direction in itself. A bewildering conclusion comes to mind. Cosmic time has only one single boundary; we might call it its "begending." This single boundary is seen as two events because our minds order the universe by human time.

The way to get out from inside a sphere is to keep on walking along a straight line. But, as the cosmology of GRT revealed, this strategy does not work for that finite though unbounded habitation that is the physical universe. There exists, however, another strategy. It is to get out of the physical universe, as it were, along a scale of complexity. This is the path nature has taken. The oldest possible fossil and largest possible object— the physical universe—with its primitive temporalities was thus left behind as life, the mind of man, and people in societies began to create new temporal realities with degrees of freedom to which nothing in the physical world can correspond.

Entropy and Time

Sir Arthur Stanley Eddington, distinguished British astronomer and natural philosopher, has sought to identify the sources of time's arrow (his coinage) among physical processes. Writing in 1928, he remarked that, "there is only one law of nature—the second law of thermodynamics— which recognizes a distinction between past and future. . . ." This law is responsible for time's arrow, the one-way property of time that has no analogue in space and that is "vividly recognized by consciousness."[16]

Eddington's clear and concise claim for the thermodynamic origins of our experience of time's passage has become an all but unquestioned dogma of the study of time in physics. It has had its uses, but by encouraging a knee-jerk metaphysics that identifies reality exclusively with whatever may be explored through physics, it has by now done more harm than good to the study of time. A close look at its tenets, sketched in this section, reveals that the relationship between the second law of thermodynamics and the experiential passage of time is spurious. As living beings, thermodynamics supplies us with two equally valid but opposing "arrows." Our consciousness recognizes neither of these trends in itself, but rather the opposition between them.

16   Eddington, *The Nature of the Physical World* (Ann Arbor: University of Michigan Press, 1958), pp. 66, 69.

Chicken soup, entropy, order, and information

The concept of entropy is introduced in most of the popular literature of time in physics through a mystic haze of implications, as if it were a concept obviously beyond the ken of the reader. It is not, provided that he or she can tell a bowl of hot soup from a cold igloo.

Prepare a bowl of chicken soup and place it, while still piping hot, in the sun on a summer's day. It will cool to the ambient temperature of, say, 27°C. In the process, it will have given up some of its heat energy to the environment. Next, take it down to the basement. After a while, more of the soup's original heat content will have been given up as it cools to 12°C. The next stop is the refrigerator (4°C) and the freezer (−11°C). Keep the soup there until the Minnesota midwinter night when you may put it on the porch (−30°C). Finally, give it to an astronaut friend and ask her to leave it in outer space, where the temperature is near the absolute zero of −273.16°C. It could not be cooled further because nothing can be colder than absolute zero. Throughout this experiment heat always flowed from the hotter to the colder body, never in the other direction.

The heat content of bodies can be used to do work only when heat flows. For instance, the heat of flames flows into the water to boil it, to make steam, to generate electricity for the lifting of an elevator. The heat of a scotch on ice flows into the ice; the work it does is the warming of the ice, the desired output is the cooling of the drink. But since heat by itself flows only from hotter to colder regions, if the heat content of a body is to serve as the source of heat energy, there must always be a heat sink. This is always possible to have, except at absolute zero temperature.

To help organize the facts of heat flow into a system of understanding, let us think of a more precisely describable system than scotch on the rocks. Imagine a box with walls that do not permit heat, matter, or energy to pass through them in either direction. In thermodynamics, which is the science of heat, such a system is said to be thermodynamically closed.

What happens to heat in a thermodynamically closed system is governed by well-known laws, such as the principle of conservation of energy. This law asserts that the total energy inside the box will remain constant. Imagine next that there is a hot object and a cold object in the box. Heat will then flow from the hot to the cold object. That heat flow may be used to generate electricity to do work, such as making a robot hop up and down and clap its hands. After a while, the hot and cold bodies will reach the same temperature, the heat will stop flowing and the robot will stop hopping and clapping. The box will contain the same amount of energy as it did before, but the nature of that energy has changed. It has become useless for doing any further work.

Now imagine that the box is large enough to enclose the earth, with its atmosphere, pastures, cows, tobacco auctioneers in Virginia, and the Paris Metro. When the inside of the box reaches thermal equilibrium, all of its free energy will have been used up (dissipated), and we will stop hopping and clapping even though the total energy content of the box remained the same.

In both examples, the free (available) energy has changed into bound, latent (unavailable) energy. This remains true for all thermodynamically closed systems, regardless of size.

In the mid-nineteenth century, the German physicist Rudolf Julius Emanuel Clausius constructed an equation to describe this fact. For use in his equation he invented a measure for that part of the heat energy which has become unavailable to do work. He called the measure entropy, from a Greek word that means "change." Since the amount of free energy in a thermodynamically closed system may remain constant or decrease, but may never increase, the entropy of a closed system may remain constant or increase, but never decrease. The principle of entropy increase for closed thermodynamic systems is the earliest formulation of the famous second law of thermodynamics.[17]

In the 1870s the Austrian physicist Ludwig Boltzmann showed that the second law is one of statistics and may be stated in terms of probabilities. In its new guise the law said that a closed system always changed from a more to a less ordered state, because the less ordered state had a greater probability to have come about. A neat room represents a lesser proba-bility than does a messy room, because if one begins to throw things into a room randomly, it is much more likely for the room to end up in a mess than neatly ordered. The second law states that a neat room left to its own devices can only become disordered; a disorderly room will not clear itself up.

With the advance of information theory, as developed by C. E.

17　Entropy is not measured on a scale with a fixed reference, as is, for instance, tempera-ture. Entropy is measured only by the changes in its magnitude, it can increase or decrease by so much. Entropy change ΔS equals the ratio of the increase in bound energy ΔQ to abso-lute temperature T, $\Delta S = \Delta Q/T$.

Shannon, E. T. Jaynes, and others, the usefulness of the concept of entropy was again extended. The second law of thermodynamics was shown to demand that the information content of a closed system remain constant or decrease, but not increase. A closed system can forget information, but cannot create any.

Now that we know all that we always wanted to know about entropy but did not dare to ask, we will assume that it measures certain scientifically objective rather than subjective aspects of thermodynamical, statistical, and information-processing systems and shall turn to the issue of entropy increase and our experience of time.

The arrow, the arrow, where is the arrow?

We learned of three different ways for measuring the one-way change of a closed physical system governed by the second law of thermodynamics: by its entropy content, which can only increase; by the probability of its state, which can only change from a less probable to more probable one; and by its information content, which can only decrease. Any of these methods could, in principle, be employed to determine the direction of time's arrow, using Eddington's program. Here are the instructions: (1) Make two entropy measurements on the largest possible object, the universe, at two different times. The larger entropy indicates a later instant. (2) Calculate the probability of two of its observed states; the state of greater probability must have been the later one. (3) Calculate its information content at two instants; the smaller number identifies the later one.

Neglecting the fact that to make measurements at two different times one must already know what time is, and also the difficulties of making the necessary measurements, the proposal itself sounds simple and convincing. But would this enterprise accomplish what it is supposed to do? I do not think so.

In his 1974 book, *The Physics of Time Asymmetry,* P. C. W. Davies considered a particular closed, unbounded model of the universe, one that he believed was likely to correspond to the real world, and gave reasons why such a universe may be thought of as a black hole. He also showed how black holes can acquire infinite entropy at the expense of gravitational potential. That is, how—if first compressed to a certain threshold density—black holes will collapse without limit to a state of infinite density and entropy. He concluded that the "universe as a whole has

unlimited capacity to increase its entropy" and that "The origin of all
thermodynamic irreversibility in the real universe depends ultimately on
gravitation."[18]

The passage of time, then, is a conscious awareness of the increase of
cosmic entropy, mediated by the gravitational behavior of matter.

But the increase in the entropy of the universe also demands that we
assume the existence of complete disorder (initial randomness) of the
universe at $t=0$ and the subsequent ordering of the world between that and
a later date. Thus, we learn from Davies' 1980 book *Other Worlds* that

> To explain where the ultimate cosmic order came from, and hence
> account for [the] distinction between past and future, it is necessary to
> consider the creation of the universe—the big bang. The cosmic
> structure which emerged from the primordial furnace was highly
> ordered, and all subsequent action of the universe has been to spend
> this order and dissipate it away. Plenty remains but it cannot last
> forever.[19]

It must then necessarily follow that for a period after the singularity,
entropy had to be decreasing, and order and information increasing. If
what we recognize as the passage of time does have its roots in universal
entropy increase then, for a while, time had to flow backward. Whatever
that might mean, the situation is rather awkward.

In spite of such problems, the metaphysical belief that identifies cosmic
entropy increase with the sources of our sense of time remains strong.
Thus, in a 1986 paper, "The Arrow of Time," Davies once again
emphasized that, "We can regard the expansion as producing the vital dis-
equilibrium that gives rise to the arrow of time."[20] Specifically, the
expansion keeps space cold (for the same reason that the expansion of
freon gas in our refrigerators keeps the refrigerators cold, as already
mentioned) while it also permits the accumulation of new structures, such
as hot stars. The flow of heat from these sources to the heat sinks provides
the limitless entropy increase (disordering) of the originally ordered
universe. Once again, in the spirit of Eddington's tenets, that entropy

18   Davies, *The Physics of Time Asymmetry* (Berkeley: University of California Press,
1974), p. 109.
19   Davies, *Other Worlds* (New York: Simon and Schuster, 1980), p. 198.
20   Davies, "The Arrow of Time," *Sky and Telescope,* September 1986, p. 241.

increase should be regarded as the ultimate source of our sense of passing time.

But there are other difficulties with the entropic theory of time's passage. To see what they are, we turn to Edward R. Harrison's solid and authoritative book *Cosmology: The Science of the Universe*.[21]

Harrison points to the broadly accepted view that the number of photons-per-baryon in the universe may be used as a measure of the entropy of the universe.[22] This cosmological parameter indicates the ratio of energy in disordered states to energy in ordered states. Baryons are seen to be ordered (organized) as compared with the speed-of-light particles, such as photons. There are between 10^8 and 10^9 photons to each baryon: the universe is made overwhelmingly of light, and "underwhelmingly" of matter.

When the universe is considered as a whole, the second law of thermodynamics is manifest by the entropy-generating processes in the stars, mentioned earlier. They release nuclear energy at temperatures of millions of degrees and radiate it into the heat sink of cold space. The large photon-to-baryon figure demonstrates that most of the entropy of the universe is already in its most disorganized state. Harrison remarks that

> Some decades ago, when scientists discussed the universe, they would foretell with bated breath the eventual death of the universe, noting how everything would fade and die, and entropy would rise inexorably and attain its ultimate value. We now realize that the heat death has already occurred; it happened long ago, and we live in a universe that has very nearly attained its maximum entropy. [P. 274]

Furthermore, Harrison maintains that "as we go back [in time], the total entropy remains constant . . ." (p. 356). But if the entropy of the universe has remained constant for most of its life, then entropy change could hardly be thought of as the ultimate source of our sense of time, even in principle.

Growth and decay—decrease and increase of entropy—are simultaneously present in the universe. Let their coexistence be represented by the shaft of a headless arrow. We are entitled to think about this shaft as

21 Harrison, *Cosmology: The Science of the Universe* (Cambridge: Cambridge University Press, 1981).

22 Baryons are heavy particles: protons, neutrons, hyperons, and their antiparticles.

two arrows in one, pointing in opposite directions, thus offsetting the directional preference that a single arrow would represent. (This kind of reasoning is used in mechanics, such as when physicists say that a body is at rest because it is subject to two equal and opposing forces.)

Since we could arbitrarily attach our sense of time to either "arrow," neither may be considered the source of our sense of time's passage. The physicist's t remains consistently eotemporal in all branches of physical science.

Figure 41 is a visual metaphor for the two opposing entropic processes of the physical world. Together, they make cosmic time eotemporal.

Life: The unresolvable conflict between loss and store

Some opposing processes of the physical world are recognized by the human observer of the seashore, along which the bard beheld

> the hungry ocean gain
> Advantage on the kingdom of the shore,
> And the firm soil win the watery main,
> Increasing store with loss and loss with store.
> [Sonnet 64]

But the physical world manifests no preference for either store or loss, only living organisms do. They assign different values to them in terms of their respective usefulness for survival. The coexistence in inanimate nature of contrary trends—formalized in thermodynamics through such variables as entropy or information—evolved, in living organisms, to certain unresolvable, creative conflicts that cannot be defined except by reference to the organic present and to intent. They are conflicts because they express themselves in many forms of hunger and in the kind of tensions that are unknown to the physical world. They are unresolvable because those tensions cease only with the death of the organism. And they are creative because each step taken to lessen the tension leads to new and unpredictable structures and functions, as I previously related in connection with the evolutionary widening of the spectrum of organic cycles and the office of moral codes.

What I have just described is the leitmotiv of the theory of time as conflict, of which the hierarchical theory of time is one aspect. It is of no further interest to us here. However, in a chapter that concerns itself,

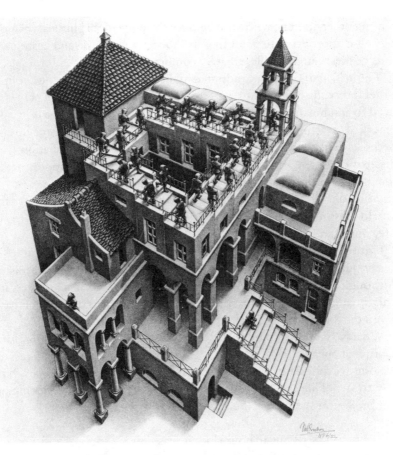

FIGURE 41   GROWTH AND DECAY PROCESSES IN THE THERMODYNAMICS
OF THE UNIVERSE. Each of the monklike figures is perceived as walking either
upstairs or downstairs, suggesting a definite direction to the steps: this way is up,
that way is down. The distinction between up and down vanishes, however, if the
circulation of the virtual movement is closed and the paseo is considered as a whole.
We then see that the activity produces no net change either up or down. Likewise,
the physical universe is governed by laws that demand the continuous increase of
entropy (disorder). But for various reasons discussed in the text, the universe also
generates negentropy (decreases entropy). In fact, we live in a universe whose total
entropy has been nearly constant since the early days of creation. Since one may
attach the sense of passing time to either of the two entropic processes, neither of
them could be its source. The physical universe at large is eotemporal. Illustration
is M. C. Escher's "Ascending and Descending," (1960). ©M. C. Escher's Heirs,
c/o Cordon Art, Baarn, Holland.

among other matters, with the origins of the universe in physical cosmology, it is appropriate to point to what I believe are the distant cosmic origins of these unresolvable, creative conflicts.

The instant of Creation amounted to the simultaneous coming about of orderability, disorderability, and the conflicts between them. There is no need to assume that out of the big bang a highly ordered universe emerged, bearing gifts to be dissipated throughout cosmic history. It is necessary only to recognize an opposition between the information-dissipating and information-generating processes that were appropriate to the physics of the early world. Then, in the course of evolution, new conflicts of increasingly more sophistication emerged and gave rise, in the world of primary concern to us, to the unresolvable, creative conflicts of life, mind, and society.

Figure 42 portrays the unresolvable conflicts between growth and decay, judged in terms of the autonomy and intent of a living organism. Although the message of the drawing may be extended to all species, it is only in humans that the unresolvable, creative conflicts of the mind have given the mute struggle of matter its powerfully articulated voice.

What is "vividly recognized by consciousness"—to repeat Eddington's claim that opened this section—and gives rise to our sense of passage is not the one-way increase of entropy in closed thermodynamic systems, but the stresses of life and mind due to the copresence of losing and storing, in our biological and mental life.

> 'tis but an hour since it was nine,
> And after one hour more 'twill be eleven,
> And so from hour to hour we ripe and ripe,
> And from hour to hour we rot and rot,
> And thereby hangs a tale.
>
> (*As You Like It*)

What the Stone Cannot Say about Time: Reflections on Time and Number

The art of conversing with stones, stated the opening sentence of this chapter, is called physics; the language spoken is mathematics. The immense strength of physical science comes from the power of number and measure. Whatever cannot be said through number and measure or

FIGURE 42 THE CONFLICT BETWEEN THE ENTROPY-DECREASING AND
ENTROPY-INCREASING ASPECTS OF THE LIFE PROCESS. This is one of Käthe
Kollwitz' dark visages of the human predicament. She called it *Death and Woman*.
A person with a different temperament might have called it *Child and Woman* by
focusing on the creative rather than the destructive aspect of life. Both perspectives
are simultaneously true: life consists of an unresolvable conflict between growth
and decay, judged in terms of the autonomy and intent of the organism. Reproduction, courtesy Galerie St. Etienne, New York.

through symbols representing them, and supported by their logic, cannot be said through the exact sciences. "The abstract mathematical idea of time as a geometrical locus," wrote G. J. Whitrow in his masterful work, *The Natural Philosophy of Time*, "is one of the most fundamental concepts of modern science."[23] What kind of limitations does this impose upon the authority of physics to speak on the nature of time?

In a paper entitled "Singularities and Time-Asymmetry," the Oxford mathematician Roger Penrose set out to find an answer "to one of the long-standing mysteries of physics: the origin of the arrow of time." Fifty-four pages of close reasoning and 114 bibliographic references later, all of them from physics, he concluded that,

> some readers might feel let down by this. Rather than finding some subtle ways that the universe based on time-symmetric laws might nevertheless exhibit gross time-asymmetry [show an arrow of time], I have merely asserted that certain laws are not in fact time-symmetric —and worse than this, that these asymmetric laws are not yet known![24]

This is an honest declaration of the metaphysical belief that has bedeviled the study of time in physics: come what may, the sources of "time that takes survey of all the world" *must* be found in and through physical science. What can induce so many brilliant intellectuals, making up the elite corps of modern physics, to maintain this indefensible article of faith? There are three powerful reasons.

The mass of living matter is negligible in comparison with the mass of the inanimate universe. It is easy to overlook the fact that the dog barking at the moon has freedoms that the moon cannot have, making the dog much superior to that pale body with sextillion times the dog's weight.

Western ideals are those of inorganic naturalism, a preference that contrasts, for instance, with the organic naturalism of China. For

---

23   Whitrow, *The Natural Philosophy of Time*, 2d ed. (Oxford: Clarendon Press, 1980), p. 174.

24   Penrose, "Singularities and Time Asymmetry," in *General Relativity*, ed. S. W. Hawking and W. Israel (Cambridge: Cambridge University Press, 1979), pp. 581, 635.

the Chinese, as for Plato, the world is seen as alive and even matter is preferably understood as having a measure of life. For the Chinese, the most exulted form of knowledge has been history; for the West, number and whatever may be done with number. If our own civilization were life-oriented rather than thing-directed, we would keep time by the migration of storks and by the rate at which fruit dries and not by ticks and tocks. Our leading metaphors for time would not be geometrical, such as a time-line, but would be organic, such as the Yggdrasil tree of Norse saga. For the sources of time experience we would consult not scientists of matter but those of living trees.

On the scale of values sanctioned by the success of science and technology, a statement about the nature of time is unlikely to be judged significant unless it can be tested for its truth or falsehood by measurement, expressible in number.

I spoke earlier of the long-standing Western love affair between number and truth: for Pythagoras, numbers were the ultimate reality; Plato's world was governed by geometrical rules; St. Augustine maintained that number and wisdom were the same; for Kepler, number had been in the mind of God from the beginning.

The German sociologist Max Weber remarked early this century that the origins of modern natural science should be sought in the readiness of Protestant asceticism to understand the world on a mathematical basis. The favorite sciences of all Puritan, Baptist, or Pietist Christians, he added, were physics and those natural sciences that used mathematics. Protestantism always preferred the most parsimonious expressions of feeling and thought, and mathematics is surely the thriftiest language there is. Protestant ethics, including parsimony in reasoning and representation, have been built into the scientific method and have become an arbiter in the scientific judgment of truth and untruth.

In sum, the belief that answers to all important questions about the world, such as those concerned with the nature of time, must be found through exact science—through number and measure—is not in itself anything "scientific" or, by any logic, necessarily correct; it is, instead, a socially conditioned dogma. It is based on the firm historical belief that mathematics is the queen of the sciences, that it forms a complete (or completable) self-consistent set of metaphors for all natural things and processes.

★

What is number? What is mathematics, that structure of permanent and unassailable logical truths that some people hold to be independent of the world of the senses?

In Umberto Eco's highly cultured, best-selling novel *The Name of the Rose,* a work in the history of science disguised as a thriller, Brother William of Baskerville—modeled on William of Occam, early fourteenth-century pioneer of scientific thought—is in pursuit of a shapeless evil that resides in a Library. The Library seems impossible to enter or, if entered, to leave, because it is a labyrinth. But Brother William has some ideas on how to decipher the message that is the labyrinth by using twentieth-century semiotics, the science of signs and symbols. While holding forth about his plans, he identifies the sources of mathematics.

> Mathematical notions are propositions constructed by the intellect in such a way that they function always as truth, either because they are innate or mathematics was invented before the other sciences.[25]

Mathematical rules are tools for survival, distilled experiences of untold generations of living things, accumulated judgments of what may be considered permanent in nature: "2+3=5," or "the ratio of the circumference of the circle to its diameter is independent of its size." How else could one account for the unobvious fact that the orbits of spacecrafts are ellipses, first identified around 350 B.C. by the Greek philosopher Menaechmus, who obtained them by slicing a cone with a plane—and judged his findings interesting enough to tell others about them?

Mathematics has been equated to the reasoning of nature itself. Its arguments have been identified with the language of Logos, the eternal controlling principle of the universe. It has also been believed through millennia that any question clearly statable in mathematical terms may also be unambiguously answered in the same terms, though we may not necessarily know what that answer is. Modern mathematics revealed that this is not the case.

In 1930, at the age of twenty-three, an Austrian mathematician named Kurt Gödel wrote a very technical paper, "On Formally Undecidable Propositions of Principia Mathematica and Related Systems." (The

---

25   Eco, *The Name of the Rose,* trans. William Weaver (New York: Harcourt Brace Jovanovich, 1983), p. 215.

reference in the title is to Bertrand Russell's monumental *Principia Mathematica*.) Gödel showed that it is possible to formulate mathematical propositions (pose questions to be answered) whose truth or falsehood cannot be decided (the question cannot be answered) using the same language of mathematics in which it was posed. The proposition may, however, be decidable (the question may be answerable) in the language of a higher-level formalism.

In everyday terms, it is always easier to play a game according to learned rules than to explain the rules to someone else. It is easier to speak a language, even correctly and precisely, than to teach someone how to speak it. It is easier to know what time is, and act consistently with that gut-level knowledge, than to explain it to someone. The reason is that to describe what one does, on whatever level of sophistication, it is always necessary to command a language of somewhat higher sophistication.

In his superb works on the theory and history of mathematics, Morris Kline has described Gödel's epochal discovery as the loss of certainty in mathematics. What for two and a half millennia was judged as a set of unambiguous rules was found to contain undecidable and therefore ambiguous propositions. The ambiguity may only be removed by using a higher-order math, which has its own unreducible ambiguities.

As mathematics keeps on developing, it will surely produce newer levels of knowledge, corresponding to the expanding reality of its creators. It is my guess that by the time mathematical methods for the handling of biological, mental, and social phenomena peculiar to those levels become available, the cherished certainty of everything mathematical will have been replaced by a system of indeterminacies. These new descriptions would recognize and formalize those degrees of freedom of the higher integrative levels that biologists, psychologists, sociologists, and historians already consider a given in their subject matter. Such a development would make the "physics envy" (my term) that now informs quantitative biology, as well as much of psychology and sociology, an example of outdated professional jealousy.

If mathematics is indeed a distilled, abstract representation of what our minds identify as permanent in nature, then something in nature ought to correspond to Gödel's discovery. It does.

Each of nature's stable integrative levels has its peculiar logic, its laws, or, as it is sometimes called, its "language." Here "language" stands for appropriate means of communication. Scientists must be careful in the ways they use the language, the logic of the process they work with. Otherwise, their reasoning, appropriate though it may be to one level, may lead them to conclusions different from what mother nature actually does on another level.

For instance, in the logic of daily experience, something is either a watermelon or the rhythm of flamenco. (Does this sound familiar?) But as we learned, in the prototemporal world, this logic does not hold. An object may be (1) a watermelon, (2) a flamenco rhythm, or (3) a watermelon and a flamenco rhythm, all at the same time.

In the logic of our life experience, an effect always follows its cause: make love now, deliver little person later; take poison yesterday, pay with life today. The logic of the eotemporal world does not contain such directionality; in that umwelt all processes are reversible, the appropriate equations permit the exchange of causes and effects.

This returns us to the story of the prisoner's paradox told in Chapter 1, in the section entitled "Ideas about Time." The lawyer's reasoning went from the future to the present: "If not on Saturday . . . if not on Friday . . . if not on Thursday . . . if not today. . . , then it cannot happen at all." This is an acceptable logic for an eotemporal world, such as that of the physical universe. That integrative level is the eminent domain of mathematics because logical implication—a relationship among mathematical propositions—does not respond to the passage of time. Although it is our skill in the use of numbers that makes time measurement possible, no clock in itself can tell us what is to be meant by the now, the future, or the past.

But for living systems, time has a direction. In the life of the prisoner it was not possible to reach next Sunday without living through Monday, Tuesday . . . Friday, and Saturday, in that sequence. In mathematical logic one may validly reason both forward and backward in time, but life and mind create and demand freedom of choice, and therefore it is not possible to live backward in time.

The good fortune of the prisoner was that he did not know that the temporal logic of the stone is not valid for the temporal logic of life and mind, and therefore could dream of freedom until the day he died.

For the complete description of the temporality of each organizational level, it is necessary to be able to look back at it, as it were, from the increased wealth and broader world of the next higher organizational level. The limitations of the child may be felt by the child, but they may be completely and accurately described only by someone who does not have those limitations.

Let me return to "one of the long-standing mysteries of physics: the origins of time's arrow." Whether or not the passing of life is a mystery is a matter of opinion. I honor both views. But the "arrow of time" is definitely not a problem that physics can solve. It is minimally a life phenomenon, and therefore its study belongs in biology, psychology, sociology, and the arts and letters.

Stones can tell stunning tales about the time of a universe too alien, too vast, and too inhospitable for life. But their reality does not include the passing of time. There is no reason to be concerned, however. Humans can tell more interesting tales and sing richer songs than the rules of the universe of stones ever could.

FIVE

Ask my song

MAN THE MEASURE AND

MEASURER OF TIME

In this chapter we circle back to the idea and experience of time—the theme of Chapter 1—because that experience and that idea are undergoing profound revolutionary changes. The role of man as the measure and measurer of time is being subsumed among the prime concerns and roles of society as the measure and measurer of time. Subsumed among, but not eliminated, just as ages ago the role of organismic timekeeping was subsumed among the tasks of the mind as the senior timekeeper, but not eliminated.

It will be recalled that neural control of organisms evolved as a part of the development of different life forms, but only after the nervous system crossed a certain threshold of complexity—that represented by the human brain—were mental capacities born and with them, the new reality of noetic time.

Likewise, society and its individual members have been evolving together, but only after a significant segment of the people on earth have reached a certain level of social complexity may the new integrative level of the global-sociotemporal arise. I believe that the global restlessness of our epoch signals the creation of that new integrative level, suggesting that we have reached what I have been calling the anthill threshold.

There is no stable platform, no intellectual discipline or sociopolitical island independent of the turmoil of our days, from which the transformations of the contemporary world could be conveniently observed, let alone reliably reported. The global family of man is both the observer and the observed, the creator of the new temporality and its servant. The best way to proceed, therefore, is to trace the background of the ongoing transformations in the social experience and idea of time, do this in terms of its major component parts, then draw such conclusions as appear warranted.

The background sketch I wish to draw in this chapter is based on the opinion that feeling is older than reasoning, time felt is older than time understood. It is possible to short-circuit reason, and react to time's passage without reasoned control in ways consistent with one's emotions. But it is not possible to short-circuit emotions and behave according to reasoned ideas about future, past, and present without the continuous presence of an emotional underpinning. *Homo sapiens* is only superficially a reasoning animal; basically he is a desiring, suffering, death-conscious, and hence time-driven creature. In this respect, the men and women of the anthill threshold are no different from their early forebears.

Therefore our voyage on the sea of contemporary turmoil will be a continuation of the journey that we began by watching my hairy ancestor prepare to fight the frumious Bandersnatch. He, and all of us since, have only been working out the conflicts between what we may individually and collectively imagine and what is possible, using our ideas of time as our ordering principles.

The first section of the chapter demonstrates through selected examples how the temperamental, emotional dimensions of time—time felt—are always present beneath reasoned expressions of the two great cultural continuities: artistic creation and political behavior. The second section identifies some of the hallmarks of a time-compact society; the closing section speaks of a number of problems unique to the nascent global sociotemporality.

Descent and Ascent: The Fable of Two Women

This section is about time travel. Not of the science-fiction kind that pretends to go into the future and the past, but of the kind we all actually do. As we know from the discussion of nootemporality in Chapter 3, the hierarchical organization of the human brain makes us continuously travel among different assessments of reality. The different ways of perceiving time are parts of those assessments; they are always present in our mental life. They make for an infinitely rich storehouse from which people select the guidelines of their behavior, consciously and unconsciously, from instant to instant. I want to introduce my discussion of the mental roaming in that storehouse of temporalities by a fable of two women.

One of the women is known only as the Wife of Lot. She and her family were spirited out of Sodom by two divine messengers, so as to escape the fire and brimstone that was to be visited upon that town because of its gay life. In return for the favor, she and her family were cordially requested not to look back. But like Orpheus in another story, Lot's wife looked back. She was punished, the Bible says, by being transformed into a pillar of salt. I imagine her changing from a woman of mind and life, the mother of two daughters, to a living vegetable, a dead body, an amorphous rock, and finally to dust. Hers was a journey of descent.

The story of the other woman begins with the sculptor Pygmalion, who fashioned out of stone a milk-white image of a beautiful maiden. Later he prayed to Aphrodite to give him a wife as beautiful as his marble virgin. Aphrodite was so moved that she brought the statue to life as Galatea, who later bore a son and daughter to Pygmalion. And to herself. I imagine cosmic dust coalescing into shapeless marble, gaining human form, and coming alive as a woman of good mind and body and a member of her society. Hers was a journey of ascent along the integrative levels of nature.

As we concluded toward the end of the preceding chapter, "from hour to hour we ripe and ripe, / And from hour to hour we rot and rot," and we do the ripening and rotting simultaneously. For this reason, each of us is a Galatea and a Wife of Lot in the same person, forever descending and ascending. During these descents and ascents we visit, as it were, then report to ourselves and sometimes to others on the archaic and modern temporalities of our minds.

Each of the temporalities crossed during these journeys has its characteristic *mood*. The atemporal mood is the suffering of the schizophrenic who feels the pull of chaos and panics for the loss of nootemporal reality. The prototemporal mood is the feeling engendered by a Jackson Pollock painting, with its incoherent islands of coherence, or by listening to the talk of an autistic child. The eotemporal mood is the oceanic feeling of Freud, a sense of directionless time. The biotemporal mood is the happiness or fright of an hour that has no distant futures and pasts. The nootemporal mood is the reasoned feeling for human reality in its fullest.

As our minds invest their psychic energies now in this level, now in that one, the mood most loaded becomes predominant, though seldom to the total exclusion of the others. Creative people seem to be instantly ready to experience these moods and to visit the different temporal assessments of reality present in their minds. We are going to survey the dynamics of such travels among the moods of temporalities, using illustrations from art and politics, those two closely related continuities of social life.

Music and tragedy: Mobilizing all the temporal levels of the mind

Music and tragedy are unique among the arts in being able to address directly the organic, mental, and social presents. Through them, they modulate the moods of time felt and speak to our understanding of time. It is a complex process, but one with manageably clear structure. Let us consider music first.

The spectrum of biological clocks, as I explained in my discussion on biotemporality in Chapter 3, is 78 octaves wide, that of audible sound is 20 octaves. Sound in itself does not become music, however, not any more than oscillating biochemical clocks become life, unless coordinated from instant to instant by a suitable organizing principle. In the life process, those principles are the stable laws of biology. For music they are the unstable conventions of aesthetics. They change from epoch to epoch, place to place, and even person to person. But unless there are such rules, one can only speak of noise and not of music.

The experience of music—the *musical present*—is created by allowing certain sounds to be heard simultaneously while others not, in analogy to the organic, mental, and social presents. Variations in pitch, rhythm, melody, and stases then help music integrate itself into the already existing hierarchy of presents. Music addresses us biologically through the organic present (as in a march or a dance), mentally (by creating musical memories, expectations, and associations), and socially (musicians, or at least their music, and audience must be present simultaneously).

Because music is the art of sound, it joins spoken language in the creation of selfhood. What spoken language is for identity understood, music is for identity felt. Music, as poetry, helps define the emotive self and its purpose, hence music anticipates—and thereby helps create—the future. Tunes identify political movements or geographical regions just as much as spoken names do: a few bars of "Dixie" stand for the identity, history, and way of life of the antebellum South.

Experimentation with musical form has been a significant part of music history. As is true for favored forms of poetry and prose, favored musical forms correlate with the views of different social groups about them- selves, with their collective temperaments, and with the ways they like to do things. The instruments also illustrate the age and place of their use. A successful German band, the Einstürzende Neubauten (Collapsing Highrises) use power tools, cement mixers, and metal grinders. One can make music with anything, they say. Indeed so. The only element that cannot be eliminated is the musician. In machine composition, the musi- cian hides himself in the software.

Some contemporary composers employ noise as the rule of harmony, play aleatory (chance) music, and use temporal elisions to suggest mysterious meanings or cause tension. The popularity of these new yet ancient forms of making music speaks of a desire to descend with the Wife of Lot into simpler forms of reality than that of our age, perhaps in the hope of finding a counterpoint to the brave new world.

What music does abstractly by musical sound alone, tragedy does con- cretely through a story and the spoken word. The unfolding of tragedy demands uninterrupted reflections upon the past, the future, and upon the choice among different paths of action thinkable in the present. These elements, skillfully combined, create what has been called the *tragic present*. Tragedy demands continuous travel among the levels of reality perceived by our minds, a restless journeying with the Wife of Lot and Galatea. The moods of time are ceaselessly evoked and are intricately mixed: we feel the terror of chaos, the call of continuity, the demands, pain, and satisfaction of being alive, and the predicament of being able to think in terms of noetic time.

A sad or heavy story, even if profound and even if it may properly be called tragic in everyday language, is not necessarily a tragedy in the richest meaning of this classic form of drama. The tragic vision involves a series of deliberate choices in the service of a nontrivial idea that the hero or heroine defends against overwhelming odds; there must be something beneath the story that is larger than its manifest content. As the story progresses, the available choices narrow until there is no choice but death.

The issues involved in a tragedy and the manner in which the events unfold must suggest cosmic or historical dimensions, ideals of order, dis- order, justice, or injustice. The tragic hero is usually a loser and yet a winner, for his fate inspires people and through them it changes the direction of social motion. Animals may be abused, they may lose their lives defending their offspring, but only humans can make a series of deliberate choices in support of an idea, a symbol they created by distilling what they judged permanent in their experience of time. Hamlet expressed it all in two lines:

> The time is out of joint: O cursed spite
> That ever I was born to set it right!

During and since the nineteenth century, the form of tragedy came to include the novel: for instance, Thomas Hardy in England, Hawthorne and Melville in America. Later it became a vehicle for social protest. There are many examples of great tragedies in contemporary literature: those by Eugene O'Neill, Tennessee Williams, William Faulkner, and Samuel Beckett. Yet after having been the most distinguished and uniquely Western form of drama for over two millennia, tragedy has ceased to be popular with theatergoers, especially in the United States. The change suggests a narrowing of temporal horizons.

By its very nature, tragedy depicts a world in which final victory over evil is impossible. This is an unpopular view for a civilization that dogmatically maintains that all problems can be solved. Be it hunger, crime, war, or cancer, given sufficient funds, time, and scientific research, the human condition is perfectible without limit. Endorsement of delayed satisfaction as a way of life was replaced by the promise of immediate satisfaction through correct problem solving. But tragedy assumes that the human condition is not a series of problems, each of which has a correct solution if we but find it, but a predicament that has no correct solution, a chronic unresolvable conflict between what ought to be and what is.

The consequent absence of the tragic sense among the vast masses of people and among their leaders has made Fate into a clown, goodness into smiling satisfaction, and evil into a petty devil. William Styron remarked in his novel *Sophie's Choice* that

> the evil portrayed in most novels and plays and movies is mediocre if not spurious, a shoddy concoction generally made up of violence, fantasy, neurotic terror and melodrama.[1]

---

1   Styron, *Sophie's Choice* (New York: Bantam Books, 1980), p. 179.

The fall of tragedy as a popular form of drama is an example of what evolutionary biologists call preadaptation, meaning changes in the behavior of an organism that in hindsight will be seen as having contributed to some important evolutionary change. In biology, preadaptation is a strange idea because natural selection works under present pressure only. But it makes sense in social science, because people can and do adapt to changes projected into the future.

In the case of the unpopularity of tragedy, preadaptation consists of the emerging, universal belief that an individual cannot be powerful enough to fight evil alone and that there is no such thing as evil (or, conversely, good) to begin with. There are only difficulties of the instant: the explosion of Challenger (an O-ring), cyanide in Tylenol (a nut, a plot), unemployment (the Government), and so on. Replacing the tragic hero by the problem-solving expert signals a narrowing of the social interest in time from history to the present. More will be said about the increasing belief in the irrelevance of history in the following section.

Individual roundtrips to the depths: The ecstasies of timelessness

When a person describes a feeling as one of "timelessness," he is making a comparison, as I explained earlier, between the readings of two of his inner timekeepers. He says that compared with his waking, conscious experience of passing time, the feeling he now has misses something: perhaps continuity, perhaps direction, perhaps broad enough horizons.

These kinds of feelings have been recognized by all civilizations. The Greek word for them was *ekstasis,* which means "to cause to stand outside one's self." Calling them the ecstasy of the dance, of the forest, of the bower, of the mushroom, and of the chalice, I will describe five common experiences that are often said to invoke feelings of timelessness. We will see that in none of them does timelessness mean a total absence of time (such that everything would happen at once) but only a recognition that the time of those experiences, as compared with the time of the fully conscious state, is somehow incomplete.

Rhythmic motion tends to produce the *ecstasy of the dance.* Dancing to a regular beat focuses the dancer's feelings upon the beat. A steady bump-bump-bump has no preferred direction in time, just as the ticks of the clock do not; the world of the beat is eotemporal; it is the absence of temporal direction that the dancer notices. Ordinarily we live in continuous tension caused by our awareness of passing and a conscious or latent image of eternity. In the ecstasy of the dance that tension lessens; the dancer's identity loses its definition as he or she becomes one with another, with the world of stars, and with all other people, and enters a never-never land. It is a return to childhood when time seemed infinite, and only the present existed. The old Shaker hymn says it well.

> Come life, Shaker life,
> Come life eternal;
> Shake, shake out of me
> All that is carnal.

The Shakers (Figure 43) believed that in the ecstasy of the dance their minds could foretell the future. The belief reflected an intense feeling to which the validity or falsity of any actual prophetic prediction was irrelevant; it was their way of describing a world in which time had no direction. If prophecies were made, the group was surely ready to modify the predictions ex post facto, if that became necessary, so as to make them fit the facts and thereby justify the ecstasy of the dance as a valid social act. I am not thinking of deliberate falsification but of the remarkable readiness of the mind to adjust reality to fit its desires.

The religious experience of the whirling dervishes of Sufism was similar to that of the Shakers. It was a search for releasing tension, as was the mass hysteria of screaming, weeping, and fainting generated by the rock 'n' roll in the 1960s. The rhythmic motion of apes under stress suggests that the ecstasy of the dance has a strong biological component shared with other species. The walk of camels is also rhythmic, although as far as one can guess, they do not use it to lessen their existential tension. It is man who can keenly feel it and even write about it, as did Nikos Kazantzakis in his *Report to Greco*.

> The camel's sure, undulating rhythm transports your body, your blood takes on the rhythm of the undulation and, together with your blood, so does your soul. Time frees itself from the geometric subdivision into which it has been so humiliatingly jammed by the sober, lucid mind of the West. Here, with the rocking of the "desert ship,"

FIGURE 43   THE SHAKER EXPERIENCE OF TIMELESSNESS. The illustration
shows the Shaker ritual of dance, a form of religious ecstasy. The rhythmic motion
focuses the psychic energies on realities older and more primitive than those of the
nootemporal world. In the archaic umwelts time is almost entirely that of the
present, with only a slight haze of the immediate future and past. The collective
descent with the Wife of Lot into the netherworld of older, simpler temporalities
may help refresh a person by relieving the burden of long-term concerns. This
lithograph of uncertain origins was made by an unidentified artist. Courtesy,
the Henry Francis du Pont Winterthur Museum, Winterthur, Delaware.

time is released from its mathematical, firm-set confines; it becomes a substance that is fluid and indivisible, a light, intoxicating vertigo which transforms thought into reverie and music.

. . . I began to understand why Anatolians read the Koran swaying to and fro as though on camelback. In this manner they impart to their souls the monotonous, intoxicating movement which leads them to the great mystic desert—to ecstasy.[2]

This is a poetic description of what the study of time calls, in prosaic words, the eotemporal mood.

The *ecstasy of the forest* is similar to the ecstasy of the dance. But the feeling of directionless continuity is achieved not by focusing on ceaseless, rhythmic change, but upon eternal rest. It is the sense of the lonely eminence of the Rockies. of a continuous now where future and past are rolled together into an eternal present. Here, too, the tension of selfhood lessens as the wayfarer feels himself integrated with the eternity of the forest.

At the celebration of Max Planck's sixtieth birthday, Albert Einstein stated it this way.

I believe . . . that one of the strongest motives that lead men to art and science is escape from everyday life with its painful crudity and hopeless dreariness, from the fetters of one's ever-shifting desires. A finely tempered nature longs to escape from personal life into the world of objective perception and thought; this desire may be compared with the townsman's irresistible longing to escape from his noisy, cramped surroundings into the high mountains, where the eye ranges freely through the still, pure air and fondly traces out the restful contours which look as if they were built for eternity.[3]

The language and setting of Einstein's thoughts are those of German romanticism, much of which was swept away by World War II and air pollution. What remained were the theory of relativity and an assurance from Einstein that time was or ought to be unreal.

2   Kazantzakis, *Report to Greco* (New York: Simon and Schuster, 1964), p. 261.
3   Einstein, *The World as I See It* (London: John Lane the Bodley Head, 1935), p. 124.

The contemplation of unchanging vistas relieves the mind of the travail of individuation. After a person succeeded in "standing beside himself," he or she may rejoin the cutting edge of organic and cultural evolution and resume his responsibility for choice.

In the *ecstasy of the bower,* better known as sexual intercourse, the participants experience the ecstasy of the dance followed by that of the forest. At the end of the dance phase, the decreased self-awareness, supported by a complex physiology, has hallmarks of sleep into which the participants are likely to fall. In the semiconscious mind, Eros may then give rise to Agape, to the light that shines in the darkness steadily and timelessly. Pablo Neruda spoke of that light in beautiful cadences.

> Kiss after kiss, I recover your little infinitude
> rivers and shores, your body's diminutive clan,
> the genital spark, made dear and delectable,
>
> that races the delicate pathways of your blood,
> breaks up from below in a gout of nocturnal carnations
> unmaking and making itself, leaving only a glow in the dark.[4]

A large variety of natural or synthetic drugs are known to produce feelings of "timelessness." These drug-induced experiences may be called the *ecstasies of the mushroom,* in honor of fly agaric, a mushroom sacred to some cultures. The aftereffects of the "mushroom" may be anything from a mild hangover to psychotic seizure. Alcohol goes under the same heading. Its power of numbing the senses and lessening the drinker's awareness of passage made it a favorite weapon of slave traders in their time and of shady characters in our own. Heroin used to be a weapon of oriental militarists and remains one in international politics; it is also a deadly commodity exported by some countries in need of hard currency. Marketing timelessness has always been a profitable enterprise.

Sniffing cocaine and descending with Lot's wife into the atavistic mind is the story of Robert Louis Stevenson's classic *The Strange Case of Dr. Jekyll and Mr. Hyde,* written a century ago. The two names are those of the same person with two, increasingly different, personalities. The

4　Neruda, Sonnet 12 in "A Hundred Love Sonnets," in *A New Decade* (New York: Grove Press, 1969), p. 61.

powers of the degenerate Mr. Hyde have grown with the sickliness of the good Dr. Jekyll, says Stevenson, who is said to have been an addict himself. The good doctor comes to think of his alterego as something

> not only hellish but inorganic. This was the shocking thing; that the slime of the pit seemed to utter cries and voices; that the amorphous dust gesticulated and sinned; [Mr. Hyde brutally murdered several people] that what was dead and had no shape, should usurp the office of life.[5]

This is an expression of disgust by the higher reaches of the mind contemplating the archaic nature of its own ancestry, with which it shares a head. A crocodile, behaving as Mr. Hyde did, would not be said to have usurped the office of life, and being slimy is not derogatory to the slime mold. While Dr. Jekyll still thought in terms of human time, with its responsibilities of choice, Mr. Hyde, the lower form of Dr. Jekyll himself, having come out of hiding (the name is deliberate) craved only for the immediate satisfaction of his primitive drives.

On a Holy Thursday a number of years ago, I heard the monks of the Benedictine Abbey of St. Pierre de Solesme, on the Selle River near Cambray, France, sing the medieval hymn, "Ubi Caritas et Amor, Deus Ibi Est" (Where charity and love abide, God is also nigh). I felt at one with the God of Christ and understood the quiet ecstasy of monkish life: the working for a final reality that had no directed time. A good name for this kind of "timelessness" is *the ecstasy of the chalice.* Although the name implies the Catholic belief in the actual presence of Christ's blood and body in the wine and host, the metaphor is well suited for many forms of religious ecstasy.

The pagan version of same may be heard in the orchestral prelude to Richard Wagner's opera *Das Rheingold,* which is a musical rendering of Teutonic cosmogony. The Rhine River wells up from its primordial sources to carry, with irresistible force, the fates of gods and of godlike men and women toward their preordained decline and ultimate fall. The music conveys a feeling of one-ness with the struggle and demise of the

---

5   Stevenson, *The Strange Case of Dr. Jekyll and Mr. Hyde* (New York: Dodd, Mead and Co., 1961), p. 68.

gods; it is a pagan religious experience, a visit to a world where time has no direction.[6]

Feelings of timelessness have often been thought of as peeps into a world that is superior to and is more advanced than ours. It seems to me that they are peeps in the opposite direction. They amount to the revisiting of childhood realities, or even those of our ancestors, where long-term futures and pasts and the burden of choice were not yet known. Life meant only existence in the organic present. Such visits may be deadly or inspiring or both; the decision whether or not to descend is itself a burden of noetic time.

Collective roundtrips to the depths: The enthusiasm of nations

The social capacity for supporting plans of communal action may be called the *enthusiasm of nations*. In Greek *entheos* meant "having the god within," also, living without aging, as did the gods of Mt. Olympus. The English version of the word means passion. This subsection examines collective passion, seeking the "timeless."

To regard collective passion as necessarily destructive is to overlook its role as one of the elementary forces behind the building of civilizations. The goals of communal emotions, whether inspired and great, evil and reprehensible, or even outright stupid, always involve dedication to something imagined, something passionately desired but not yet at hand.

By convincing people, by cajoling, by terror, by shrewdness, by force, by intelligence, or by charisma, each of the twenty-five leaders shown in figure 44 was able to enlist others to do his or her bidding. Let these people represent the infinite variety of preferences in values, behavior, and manner of doing things.

Is there anything common to the enthusiasm of nations beyond what I have listed as cajoling, terror, charisma, and so on? I believe there is. In democracies, the selection among possible goals, ways, and means is done by popular vote. But historically, decisions have always been made by the use of the collective violence of wars and revolutions, and it is through

---

6　The imaginative reader may extend this list at his pleasure. For instance, ancient Chinese and Japanese legends speak of the power of the Go game to suspend time for those who are absorbed in playing it. That experience may be thought of as an example of the *ecstasy of pursuit*.

them that the forces which move the enthusiasm of nations may be most clearly seen. How do these forces involve the trips down to and back from primitive temporalities? Figuratively and not so figuratively speaking, by song and dance.

Claude Joseph Rouget de Lisle was a French captain of engineers and a royalist. In 1792 he composed a march, music and verse, in support of an unpopular war the king of France was waging against the Prussians. De Lisle called his march "The War Song of the Army of the Rhine." Then the march changed sides, like Lilli Marlene did in World War II. For, in spite of its royalist intent, the emotions mobilized by the march corresponded to popular sentiments that were not for any war on the Rhine but for internal revolutionary change. It gave vocal expressions to people's hopes and, by so doing, helped realize them.

The march was sung by volunteers from Marseilles on their way to join the Parisian Jacobins, a radical group advicating egalitarian policies. Singing helped the volunteers forget their hurting feet, missed lovers, and no doubt lice, constant companions to humanity until lately. The march became known as "La Marseillaise" and, in 1795, the French national anthem.

In 1830 there was yet another revolution in France, ending in the victory of the bourgeoisie over the aristocracy. In July of that year, in Paris, the composer Hector Berlioz conducted a makeshift military band in playing the march. The band was in the shop of a haberdasher on the second floor of a building, with its doors open to a balcony that overlooked a square. The people, wrote Berlioz in his *Memoirs* were "men, women and children . . . hot from the barricades. . . ." But the crowd remained silent; this was not what he expected. After the third stanza he could not contain himself and yelled, "Confound it all—sing!"

> The great crowd roared out its "Aux armes, citoyens!" with the power and precision of a trained choir. Remember that the arcade leading to the rue Catherine was packed with people, as was the arcade leading to the rue Neuves des Petits-champs and . . . that these four or five thousand voices were crammed into a resonant space bounded by the clapboard shutters of the shops, the glass on the roof and the flagstones of the arcades. . . . Remember the effect of this stupendous refrain. I literally sank to the floor, while our little

"Why vote? Politicians are all the same."

It's a weak excuse for not voting.
Though it's far easier to look back at what has been than to chart the course of what will be, history has proven that all politicians are not quite the same.
And the countries they lead are never quite the same again. YOUNG & RUBICAM INC.

Vote Tuesday, Nov. 6

FIGURE 44   MOSAIC OF THE ''ENTHUSIASM OF NATIONS.'' To different extents, all political, military, and spiritual leaders have the skill for engendering in their followers a feeling of enthusiasm. Often, but especially when it can express itself in collective behavior, that enthusiasm is experienced as an altered change of awareness, a descent into an archaic temporal reality where the tension of responsibility for choice lessens, because the definitions of future and past become hazy. Considering the plurality of ways and means that human imagination continuously creates to help guide behavior, it would be difficult to assure collective social action were it not for the human predilection for losing the burden of hard-earned selfhood. Courtesy, Young and Rubicam, Inc.

band, aghast at the explosion it had provoked, stood dumbfounded, silent as birds after a thunderclap.[7]

Few people can listen to the "Marseillaise" and not feel like marching from here and now to a timeless world of eternal victory, there and then. Wherever. It is not possible to play the music on a printed page, but we may attend to the words; they are the Yang to the Yin of music. Each line is loaded with emotions, each encourages descent into the prerational world of childhood and, even deeper, into the archaic realities of the mind. Each line pushes the singing marcher a step closer to "having the god within." Here are the first four lines.

> Allons enfants de la patrie!
> Le jour de gloire est arrivé!
> Contre nous de la tyrannie
> L'étendard sanglant est levé. . . .

"March, children of France!" You are a child without personal identity or a past; indistinguishable from the other children of France.[8] "The day of glory is at hand!" Only the shining emotions of today are real, we live in an eternal now. "Against us, the tyranny." The tyrant's name is not given, the enemy is Tyranny. "The blood-soaked flag has been raised," deserves special attention.

When I spoke earlier about collective identity I remarked that it had to be represented by a symbol; even a clan is too complex to be thought of in terms of anything other than a name, a tune, a coat of arms, or a flag. For Longfellow, it was the "banner with the strange device, Excelsior!" There is the burning cross; the swastika; Mao's Little Red Book; the coat-hanger for freedom of choice; a donkey for the Democrats; and a pink carnation for weddings because it has the color of young flesh, as its name clearly says. These are some of the accoutrements of the species whose members use symbols to help assure the continuity of their society through time.

"The blood-soaked flag" communicates directly with our archaic

---

7   *Memoirs of Hector Berlioz,* trans. and ed. David Cairns (New York: Knopf, 1969), pp. 132–33.

8   From a recent decision of the United States Supreme Court concerning the uniform appearance of members of the armed forces: "Uniforms encourage a sense of hierarchical identity by tending to eliminate outward individual distinctions except those for rank."

selves. As a means of communication among living things, the sight and smell of blood is as old as the age of the circulatory system. It first appeared in the earliest invertebrates 400 million years ago as a colorless plasma; it became colored in vertebrates. Many people faint at the sudden sight of blood, and if sufficiently weak, most people will. Fainting is the human body's way of short-circuiting the conscious mind: "I have enough trouble as it is, do not interfere!" Animals do not faint at the sight of blood, nor do they drink it symbolically. A good jackal would enjoy the Fountain of Blood in the Ayatollah Khomeini's Iran, erected to encourage his subjects to die in battle with enthusiasm. Let the "Marseillaise" stand for the ways of song-and-dance that help communities descent collectively into archaic worlds of time.

The French Revolution was a trip down into the depths with the Wife of Lot as a *tricoteuse,* followed by an ascent with Galatea as *Liberté,* the bare-breasted French figure of freedom. The Reign of Terror was followed by the comet-like rise of Napoleon, the revolutionary war was replaced by the Napoleonic wars, the focus of enthusiasm shifted.

The goals of the French Revolution were inspired by those of the American Revolution; in its turn, the French events inspired other people and nations across Europe and the world. They moved Beethoven to compose his Ninth Symphony, declaring the brotherhood of all people in music and words as stirring as those of the "Marseillaise." And thus the collective roundtrips with the two women have been going on from generation to generation.[9]

Beyond music and verse, the most powerful single tool to create and guide the enthusiasm of nations had been the word itself. In the advanced lands of our age, it is known as the art of oratory. A century and a half ago Daniel Webster debated the Devil and John Calhoun debated everybody. The man whom many consider the greatest orator of our century was born with a congenital lisp and stammer. But Winston Churchill taught himself a magistral command of the language: his gift of mixing the rational with the emotional and delivering it all in an accomplished literary English has not been matched. But then came film and television oratory, which

---

9   Those who prefer an example closer to home may go through a similar exercise in reasoning, using the Civil War lyrics, "John Brown's body lies a-mould'ring in the grave But his truth goes marching on. . . ."

demanded the tailoring of word and message not by any moral or intellectual standard but primarily by its raw emotional appeal to the largest possible audience.

The pioneer of the photo-opportunity oratory was Adolf Hitler, past master of organizing public frenzy, whose skill has been studied in more detail than that of any other contemporary leader. In a biography entitled *Hitler,* Joachim Fest gives an account of a typical collective descent into the "timeless." Hitler, writes Fest,

> delivered one of his passionate speeches that whipped the audience into a kind of collective orgy, all waiting tensely for the moment of release, the orgasm that manifested itself in a wild outcry. The parallel is too patent to be passed over, it lets us see Hitler's oratorical triumphs as surrogate actions of a churning sexuality unable to find its object. . . . His oratorical discharges were largely instinctual, and his audience, unnerved by the prolonged distress [of waiting] and reduced to a few elemental needs, reacted on the same instinctual wavelength.[10]

The crowds were willingly seduced to become his lovers in the collective ecstasy of a political bower. The rhythmic chants of "Sieg Heil!" made people shake in the presence of their god as the Shakers shook in the presence of the Lord. But whereas the Shakers' dance was a ritualized descent and ascent, the journey of Nazi Germany was only a descent and the best demonstration in modern history of the diabolical in man. Hitler's war laid Europe to ruin, buried its feudalism for good, and pushed that continent, willy nilly, into the democratic struggles of the twentieth century, in a demonstration of the Wife of Lot—Galatea dialectic.

Ferocity of enthusiasm, the fire that destroys and builds, is hardly limited to the French and the Germans. China's youth also felt the timeless "god within" when Mao whipped up their emotions in his anti-cultural revolution of vast dimensions. The list of examples could go on and on.

Instead of listening to magicians of political power, let me turn to the words of a sensitive Englishwoman, convinced pacifist, feminist, and accomplished writer. In her *Testament of Youth,* Vera Brittain recalls her stay on Malta as a nurse with the British Army during World War I.

---

10   Fest, *Hitler* (New York: Harcourt Brace Jovanovich, 1974), p. 323.

I may see the rocks again, and smell the flowers, and watch the dawn sunshine chase the shadows from the old sulphur-colored walls, but the light that sprang from the heightened consciousness of wartime, the glory seen by the enraptured ingenious eyes of twenty-two, will be upon them no more. [The] world for all its excitement of chosen work and individualistic play, has grown tame in comparison with Malta during those years of anguish.

[The] causes of war are always falsely represented . . . but the challenge to spiritual endurance, the immense sharpening of all the senses, the vitalizing consciousness of common peril for a common end, remain to allure those . . . who have just reached the age when love and friendship and adventure call more persistently than at any later time. The glamour may be the mere delirium of fever . . . but while it lasts no emotion known to man seems as yet to have quite the same compelling power and enlarged vitality.

I do not believe that . . . any Disarmament Conference will ever rescue our . . . civilisation from the threatening forces of destruction, until we can somehow impart to the rational process of constructive thought . . . that element of sanctified loveliness which, like sunshine breaking through thunderclouds, from time to time, glorifies war.[11]

These are beautiful and subtle lines. They paint a lyrical image of the unresolvable conflict between the archaic concentration on the present—the excitement of the chase—and the long-term plans and memories of the noetic mind. Vera Brittain wrote her book as she was approaching forty. Would she have preferred to be twenty-two again, with the world at war? The lines do not say.

Driven by its knowledge of time, our species is an unruly lot.

Since our species left behind its animal competitors, it has been burdened by being able to dream without thinking, but not to think without dreaming. Galatea and the Wife of Lot run a shuttle service between the advanced and simpler forms of temporal realities present in the mind; they help maintain the unresolvable conflicts that have been driving people to become so frighteningly creative and destructive.

11   Brittain, *Testament of Youth* (El Cerrito, Calif.: Seaview Books, 1980), p. 281.

The Time-Compact Globe

The thesis of this and the following section is as follows. Society and its members have evolved together: nootemporality and sociotemporality have complemented each other throughout history. This relation is likely to change, however, if a significant portion of the earth's population reaches a certain level of social complexity. Thereafter, the global social-ization and evaluation of time will subsume the office of the individual as the measure and measurer of time. In this and in the following section, I give reasons why I believe the necessary level of complexity has been or will soon be reached around a time-compact globe and outline some of the problems of transition, as I see them.

The basis of the multiplicity in the modes of the socialization and evalua-tion of time—a plurality that has been with us since the beginning of his-tory—is being rapidly whittled away as economic needs and political necessities force mankind into adopting a shared rhythm of labor, similar methods of industry, and identical reasoning in science. Unlike religions, philosophies, and political ideologies, which do not tolerate each other though as a species of thought allow for vast variations, the methods of in-dustrial production allow only for minor variations from place to place, scientific reasoning allows for none.

As a result, a new and distinct integrative level seems to be in the making: that of a time-compact global society. This section discusses a number of changes that form a part of the emerging new temporality.

The global present, the perceived irrelevance of history, and the greying of the calendar

The width of the social present depends on the speed at which those messages that make collective action possible may be exchanged. We saw how, in 1784, the Bath mail coach narrowed the London–Bath social present from eighty to twenty hours. On July 4, 1826, on the fiftieth anni-versary of the adoption of the Declaration of Independence, the second and third presidents of the United States—John Adams and Thomas Jefferson—died in Quincy, Massachusetts, and Monticello, Virginia, respectively, only hours apart of each other. Their simultaneous passing did not become known until many weeks later. In the 1920s and early 1930s the "Amos 'n' Andy" show stopped the country at 7 P.M. Chicago time with the slogan "Brush your teeth with Pepsodent." In the 1990s the televised wedding of a Pope to a Chinese princess would stop much of the world instantly. In case of a star war, events would happen so rapidly that all decisions would have to be made by computers, bypassing the President. This is a story of a continuously narrowing social present.[12]

In the United States, the Baby Bell companies operate a total of 1.2 billion miles of wire and 1.5 million miles of microwave links. To these must be added the unknown mileage of intercompany, military, and illegal links. Worldwide there are over 1.3 billion radio receivers, not including those of commerce, industry, and government.

The increasing communication density is necessary to assure that events that should happen simultaneously do, while others that should not, do not. We have already encountered these requirements as the constituting features of the organic, mental, and social presents. In the case of organisms and the brain, the rules of do's and don'ts are set by the laws of biochemistry and physics. They are not negotiable. The principles that could hold the many social presents of the globe together in a *global present* are now being negotiated. A red phone in the White House and, I assume, a white phone in the Red House are for publicity, but they do make the point.

A system that may serve as a prototype for an eventual worldwide communication network is the Information Network System of Mitaka, a suburb of Tokyo. It is designed to integrate into a single web all electronic communication means within its territory. It includes video phones, sketch phones, facsimile transmission, phones that show the caller's number, and conference circuits; it provides connections to department stores, investment brokers, and possibly to marriage brokers, dating services, and computerized fortune tellers as well.

When multicellular organisms began to develop from single cells, some of the cells had to evolve new behavior. Henceforth they could not move: food had to be carried to them, waste products had to be carried away

---

12   In the following paragraphs I deal entirely with technical communication links. But the narrowing of the global present may also be observed in the biological domain. The rapidly increasing rate at which human gene pools are mixed—a process of biological homogenization—had been commented on by sociobiologists. The deadly disease AIDS (acquired immune deficiency syndrome), of which a few years ago only a handful of people had even heard, is expected to claim a hundred million victims around the world in the short space of a decade.

from them, and their functions had to be coordinated. In return, as it were, the multicellular organisms acquired dimensions of freedom unavailable to unicellular organisms.

Society is not a living organism—that idea was left behind in the nineteenth century—but its functions do bear resemblances to one. In this case, consider the stay-home paradise of people having an efficient information network. They may conduct their business from their living room, join the cottage industry of computer-operating fathers, receive food at home, and, if ill, get diagnosed by wire. They need not go anywhere at all but may watch their favorite show on television while at the computer keyboard. The individual will have become dependent for his survival on the information network. He will have to worry only about the present; long-term plans for the future and learning from the past will have become the task of specialists.

We are not there yet, but the machinery for transferring the responsibility for planning and memory from the individual to the community is already in place across extended regions of the globe. The large social institutions of the West, for instance, hold an immense amount of liquid data on their countries' citizens. Through that data, Big Brother is ready to monitor the present behavior and opinions of one and all, and use the information to design future paths of actions to which the governed have or have not agreed. In totalitarian lands, this is done as a matter of government policy; in the free world, it is being developed in the name of market surveys, public opinion surveys, credit checks, transportation needs—you name it, the local merchants or Town Halls have it. We are witnessing, I believe, preparations for the merger of the many social presents into a common global present.

Let me turn now to another element in the making of a time-compact globe, the *perceived irrelevance of history*.

In 1278, Roger Bacon, early pioneer of scientific thought, was condemned to prison by his fellow Franciscans for suspected novelties in his teachings. A steady appeal to the past and the assurance of continuity with history were judged necessary ingredients of all valid opinions about anything. The pendulum has swung to the other extreme: a steady appeal to the present and assurance of usefulness for the immediate future are judged necessary ingredients of all opinions worth listening to. Enrollments in the history departments of universities may go up or down, but a broadly held belief in the irrelevance of history remains and is becoming further elaborated. This narrowing of the temporal horizons has far-reaching consequences.

Pamphlets about issues of immediate concerns are known to have been published in ancient Rome. But manuscripts and, later, printed books have generally dealt with matters of lasting interest. No one wanted to publish a book that would not be around for a long time. Today's mass-market publishing demonstrates an opposite policy: the emphasis is on the best seller now, not on the long seller for decades and centuries to come.

In the domain of public policy, we are shifting resources to the gratification of current needs without much thought as to who, when, and how we will pay the price for it. In the economic world, it is the well-to-do who make the loans and collect the interest, hence ensuring the transfer of wealth from the average citizen to the rich, surely to the joy of Marxist historians, who may claim that they have told us so. In the ecological dimensions of our lives, the problems of shortsightedness are too well known to be repeated here. Interest in science-fiction futures is not the same as acknowledging a present responsibility for the future and expressing it in a willingness to make sacrifices in the present, for distant goals.

The reading public has a great hunger for chronicles—for real or imaginary stories of past events retold—and to satisfy that hunger, we have chroniclers. Some are quite good, others quite bad. But there is little feeling for history as the matrix in which individual and social life unfold day after day. Human reality is fundamentally historical; an ignorance of history deprives people of their sense of continuity and therefore of the elementary tools necessary for an understanding of what is happening to them and to others, and why. To be thus insensitive amounts to losing one's social and cultural identities. Dramatic chronicles sell well, but do not replace the important messages that come from the serious study of history, such as the recognition of the extreme complexity of the past, the nonexistence of single-valued answers, and the realization that the past is much more of a creation than a discovery. And, if the past is so ambiguous, why should the present, growing out of that past, or the future, growing out of the present and the past, be less ambiguous, less multi-valued, or less complex?

A civilization that focuses on the present and the immediate future as the only significant parts of time gets dislocated in history. For industrialized countries, the effects of such dislocations are profound but subtle, as I shall try to show in the following subsections. For many Third-World countries, the effects are profound and not at all subtle. These countries have been jettisoned across several historical time zones into a pastless present and suffer the cultural equivalent of jet lag. Latest-model computers adorn the governmental offices of countries where women are still circumcised to make them faithful wives and the earth is still fixed in the center of the universe.

But . . . if I were asked to help make the people of the world receptive to the demands of a time-compact order, I would first try to make history appear irrelevant. It is much easier to secure cooperation among people without an understanding of history than those with many and usually antagonistic histories.

The French Republican calendar was the first modern plan to provide a new rhythm for social life based entirely on the needs of a new society, as perceived by the creators of that calendar. It was, as we saw, a miserable failure. But we witness in our age a revolutionary new recasting of the calendar: profound, silent, and irreversible. It is done without governmental action or fanfare or even much public attention.

The *greying of the calendar* is the smoothing out of differences between day and night, the disappearance of distinction among the days of the week and among the seasons of the year. In urban regions of the United States, many stores remain open twenty-four hours a day, every day of the year save perhaps Thanksgiving and Christmas. Sunday shopping has become a way of life for millions. Food stores, laundromats, gas stations, diners, and machine tellers are ready to serve around the clock, around the calendar, all who have money to spend.[13] Radio stations in every corner of the United States broadcast all night. Then there are the night shifts, with their transportation needs, the public utilities and the permanent vigilance of the military. Forty percent of the world's population lives in cities; most of those people care little or not at all for machine tellers or rock videos twenty-four hours a day, but the cities of the United States, Western Europe, and Japan pioneer and suggest the direction of development.

13   *The Coyote Howls* is the aperiodic newsletter of the National Task Force on Prostitution. The reference in its title to that nocturnal hunter is a literary device and an acronym (for Call Off Your Old Tired Ethics). Otherwise, there is as much prostitution going on in cities during the day hours as during the night. The disappearance of distinction between day and night goes both ways.

The sociologist Murray Melbin has noted that nighttime social life in urban areas resembles the social life of former land frontiers and named this phenomenon the colonization of time. As on land frontiers of old, social contacts at night are still sparse, which pleases the pioneers of the night, people who are likely to be loners. Frontier settlements were isolated, as are the night stores today, pockets of activity in a sleeping and perhaps hostile environment. But as with earlier land frontiers, the night-time world is on its way to becoming the habitation of all.

Beneath these ongoing changes, our biological rhythms, notably the circadian cycle, have remained entrained; it would take drastic measures to replace it with another rhythm. But the circadian cycle need not be in phase with the sun. Earlier I gave an example of the separation of another biological rhythm from its astronomical origins. The menstrual flow in women, originally probably in synchrony with lunations, became independent of the moon and randomly distributed as social life itself achieved independence of the moon's phases; but the synodic rhythm remained entrained. Likewise, the circadian rhythm of individuals is now shaking itself loose from sunrise and sunset and is on its way to becoming independent of its astronomical origins, while the circadian cycle itself remains entrained. The social and industrial need to have people available for working at night is sufficiently important and widespread to have attracted the attention of the Scientific Committee on Night and Shiftwork of the International Commission on Occupational Health. Sociologists, biologists, psychologists, and business organizations meet regularly to discuss the many-sided problems of the colonization of the night.[14]

If I were asked to make the people of the world receptive to the demands of a time-compact global order, I would suggest the greying of the calendar and the colonization of time as two preparative steps. For it is much easier to secure cooperation among people who have no calendars than among people with different calendrical traditions, maintained among other reasons—as we have seen—for the purpose of securing distinct group identities.

---

14   This "Shiftwork Song," sang to the tune of the German carol "Oh Tannenbaum," is from a recent issue of *Shiftwork International Newsletter,* published in Edinburgh, Scotland. "Oh shiftworker, oh shiftworker, We find you napping all the time . . . You have to watch the control room, You have to stop the plant go boom! Oh shiftworker, oh shiftworker, please compensate at home."

*Surplus people, birth and death control, and changes in the
texture of private time*

The needs of the world's people for food, housing, goods, health care,
and services, judged in terms of existing shortages and expectations,
outrun by far what the existing economic and political systems could
produce, deliver, and distribute. Even if the productivity and distribution
means were available, it is doubtful whether the earth could keep on
providing the necessary raw material at a sustainable rate. Our ecological
problems are examples of the coevolution of species with their environment. Evolution is always coevolution, but when the pasture is large, the
cows few, and the long-time effects irrelevant, environmental changes
may be neglected.

The population explosion, the advances of technology, and the finity of
the earth's resources have made environmental issues non-negligible and
have begun to force upon us birth and death control. In their turn, these
measures have been altering the texture and boundaries of private time.
The details are complex, the answers many-valued, but the major issues
may at least be sketched and the way they bear upon the socialization and
evaluation of time discerned.

Recently thirty-five heads of state of countries representing more than half
of the earth's population ran a full-page statement in the *New York Times*
to point out to ignorant readers what informed readers already knew.
Each week one and a half million more people are born than die. At this
rate of growth, by the end of the century the world's population will
approach 6.2 billion, with a weekly net gain of 1.8 million. Ninety percent
of that increase will take place in the Third World. To absorb the new en-
trants into the labor force, the world economy must create 200,000 more
jobs per day. Have you seen a million new job openings advertised during
the last five days?

One remedy is an increase in productivity, so that more people may be
fed, clothed, and housed. But no responsible economist, social or political
scientist, or statesman would judge the necessary increase in productivity
feasible. We have passed the point at which the invention and manufacture
of better mousetraps were sufficient. The problem now involves complex, interdependent changes among rates of population growth, the

designs of political and commercial forces, climatic changes brought about by industrial and agricultural practices, and the decrease of biological diversity related to all these variables.

Attending to the issue of biological diversity, I want to note that several near-simultaneous mass extinctions of species are known to have taken place during the history of organic evolution. Not all their causes are known but since there were no people around, they could not have been manmade. This time they are, and we are one of the threatened species. What kind of mutations does this manmade environmental change favor? Perhaps it favors fish that live on pollution, insects that thrive on radiation, or advanced forms of mammalian life that favor very closely-knit social behavior. I do not know. But the changes make for a more homogeneous and therefore less creative biosphere. It is ironic that the deterioration in the quality of life that must accompany the homogenization of the biosphere is an unwanted side-effect of the increased productivity intended to improve the quality of life.

An alternative remedy to the population explosion is the use of contraception. Fertility control in the animal kingdom is an important part of population dynamics. Mammals, for instance, have four major mechanisms that help them space the birth of their young: varying the age of sexual maturation, controlling fetal and embryonic deaths, limiting the periods in heat, and, in a number of primates, lactational infertility.

The first two of these are present in man; limited periods of heat are not. Being able to deliver children the year round has the advantage that, at any one time, only a small percentage of females are unavailable for joining the labor force or caring for children. Also, in both sexes, the continuous availability of libidinal energies can fire creativity and destructiveness regardless of season. The fourth control mechanism, lactational infertility, exists in humans but as a natural fertility control it has fallen victim to bottle feeding, a necessity for many working mothers. To overcome this disadvantage of bottle feeding, it has been seriously suggested that baby formulas marketed in the Third World be packaged with contraceptives.

In the People's Republic of China, where stick and carrot are both employed to hold families to one child only, the policy has not been successful outside of the cities. Party or no Party, Chinese peasants are as hardheaded as their counterparts elsewhere and believe that having children (specifically more sons) means more working hands and more security for old age. This belief, held by all older agricultural societies, survives in the industrial West in the objection of the Roman Catholic Church to birth control on theological and moralistic grounds.

In both China and the West, however, resistance to contraception is being ground down, suggesting the eventual broad acceptance of birth control—which in itself will lead to other dilemmas. For instance, if the one-child drive succeeds in China as planned, by the end of this century 40 percent of China's population will be over sixty years of age. At that time, the one-child policy would have to be abandoned because there would not be enough working hands to support the elderly population; such a step would open the way to population growth and an untenable population density. One possible alternative would be to put people to death when they reach a certain age. The resurgence in modern China of the ancient practice of female infanticide may be the harbinger of a policy of a controlled decimation of the population.

And there are other problems. No one knows, for instance, the long-term effects of having millions of children grow up without siblings and, in a few generations, without uncles and aunts. The character-forming role of the family will have to be taken over by further socialization of child raising—in an already socialized land.

In China and elsewhere, the broad use of contraception will also lead to having many more childless couples than would otherwise be the case. Again, no one knows the long-term effects on individuals and on society of this new configuration in human affairs. For the moment not having children is the peacetime equivalent of a soldier dying in a war: in both cases, the biological continuity of the individual is cut short for the benefit of the community. But having no offspring has been an unacceptable way of life through the 200-million-year history of the mammals. Organic evolution has always striven for maximum reproductive potential: the female body evolved to deliver children, the male to fertilize millions of eggs. The male body could not care less whether any of its sperms grow into children, but the male psyche, and the female body and psyche do. Taking the pill or undergoing a vasectomy is a conjugal act. Contraception has thus moved into the marital bed and will remain there in a ménage à trois until nature finds another way to control population.

In response to the new demands and working on the organic, mental, and social levels, natural selection will have to favor the development of nonreproducing persons. This is already happening through the social acceptance of gay life. The freedom so granted permits the development of the maximum number of nonreproducing humans that genetic variations can produce. It is a method of birth control without government order, a step toward the human anthill of sociobiology.

No one knows exactly how the nature of private time will change as decisions about sex and the commitment to beget and raise children are transferred from the individual to society. But it is certain that the changes will be profound.

The coevolutionary character of life is not negotiable: the cow, the pasture, the people, their children, their brain children, and the oil shales form a single, dynamic unit. The economic and population facts of our age are also a given. Should contraception, the silent friend in the marital bed, be shown its way out the door or else prove inadequate, a different visitor will be climbing through the window, just before dawn. It will be a policeman or policewoman carrying a death warrant; birth control will be replaced or at least supplemented by death control. This unsavory idea is not a nasty figment of the imagination. After the meteoric rise of science and industry, the idea that humanity has limitations has been rediscovered.

Death control is not new: mass murder and slavery have been the ways of mankind. Antiquity abounds with examples but it is sufficient to begin with the Tartars of the thirteenth century and the Turks of the sixteenth century. They killed or carried away as slaves two-thirds of the people of southeast Europe. In the United States on the eve of the Civil War, there were 4.5 million African slaves. They had been gathered and shipped to this continent at carefully maximized profits and minimized humanity. During World War I over one million Armenians were killed by the Turks. Later, Stalin deliberately starved to death millions of peasants to make a point: small farms must be changed into more efficient agribusiness. In World War II over six million Jews were exterminated by the Nazis. Mao's anticultural revolution may be unparalleled in its numerical terror. History has been a series of holocausts.

Scientific and technological advances increased the efficiency of all know-hows: the repair of human joints out of joint and the replacement of hearts with clocks (as did earlier the Wizard of Oz). Our garden stores are full of "-cides," as in herbicides and pesticides. The coining form comes from the Latin *caedere,* "to kill." Though I see nothing wrong with killing lice, there is something definitely wrong with genocide.

Mass murder in its contemporary form was pioneered by the Nazis, who made it into an art of state. First, the victims received negative advertisement, then they were made homeless, emaciated by work and hunger, and forced to become filthy. In short, they were made murderable. Killing Jews and other "enemies of the state" was not against any law, because laws were written for people and the victims were not people. That this could happen in one of the most highly civilized nations demonstrates that collective enthusiasm, if so directed, can easily override the ethical veneer. Our archaic selves reside just beneath the surface of rational, humane behavior, ever ready to pull rank on it. Young men riding their cars on our roads at night drink their beer, then crush the cans in a macho show of strength; this makes a handsome, recyclable beer can into a piece of ugly garbage that may, therefore, be thrown out the car window.

The Nazi system was backed by an efficient and reliable Western-type bureaucracy. All such bureaucracies are secularized forms of the monastery's organized labor, working for a heaven on earth or at least to provide the best possible conditions for its citizens. In performing its task well, a good bureaucratic machine is an impressive social creation. In an essay called "Bureaucracy," written in 1916, Max Weber remarked that,

> The fully developed bureaucratic mechanism compares with other organizations exactly as does the machine with the nonmechanical modes of organization.[15]

Today the ideal bureaucracy would be compared with a computer: instructions and raw material in; tax bills, social security checks, marriage certificates, and trained people out. Had it not been for Hitler's military defeat, his bureaucracy would have eventually seen to it to erase the memory of the death camps from the data banks. They would have survived only as uncertain tales of horror, like that of the thirty villages in the

---

15   Weber, "Bureaucracy," in *From Max Weber: Essays in Sociobiology,* ed. H. H. Gert and C. W. Mills (New York: Oxford University Press, 1946), p. 214.

Ural mountains that have ceased to appear on Soviet maps since 1958 because a nuclear dump did away with them in 1957.

The Nazi death camps began as collection places for slave labor; the notorious Auschwitz camp was first run by the government to supply workers for the A. G. Farben Industrie's plant that manufactured synthetic rubber. It was a jail jointly owned by the government and the business sector. It is certain that in Hitler's mind it was never to remain a nonlethal weapon, and it did not.

As the end of the century approaches, the creation of minorities by legal means and their subsequent legal slaughter is within the power of all governments. The methods are at hand and ready, should socioeconomic conditions demand and political considerations dictate their use. Thus, in the United States, the Prison Construction Privatization Act of 1984 was designed to "unleash private capital" for the building of $1.5 billion worth of prison space for 50,000 inmates. The financial and business weekly *Barton's* commented that, "The private jail market is ripe. And it's the brokers, architects, builders and banks—not the taxpayer—who will make out like bandits." Several jails-for-profit corporations are already operating, run for the government by the business sector. In September 1985, the Corrections Corporation of America offered $250 million for a ninety-nine–year lease of the entire Tennessee prison system.

Jails for profit are a very far cry from the death camps of history. Nevertheless, they are experiments in the building of machinery for legally disposing of surplus population. The inmates of these jails are on their way to being defined as enslavable minorities, created by the unwillingness of people to pay the costs of maintaining norms of behavior that they say they espouse. Minorities need not be the wetback, the poor, or the out-of-wedlock. By the turn of the century 13 percent of the world's population will be over 65 years of age and, to different degrees, a burden on their societies.

The reader may recall that in organic evolution the improvements achieved by one species always lead to the deterioration of the environment of another species sharing the same ecological niche. For instance, the human population explosion and improvements in the human environment have made for the deterioration of the environment of many other species. The same principle also informs the economic life of societies in both peace and war and, on a time-compact globe, the social policies of nations competing for resources and useful manpower. If one nation succeeds in developing a method of extensive bloodletting without a substantial backlash from within or without, other nations having a surplus population will copy those methods for their own benefit.

Each person, whether killed by the Tartars or the Turks or executed in a utopian community because he had reached the age limit, will die resenting the arrogance and finality of death. Would this also be the case for someone born and raised in a society where death control was the democratic and necessary means for keeping the population in balance with resources and services?

Chidiok Tichborne was a sixteenth-century Catholic conspirator against the life of Queen Elizabeth. He was caught and sentenced to death by hanging, drawing, and quartering. On the eve of his execution, not having been misled by a lawyer using a logic of time inappropriate to human time, he wrote a letter to his wife. It included a three-stanza elegy, a deeply moving masterpiece. He was twenty-eight years old. This is the closing stanza.

> I sought my death and found it in my womb,
> I looked for life and saw it was a shade,
> I trod the earth and knew it was my tomb,
> And now I die, and now I was but made;
> My glass is full, and now my glass is run,
> And now I live, and now my life is done.

Pigs' lives are often counted in months to slaughter day and merchants count so many days before Christmas. Would counting one's life backward from an assigned death date become as acceptable as counting birthdays forward toward its known–unknown ending? Noetic time arose from the tension of ceaseless bargaining with death. How would the absence of bargaining change the nature of noetic time? How would the texture of private time change if everyone knew that at the age of twenty-eight he or she would be painlessly or otherwise disposed of? The answer must be worked out by the reader.

The bomb in the closet, the computer in town hall, and the global socialization and evaluation of time

This subsection is about the role that two great brain children of mankind—the Bomb and the computer—have been playing in the march toward a global socialization and evaluation of time.

THE BOMB: FOREMOST SOCIALIZER OF GLOBAL TIME

The Bomb is a mode of death control, an environmental selection pressure exerted by the created upon its creator, an example of coevolution between people and their brain children. Its potential effects are appropriate in magnitude to the 4.8 billion people of the earth equipped with computers, hammers and sickles, and hungry for food, life, and sex. Although the Bomb works by the principles revealed by special relativity theory (see fig. 37), it was made possible only through the cooperation of a distinguished group of statesmen, scientists, engineers, industrialists, the armed forces and taxpayers of a few lands, and a small army of mostly Russian and not so distinguished spies.

It is estimated that the existing nuclear weapons arsenal is equivalent to 4 tons of TNT for every man, woman, and child on earth. What would an Inuit family do if the mailman had delivered them their fair share of 20 tons of TNT (4 tons each for the husband and wife, 4 for each of the three kids), together with an invoice for $50,000, which is the family's part in the American national debt?

Perhaps the Bomb will be used and met head on by the beam of a nuclear X-ray laser. If so, the illiterates and the hungry—the majority of people on earth—together with the literate and the well-fed will look up at the remarkable fireworks and in many different languages say something that resembles the last stanza of the hymn that was in Michelangelo's mind while painting the *Last Judgment*. It was written by Thomas de Celano at the turn of the twelfth century.

> Day of tear and late repentence,
> Man shall rise to hear his sentence.
> Him the child of guilt and error,
> Spare, Lord, in that hour of terror.

Those who will have watched the show and survived it without harm, and surely there would be many, would return to their shacks, perhaps speak of the curious outcome of the first international scientific and industrial project, finish their dinner of chicken skin, maybe enjoy the ecstasy of the bower, and then begin having new plans appropriate to their temperament and tradition. Because dreaming up plans as an occupation never dies as long as there is a man or a woman around, even if patience and resignation are the socially conditioned responses in their lives. Even nightmares are dreams whose usefulness is to keep us on our toes.

On a March evening when English ships entered Boston harbor, the people were rallied by the cry, "Town born, turn out!" Repel the unwanted! Before independence from the unwanted side effects of industrial civilization may be gained, "Earth born, turn out!" would have to rally the people to defend themselves against the creative power of their minds.

But such a call against the Bomb has never carried much weight because its position on our scale of values is an ambiguous one. It is a homegrown public enemy number one, but also the defender of peace, employer directly or indirectly of millions of hands, and the logo of a civilization that holds that anything science and industry can do must be done. Nor do I believe that the Bomb will ever be used to lay parts of the earth to ruin. It is not necessary for it to be used as a weapon; it is already doing something like that just by sitting in the closet. It is like the poltergeist that lives beneath the steps and makes them creak, but otherwise remains out of sight.

The mere existence of the Einstein bombs has been forcing a shift in the nature of armed conflicts from large-scale wars to the fragmented, inexpensive warfare of organized terrorism. But since the possibility of large-scale wars still exists, the Bomb is forcing the great powers to arm, because, once again, an improvement in the weapons arsenal of one nation appears to all others as a deterioration of their own environment.

The making and chasing of this deadly Bandersnatch and its ilk have led to the war economies of our era. The Soviets have been living with it since the revolution of 1917; the United States has lived with a war economy only occasionally until the recent steady rise in military spending. The price of arming is so staggering that in the view of Theodore H. White, distinguished observer of the contemporary scene, it is likely to bankrupt all the Bomb makers. A fiscal calamity so created would favor China, the Third World, and the Soviets, because their people already know how to live in poverty.

The war economies have been promoting the socialization of time in at least two ways. First, they do so through the industrial and economic concentration necessary to keep the Bomb and its affiliates always up-to-date and ready for deployment. Second, they do so by demanding that the global present be kept as narrow as communication techniques permit, so as to help each nation better safeguard itself against an attack by another. In the United States, the consequent regimentation of labor and daily life is not glaringly evident because of the immense wealth of the nation, but it will surely become all too evident upon any substantial downturn of the nation's economy. For the Soviets, regimentation and scheduling of private life has already been written into the routines of one and all.

THE COMPUTER: FOREMOST HOMOGENIZER OF THE EVALUATION OF TIME

The computer's social position is as ambiguous as that of the Bomb: it is a friend administering to people and an enemy totally disregarding them. A good way to understand the computer as a powerful force in the global homogenization of time and values is to think of spelling.

In various documents, the name of the sixteenth-century English poet Christopher Marlowe is shown as Marlo, Marley, Marlin, Merling, Marling, and Morley. His contemporary, William Shakespeare, was baptized in 1564 as "Gulielmus filius Johannes Shakspere." In his marriage licence to Anne Hathaway he is William Shagspere, on the first page of his will he is Shackspeare, and on the third page Shackpere. People of that age were free about the spelling of words and names.

About a century after Gulielmus Shakspere, Samuel Johnson's famous *Dictionary* was published, and with it, the era of uniform spelling arrived. Henceforth, to be able to communicate and be understood correctly one had to refer to the *Dictionary*. With that step toward more precise communication, the individual lost some of his freedom while the community gained an added degree of coherence. A similar but much farther-reaching exchange of individual freedom for social coherence is taking place in our days through the use of the computer.

Computers began as number jugglers and that is still their foremost use. But computers today also spend much of their time on tasks whose inputs and outputs are not numerical, though inside the machine the world is known only by number. (Each time I press the letter K, the keyboard says to the computer: 1000101.) It was a computer that controlled the small robot that followed people and spoke to them in Japanese at the Tsukuba '85 Science Fair. If no answer came, it pleaded in English, "I want to go to England." Computers have gone to many lands.

In the advanced computer states one is continuously treated to the exhibit of faster, larger, and increasingly more sophisticated computers that prove themselves inadequate to their task, which is to decrease the traffic jam in the flow of information and data. The reason is that the rate at which information is created for the computer to organize, process, and deliver and the rate at which the results are demanded have both increased while the machine was made, installed, and debugged. An advanced industrial nation creates and demands more information than it is able to process and distribute. It can even create more computer power than it is able to educate (via software) and absorb in the information labor market, that is, integrate into society.

The situation is analogous to the increasing complexity of life forms in the course of organic evolution. Computer programs designed to facilitate the production and distribution of goods and services or the running of large organizations often need hundreds of thousands or millions of instruction lines. Unavoidably, these come with errors that need other computers for their diagnosis. As the number of instruction lines goes up, it becomes increasingly difficult and finally impossible to obtain a faultless program. Having unpredictable and perhaps fatal mistakes imbedded in long computer programs is a statistical certainty. When the computer becomes an administrator, this statistical certainty is the equivalent of an efficient bureaucracy sending the innocent to the gallows. In this case, the innocent may be a whole segment of population.

In the industrially advanced lands, millions of people spend most of their working lives in the company of computers. Through these people, every other person has computer connections: it is like living on the seashore where everyone is either a sailor or related to one.

Computer literacy (knowing how to use computers) has become as much of a prerequisite for the conduct of our daily lives as is car literacy (knowing how to drive a car), steam-iron literacy (knowing how to use a steam iron), and diesel-locomotive literacy (for those so employed). Each of these devices demands a special way of interaction. Computers demand a way of talking to them, which carries with it a way of thinking; this communicates itself to peoples' assessments of the nature of reality and to their scales of values. In the coevolution of people and their computers, the most significant step has not been that computers became user-friendly, but that people became computer-friendly. We have learned how to adapt to the ever-present computer just as we learned to cross the seas strapped into an uncomfortable seat that may be blown up at any moment.

Computers serve as the control centers of the myriad wire and radio links that maintain the global present. The responsibility for guaranteeing that it is so maintained is the role of the computer-as-bureaucrat. Bureaucracies and computers have much in common: both must work reliably by fixed laws and without ambiguity. Because they share these talents, computing machines and bureaucratic people can easily be combined. An ideal computerized bureaucracy receives its instructions (the software) and its supplies (paper, ink, data, food, weapons, medicine, people, canned sardines). It processes its input by the rules of the software and produces words, bills, checks, well-run jails, lotteries, mass-educated people, meals on wheels, art, and standards of value.

But an ideal bureaucracy and even a not-so-ideal computer cannot make value judgments and, as Doctor Johnson's *Dictionary* did for spelling, they favor uniformity in ways of thinking. I spoke earlier of the decrease in the plurality of species making for a less resilient biosphere. The homogenization of thought, not created but surely promoted by the computer, leads to a less resilient cultural, scientific, and intellectual system. It does so by narrowing the spectrum of ideas out of which appropriate responses to new challenges may be selected.

In education, computers overwhelmingly favor testing for facts because factual questions are the only kinds that may be answered more or less unambiguously. Also, for computer testing to remain cheap, it must use single-valued answers. For these and other reasons, computer testing promotes rote memory rather than the learning of inference, analysis, interpretation, and the application of broad principles. The latter tasks, necessary to keep a modern society viable, are becoming the domain of a minority of people: those at the policy institutes of the government, at private foundations, in think tanks, at the top research laboratories, at a few of the institutes for advanced studies, at some of the universities, and the occasional wild card of a genius. We are witnessing the emergence of a new class of scribes, a meritocracy described by the British sociologist Michael Young, separate from the average citizen.

Computers force their users to phrase their questions in ways the machine can accept. Concerns that ought to be formulated in terms of the wealth of historical time, such as social, political, and fiscal policies, are posed and solved in terms of immediate needs. Judgments that should be based on long-term memory and expectations, and in full awareness of the plurality and ambiguity of human needs, are replaced by calculations. It is computer mentality that makes the pollster ask, 30 seconds after the show is off the air, "Who, in your view, won the presidential debate?" then publish the results. These results influence people who, in a similar computer state of mind, bypass their own critical faculties.

In the financial world, on an average day, computers transfer over \$300 billion among banks on five continents. Computers make possible the debiting and crediting of tens of millions of accounts daily, and the investment of New York City's funds each evening until next morning. The nineteenth-century preference for long-term securities has been replaced by the six-month, one-month, fourteen-day, and one-week securities. They keep company with all the other changes that narrow the horizons of social time.

We learned earlier that identical objects can only be governed by statistical laws. We also noted that members of a set not otherwise identical (all dollar bills, all weekend drivers, all divorced males between the ages of 40 and 45) may sometimes be assumed identical for a specific purpose and found to be governed by statistical laws. With the population explosion, the homogenization of needs, and the uniformity of production and distribution methods, statistics must replace individual attention in public administration. The umwelt, the working reality of all statistical aggregates, as I also demonstrated, is prototemporal.

The trend around the world is toward an increase in efficiency for the meeting of social needs. This is best accomplished by large organizations, for they can more easily prevent duplication of efforts and can better afford the use of advanced technologies. In the area of health services, for instance, one result is that patients have increasingly become numbers and physicians trained conduits between drug and equipment manufacturers and hospitals on the one hand, and the sick on the other.

We will recall that in the probabilistic world of quantum theory, we may talk of one or another electron if there are two, but we cannot talk of "Electron 1" or "Electron 2," as if each carried a different label. For if they did, they would not be indistinguishably identical and could not be handled by probabilistic, statistical laws. In the context of social services, we may also talk about one Medicaid patient or one million Medicaid patients and think of them as so many different people, but in handling them they must be assumed identical. The behavior of large masses of people, taken to be that of indistinguishable individuals, is studied in mathematical demography through the development and testing of deterministic models of population dynamics. These models are mathematical formulas representing the probabilistic—prototemporal— components of historical conduct, that is, those actions of people that form the most rudimentary, hence most conservative, behavioral traits. The usefulness of such models is impressively demonstrated by the success of election-night predictions based on meager early returns.

However, from the point of view of the individual, entering the proto-temporal world of an efficient, computer-run bureaucracy amounts to becoming a nameless person for one or another purpose, or for all purposes. The reader may wish to reread the earlier quote from *Doctor Zhivago* (p. 179) and reflect on what the artist wanted to express in figure 45.

In most parts of the world namelessness has been a way of life; not so, at least in principle, in the West. American tradition in particular demands respect for the individual as distinct from all others. On the other hand, a fair and equitable approach in case of need, if it is to be done efficiently, leads inevitably to the erosion of distinctions among people, that is, to a statistical situation. Anyone who ever tried to get a computer mistake corrected knows the indifference of the computer-as-bureaucrat to the person. To say that it is the technician or employee handling the computer who is callous and should be responsible for attending to the clients consti-tutes an indefensible argument: there are not enough people to attend to the clients handled by the computer, that is why we have computers to begin with.

The computer and the Bomb have been forcing the superpowers to approach each other in their methods of handling matters of state and charting policy. The Bomb keeps conflicting interests from exploding

FIGURE 45   THE LOSS OF DISTINCT IDENTITIES is a necessary corollary of the
use of statistical laws in public administration. The nature of social time then also
changes because the scale of values, pegged to the statistical behavioral laws,
permits less and less deviation from the norms assumed or ordained for average
behavior. This illustration is by the Vermont artist George Tooker and bears the
title, *Government Bureau*. As a longer title it could have borne two lines from
a poem called "In Bohemia" by the nineteenth-century Massachusetts poet John
Boyle O'Reilly, who spoke of "The organized charity, scrimped and iced / In the
name of a cautious, statistical Christ." Reproduction, courtesy of the artist and the
Metropolitan Museum of Art, George A. Hearn Fund, 1956.

into a global war; it forces anger, frustration, and humiliation to take the form of an epidemic of violence. The computer hastens the rapprochement by aiding the creation of ever-larger corporations in the free world. These are steps toward the global socialization and common evaluation of time, even though the conservative ideologies that support corporate power say, and believe, they oppose it. Ideologies are becoming increasingly less significant as homogenization in the ways we do things takes hold.

In the Soviet Union, large, vast centralized institutions are maintained as matters of state policy; a corollary practice demands the maintenance of a ruling class not responsible to the people. Party officials cannot be voted out of office. In the United States, power has been increasingly concentrated in the hands of the leaders of industry, commerce, and labor, who are not held responsible to the people at large. They cannot be voted out of office for they have never been elected.

Agricultural policies by the American and Soviet governments illustrate the rapprochement poignantly.

In the Soviet Union, secrecy about everything, including agricultural production figures, is a way of life. This secrecy has matched the secretive character of the Russian soul, recasting its complaints about the harshness of life into the messianic message of Communism. That kind of darkness has arrived at the shores of the United States only with the equally messianic policies of a conservative government that has been laboring hard to introduce secrecy into the American governmental process.

In the Soviet Union in the 1930s, 25 million Russian peasants were either allowed to starve to death, sent to the gulags, or deported to industrial centers to become factory workers. This was, as I previously mentioned, Stalin's way of promoting agribusiness. In the United States in the 1980s, we have witnessed the irreversible end of the American dream of a home on the range. Not for everyone, but for the majority of Americans, gone is the respect for the land, the aphrodisiac of the fresh hay and the just-turned earth, and the lifelong love–hate relationship with a territory—the kind of local patriotism that has been with us since we became human. The changes are forced by a conservative administration that preaches individualism but recognizes as an economic aspect of the narrowing global present the fact that the world is now a single market. The nation can only be made competitive in trade if all but the largest farms are eliminated. The surplus farmers must be made into factory or high-tech workers.

The American farm crisis does not have the barbaric horrors of Stalin's collectivization mainly because the traditional attitudes of the two lands toward authority are so different. Russians demand it, Americans criticize it. Russia has a history of despots, the United States one of Presidents. And beneath it all there is the blessed wealth of the American earth that can support social experimentation without catastrophes. Otherwise, the goals of Stalin and those of a conservative American administration are identical: collectivize your systems of value and socialize your schedules.

The call for a complete global socialization and evaluation of time, demanded by the needs of a time-compact globe, cuts across all cultural, ethnic, and racial boundaries and differences in political ideology.

The central beat bank, how the eye interprets the ear, and the dilemma of Disneyland

This subsection sketches a few examples of cultural change, expressed in some popular forms of art-as-communication, and relates them to the emerging global socialization and evaluation of time.

Art emerged from behavior recognizable among animals: red is the favorite color of apes and humans. The use of pigments, presumably for ritual body painting, is believed to be 1.6 million years old; a decorative amulet found in Hungary is estimated to be 100,000 years old; hand stenciling, ceramics, and sculpting go back 30,000 to 35,000 years. There are good reasons to maintain that art, as it is known in the history of modern humans, evolved as a method of transmitting information that could not be transmitted biologically or easily shared through words.

With the increasing diversity of cultural and individual temperaments and scales of value, art itself became differentiated. It remained for our epoch to reverse that trend by producing a great deal of uniformity in popular art and enlisting it in the service of social coherence around the globe. The "enlisting" is not directed by a central body but rather is forced by the uniformity of the methods of communication, modes of industrial production, and the consequent interpenetration of commercial, political, and military empires.

The most accomplished and totally socialized form of art-as-communication are the ritualized happenings and imaginative props that make up the four Disney towns: Disneyland, Walt Disney World, EPCOT Center (for Experimental Prototype Community of Tomorrow), and Tokyo Disneyland. They magnify the ideals of contemporary America: the good life, good fellowship, no deep search for meaning, interesting gadgetry, light entertainment, and mass-reproduced art. They do so no less impressively than the medieval cathedrals did in magnifying the beliefs of their age: the difficulty of life on earth, fellowship with disturbing depths, moralistic entertainment, and original works of art inspired by spiritual ambitions.

Neither of these lists is complete or exclusive, but considered together, they convey a valid contrast. The central element in that contrast is the difference of attitudes toward time. Cathedrals convey concern with permanence and suggest the ecstasy of the forest. The Disney towns convey an image of reality as change, and suggest the ecstasy of the dance. In the Disney towns nothing is ever at rest, everything is in constant motion. The engineered happenings are unified by Walt Disney's belief that if people get the right information, they will take the right action. The right information resides in the method of presentation itself: a contagious, ruthless happiness. The reason for the happiness is an inorganic, one might say antiseptic, fairy tale spirit.

Traditional fairy tales are rooted in the experience of continuity, and they indicate this symbolically. Each is woven around a conflict important for the child and for the child in the adult. For this reason they can serve as guides to making sense of the turmoil of feelings about one's self and about the world. Good struggles with evil, good wins, and the present and future are once again tied to the past. It is not the arrival that matters, but the development, the journey, the building of continuity. The stories justify the moral values that children are told to heed and that adults try to believe.

Such early classic Disney movies as *The Three Little Pigs* or *The Old Mill* were tales in that tradition: conflicts were posed and resolved; diligence won over indolence, and foresight over present concerns. In contrast, the Disney towns are carefully planned to give no hint that there can be problems and that achievement demands sacrifice. The following is from a report in a 1985 issue of the *New York Times*.

Every day through September, 1986, "America on Parade," a spectacular Bicentennial salute, marches straight down . . . Main Street, U.S.A. in the heart of Disney World. And there, at the head of the parade, bearing drum and fife and Betsy Ross's original pennant, dressed in tricorner hat and patched with bandages, stand the three symbols of the American Revolution: Mickey Mouse, Donald Duck and Goofy. (September 28, 1985)

The world is a happy place, passing and conflict are unreal, we have all arrived in utopia, to begin with.

These cathedrals are spiritual runts of "certain unalienable Rights," including "the pursuit of Happiness," referred to in the Declaration of Independence. For Thomas Jefferson, the author of these words, happiness meant a communal sense of having assisted the good in its struggle against evil, an assurance that time had meaning. Walt Disney is no less a representative of the America of our age than Jefferson was of the America of his. The Declaration of Independence begins with reference to history: "When in the Course of human events. . . ." Walt Disney's declaration is addressed to an ahistorical society. His Audio-Animatronix figures of stylized hippos, crocodiles, and people, moved by compressed air and controlled electonically, continuously sing, talk, and dance following the criteria of the Disney ballet, which is never to keep people seeing or hearing anything longer than their powers of concentration permit without effort. The figures must be instantly legible, the disembodied, soothing voices instantly comprehensible. The only reality is the organic present.

The music of the central beat bank comes through thousands of speakers sunk into the grass or otherwise blended into the environment. They create the kind of euphoria I called the enthusiasm of nations, the collective futureless, pastless ecstasy of the "Marseillaise." It is a feeling easy to share. Every country that can afford it tries to copy the Disney method of moving people and holding their attention. Exhibitors from around the world flock to the Disney towns to study them, while the Disney management keeps the time-and-motion studies upon which its people-moving technique is based strictly confidential.

What is the basis of the universal appeal to one and all—including this writer, the son of his age—of these latter-day cathedrals, these orchestrated ballets, musical theaters, simple storytelling, stage architecture, stage painting and sculpting, real technology and real dining out?

I was reflecting on this question while in line for a show in one of the Disney towns. I was watching a pretty chaperone press a button labeled "Scandinavian" (!) to bring up a canned message for a group of Scandinavians she spotted. (The message was in Swedish.) To the left of her button-pushing hand, there was a red button labeled "Crowd Control."

The answer came to my mind in a sinister vision. I saw the young woman press that button and watched a score of colorfully dressed policemen emerge from their underground lair where they had been attending to the vacuum system that disposes of garbage and helps keep the place immaculately clean. While arresting the malefactor or clubbing him to the ground, they kept on singing their joyful song, "It's a Small World After All!"

Waking from this realistic reverie, I realized that the freedom and wealth of contemporary America has created as its ideal the utopia of our age. It is that of an orderly world, one without problems, a society where the needs of the citizen for bodily comfort are taken care of by magic invisible powers, where his demands for mental stimulation are satisfied by simple questions and undisturbing answers, and where disturbing individuals are joyfully disposed of.

The California Disneyland had a great appeal to Nikita Khrushchev. He would have been even more impressed had he known of the plans for the late 1980s to build a serious Industryland for adults, with trips through working factories where visitors could watch employees make chocolate bars, golf clubs, and stuffed animals. This dreamland is indistinguishable in its main features from the pie-in-the-sky promised by classical Communism: and that is the dilemma of Disneyland.

The marble statues of saints, the art of their days, spoke about the continuity of history. The art of the Audio-Animatronix figures of the Disney towns, dancing their restless ballet, speak of the present as the only significant aspect of temporal reality.

Another form of socialized art has also been coming into view: it is collective art with very large numbers of objects or participants.

On Memorial Day weekend in 1986, five and a half million people formed a chain across the 4,000 miles from Los Angeles to New York, in the "Hands Across America" project. It was a collective act of vast proportions, a benefit show in aid of the homeless and the hungry of

America. I suspect it will be seen as the forerunner of celebration dances or national wakes at later occasions, performed across the land to the synchronous tune of a single beat bank.

Instead of designing individual homes, we have prefabricated, standard houses and towns designed in a more or less uniform manner. Marketplaces, instead of being different from town to town, have given way to the malling of America; sculpture, instead of being made for individual display, tends toward parks, such as the Park of the Happy Streaker in Oslo, Norway, with 300 nude figures by the same sculptor. Instead of an opera house, exhibit hall, art library, and concert hall we have the Barbican Center in London, with its geometry in stone and yellow arrows like the yellow brick road in the *Wizard of Oz,* and the Lincoln Center in New York, which at least is spacious and balanced in appearance.

Instead of individual research laboratories, we have research towns and cities such as Tsukuba in Japan, Hefe in the Anhui province of China, and Akademgorod in the Soviet Union. Instead of bordellos or red-light districts, there are packaged sex tours for Japanese businessmen to different Asian capitals, replacing the traditional visits to the flower-and-willow world by guided tours of the botanical garden.

Walt Whitman heard America sing in "varied carols . . . Each singing what belongs to him and her and none else." As the end of our century approaches those songs may be heard in diners, supermarkets, trucks, and doctors' offices, on the telephone while on hold, and on ghetto and suburban boom boxes. But they are not particularly varied. Most of them are carefully engineered network music, the rhythm of a central beat bank. Videocassettes and taped music by the branches of that central beat bank are changing the texture of Indian village life. In Muslim countries, in spite of the ruthless rules of fundamentalist governments, the change goes on irresistibly. Indians, Muslims, and everyone else are trying to reconcile their differing traditions with what appears to be a likely future of a larger community or, if the reconciliation is not possible, to eliminate both past and future and attend to the present in a readaptive step to the time-compact globe (fig. 46).

Technical perfection in producing exact replicas of art objects and music is yet another contributor to global socialization. In one way it is a gain: what used to be the experience of the privileged few may now be a source of inspiration shared by the many; the desire of artists to be heard and seen by all has become feasible. In another way something has been lost: the

dignity of uniqueness. Having an exact replica in every house and apart-
ment of a once and only event or one and only object amounts to the loss of
well-defined instants in time and locations in space. Should it be possible
to reproduce Leonardo's Mona Lisa so well that even experts could not tell
the copy from the original, the Louvre in Paris would lose a part of its
identity. If the last concert of Peter, Paul, and Mary before they broke up
could be seen in a 3-D hologram and be heard with super-high fidelity so
that it could not be told from the original when seen from a distance, historical time would have lost a hallmark. Both the calendar and the cultural
map will have taken a step from being many-colored toward becoming an
average grey.

Then, there is the Book.

In *The Ebony Tower,* John Fowles speaks of a young painter who
"suffered the most intense pang of the most terrible of all human depriva-
tions; which is not of possession but of knowledge." For perhaps three
millennia, the carriers of acquired knowledge that could lessen the pangs
of hunger for knowledge, the evolving genes of civilizations, have been
the words inscribed on tablets, written on parchment, or printed on paper
and bound. In whatever form, they are known as books. Every two years,
there is an exhibition of foreign books in Moscow. Russians, deprived of
literature from the West by the fear of knowledge built into the Soviet
system, travel thousands of miles and wait in line for hours to be able to
read a few pages of books that are available to every American if he or she
wants it.

Yet the book does not seem to be the appropriate method of communication for the emerging global order of vast, inert masses.

Publishers Weekly, the periodical of the book publishing industry,
recently carried one of George Steiner's remarkable essays, titled "Books
in an Age of Post-Literacy." In it he attempted to identify the position of
the book in a time-compact society in which technology, politics, and
economic interests interact to create a new global society with its peculiar
means of communication. Steiner writes that,

> the situation is almost classically that of a Marxist analysis: the
> concentration of the marketing and the dissemination of books not
> only in a very few hands, but in hands which are politically and socio-
> logically scarcely distinguishable. Whatever the difference of style, of

FIGURE 46   ISLAMIC PRAYER WATCH, A SYMBOL OF THE TIME-COMPACT
GLOBE. The watch displays both the Gregorian and the Islamic day, month, and
year counts. Geographic locations of over 120 cities are programmed into its data
bank; for other locations longitudes and latitudes may be entered. For the location
identified, the watch will display the calendar dates as well as local time in hours,
minutes, and seconds. It calculates and upon demand displays the six prayer times
of the day. If desired, it gives a 20-second warning signal before each prayer time.

The times shown in the figure are those for Boston, Massachusetts, on September 10, 1986. The ASR is the afternoon prayer, to be said when the length of the shadow of an object is the length of the object, plus the length of its shadow at noon; the MAGHRIB or sunset prayer is recited when the upper edge of the sun just touches the horizon. ISHA, the evening prayer, is said toward the end of the evening twilight.

Also upon command, the watch can display the geographical direction to the Ka'ba in Mecca, along the great circle that goes through the selected location and Mecca, and the zone time of the location calculated from the meridian of Greenwich, England. Each time there is a delay in presenting a display because the computer is calculating, the message *Allahu Akbar* appears in Arabic and Latin characters.

The Islamic prayer watch is a contemporary descendant of the long line of intricate calendar clocks of Christendom, which incorporated the rich tradition of Arabian science of the golden age of Islamic cultures. The primary purpose of astronomical clocks, as we have learned, was to help explain the dynamics of the universe, functioning like clockwork under the tutelage of God the Father. Timekeeping as we know it today was only a secondary task. In contrast, this watch is primarily a utilitarian timing device; it does not propose to explain anything. Its computers could be easily reprogrammed to display appointments, the arrival and departure of scheduled airlines, or the crucial times of a manufacturing process. The connection of the watch to the mathematical, cosmic order of planets and stars is indirect and totally masked by its operational and social purposes.

Following the traditional role of all calendars, the Islamic prayer watch offers a temporal rhythm of life that helps secure the collective identity of its users. This task is an increasingly difficult one because of the pull of homogenization of a time-compact globe, promoted among other factors by the very technology that made this watch possible.

The alarm of the watch is an inorganic version of the call that inspired the astronomer, mathematician, and poet Omar Khayyam to write,

> Alike those who for to-day prepare,
> And those who after tomorrow stare,
> A Muezzin from the Tower of Darkness cries,
> "Fools! Your reward is neither Here nor There.

Reproduction, courtesy of Spacetronic S.A. Geneva (Switzerland), subsidiary of Dar Al-Maal Al-Islami Trust; photo, courtesy of Peter Fraboni and the Nature Center for Environmental Activities, Westport, Connecticut.

personality, of anecdote, they constitute, so far as culture goes, an almost monolithic and monopolistic vision.

(May 24, 1985, pp. 44–48)

In 1958, 72 percent of all books were sold by independent, one-store firms. In 1985, 52 percent were sold by four large book chains, and ten publishers accounted for more than 85 percent of the mass market. It is a Marxist dream come true, made possible by capitalist production and marketing methods that those who claim to be the pioneers of Marxism could neither permit nor make happen in their lands.

"Reading in the old, archaic, private, silent sense . . . may become as specialized a skill and avocation as it was in the . . . monasteries during the so-called Dark Ages," Steiner continues.

> It may well be . . . that the privately owned book, in a format such as we know it, in type (even where such type is electronically cast and composed), will become a luxury object. It will become an article for special use. . . .

What Steiner aptly calls the "airport book," the mass market paperback, is such an article. It is bought to satisfy an instant craving for something the title and the cover vaguely promise. Read or unread, it is then tossed, leaving an emptiness to be filled by the next purchase. It is the ideal book to satisfy the literary needs of the mass market customers; financially it is good for the industry that needs the profits of the "airport book" if it is to continue to publish what the best of publishers judge lasting.

Steiner's worry is understandable, considering that picture and voice have been rapidly replacing the printed word as the most appropriate means of communication for a time-compact globe. Historically, since most people could not read, the best way to relay messages has been by word of mouth: the ears read the message. Illiteracy is still very high: in the United States, 27 million people cannot read at all and, according to the report of the Librarian of Congress to Congress, 23 million adults are functionally illiterate.

"If you can't read all about it, hear all about it!" is the pragmatic call of the time-compact globe, heard on CBS radio in New York. Can't or won't. The great challenge to the survival of the book is not from the illiterate but from the large masses of the aliterate. These are men and women able but unwilling to read anything outside that required by their trade and everyday tasks. The Librarian of Congress also noted that only one-half of the American public can be called book readers, defined as people who read two or more books a year. Sales figures are misleading in assessing the popularity of literature because they are inflated by the sales of how-to and fad books, and by the pulp known in the industry as that of gore, blod, and semen. The hard facts of the market are to free lands what censors are to totalitarian regimes. These censorial facts have been pushing the publishing industry from the printing and distribution of literature and science to producing computer software and novelty items.

As if these changes were not enough, there is television in its many shapes, surely the most appropriate means of communication for a time-compact society because of its format and the way TV programming evolved. The span of attention is matched as well as managed by the rhythm of commercials and/or by rhythm and character of the script. The appeal is to the sight, and a powerful one it is, because, as we learn from Shakespeare's *Rape of Lucrece,* "To see sad sights moves more than hear them told / For then the eye interprets the ear."

The screen crosses cultural and political borders more easily than does the printed word because it employs an older mode of communication. Even frogs, horses, and hawks can see stories, though they cannot write or read about them. If I were asked to help make the people of the world ready for the demands of a time-compact global order, I would recommend communication by dancing pictures: the machinery for it is already in place.

The sociotemporality of the time-compact globe has no history to which we can turn for guidance. One can only try and discern some of its hallmarks, which was the purpose of this section. Here is a sampling, but not a summary, of what we have found:

> The merging of many social presents into a steadily narrowing global present promotes the development of a worldwide rhythm.
> The popular perception of the irrelevance of history leads to the gradual removal of people from the continuity of their different pasts.
> The greying of the calendar evens out the differences among seasons, months, weeks, and days and between day and night as far as social behavior is concerned; it thus lessens the distinctness among

different patterns of collective behavior and scales of value, to the extent these derive from different calendrical traditions.

There are changes in the texture of private time, attributable to the necessities of birth control and death control. These controls are forced upon us by population and ecological pressures. They modulate those aspects of human life that gave rise to the sense of time in the first place: securing continuity through biological issue and bargaining with death.

There is a lessening of the hold of ideologies, as expectations around the globe become more uniform. The computer hastens the homogenization; the Bomb keeps frustration at the terrorist level.

Art, born 30,000 years ago, is alive and well, and as always, it communicates values that are difficult to convey in words. The values seem to be those of a time-compact global society.

Time at the Anthill Threshold: A Turbulent Act of Creation

The thesis of this and of the prior section is that if a significant portion of the earth's population reaches a certain level of social complexity, the global socialization and evaluation of time will subsume the office of the individual as the primary measure and measurer of time. In the preceding section we have explored a number of social and cultural changes that suggest that a new kind of temporality is in fact being born.

There is yet another kind of argument that seems to lead to similar conclusions. Sociobiologists have estimated that the rate of genetic learning— the rate at which new biological structures can arise—is eight billion times slower than the rate at which the human mind can learn. Whatever the numerical value of that ratio may be, it is being rapidly diminished by a combination of genetic engineering and the continuing population explosion. The changes presage a condition when the rates of genetic and mental learning may become comparable. Genetic engineering contributes to the rapprochement by producing new biological structures and functions on demand (at least in principle); a larger gene pool does its share—very much in practice—by providing a larger number of mutations from which nature and society may select.

"You, Sir, with the readiness to sleep for three days and nights and work for three days and nights, were you engineered or were you a spontaneous mutation found by research workers in the teeming metropolis of Portage, Saskatchewan?''

When the time comes for such a question to be seriously addressed to a person, the time-compact globe will have socialized not only the machinery of its civilization—language, science, technology, and art— but also organismic life itself. The blurring of the difference between cultural and biological evolution is an aspect of the crossing of the anthill threshold.

It is very difficult, perhaps impossible, to outline the nature of time appropriate for the society of a time-compact globe, as that society is here envisaged. Certainly, it could not be done with the same precision with which we were able to trace, for instance, the hallmarks of the cyclic and aging orders of life. The problem is that to describe the temporality of a time-compact globe we must contemplate the attributes of a new organizational level, having its own peculiar language and logic, yet all we can use are the language and logic of our noetic world. We are like the fish in Rupert Brooke's poem "Heaven," contemplating the unthinkable:

> Fish say, they have Stream and Pond;
> But is there anything Beyond?
> . . . somewhere, beyond Space and Time,
> Is wetter water, slimier slime.[16]

But it is possible to point to some of the unique, time-related difficulties of a global socioeconomic system. That, at least, will give us guidance concerning the dynamics of the postulated transition.

The first subsection personifies the unresolvable conflicts of knowing time; it sets the stage, as it were. The second subsection observes the events on that stage in our epoch.

Setting the stage

Ingmar Bergman is the son of a Lutheran minister and grew up in a family seldom without tension. When he was thirty-eight years old, he made one of his beautiful motion picture ballads. It is set in medieval times

---

16   Brooke, *The Complete Poems* (London: Sidgwick and Jackson, 1942), p. 132.

because Bergman felt, as have others, that our epoch resembled the Middle Ages in its ruthless wars for eternal peace.

The Seventh Seal is the story of a young couple, Mia and Jof, and their infant child. They are traveling actors who stopped their covered wagon in a nordic village and began their show. The people of the village watch with expectation: they joke, frolic, and relax. Then, in Bergman's script,

> a rapid change occurs. People who have been laughing and chattering fall silent. Their faces seem to pale . . . children stop their games and stand with gaping mouths and frightened eyes. Jof steps out in front of the curtain [of the covered wagon] . . . some of the women have fallen on their knees. . . .[17]

A line of people approaches, led by monks; they are followed by men, boys, old men, and children, all of them beating themselves with steel-edged scourges. We hear the haunting tune of the Latin hymn "Dies ire, dies ira . . . ," which the reader may recognize as the same hymn that inspired Michelangelo in the design of the *Last Judgment*. The flagellants are declaring the end of the world, the Last Judgment is imminent. In the events that follow, all perish except the small family.

A rainstorm comes and passes. Jof and Mia crawl out of their hiding place in the wagon to "look across ridges, forests, the wide plains and the sea," writes Bergman. Jof points to the distance, toward "the dark, retreating sky where summer lightning glitters like silver needles over the horizon." On the ridge of the hill, against the sky, he sees their former companions. "They are all there. . . . And Death, the severe master, invites them to dance. . . . They dance away from the dawn and it's a solemn dance toward the dark lands," he says to a doubting Mia, who looks at him with gentle devotion. "You, with your visions and dreams!"

This, then, is the stage: Bergman's vision of mankind, torn between an awareness of passing and dreams of eternity (fig. 47, *top*).

A contemporary form of that dream of eternity may be represented by the figure of Albert Einstein. His thought reflects the Platonic belief in exact science, disclaiming the reality of time.

---

17 *Four Screenplays of Ingmar Bergman,* trans. L. Malstrom and D. Kushner (New York: Simon and Schuster, 1960), p. 123.

A contemporary form of the keen awareness of passing may be represented by the figure of Winston Churchill. His thought reflects a belief in time as the central reality in the enterprise of being human.

Einstein came from a peaceful family and attended a Catholic school from ages five to ten (fig. 47, *bottom left*). At nine he was still not totally fluent in his speech, which made his parents fear that he might be abnormal. When he was sixteen, he came upon a scientific problem that took him ten years to solve. He did so at age twenty-six, when he proposed a new theory of physics that led to a revolutionary new understanding of time in the physical world. That glorious achievement, the mind searching for lasting patterns, is also the theory upon which the Bomb, the most inglorious weapon of genocide, is based.

Winston Churchill was the son of two headstrong lovers: Lord Randolph, the son of a British duke, and Jennie Jerome, beautiful daughter of an American adventurer (fig. 47, *bottom right*). Winston was born prematurely, and very suddenly, in the coatroom of Blenheim Palace at 1:30 in the morning. His early appearance might have been brought about by Lady Churchill's dancing at the Palace of St. Andrew's, followed by a bumpy coach ride.

"Study history, study history," he told once to an American student. "In history lies all the secrets of statecraft." His favorite hymn, played at his memorial service following his written instructions, was our "Battle Hymn of the Republic": "He has sounded forth the trumpet that shall never call retreat. . . ."

The stage, the play, and the observers

In one of her sensitive and well-informed newspaper columns, Georgie Anne Geyer, writing from Paris, had this to say about the turbulence of the contemporary mind.

> It is not only beautiful and aged Western Europe that is awash in ambivalence. It is not only in Europe that the utopian ideologies of the past are being rejected.
>
> The questions being asked in Eastern Europe are not that different from those being posed in the West: How do we live without faith or ideology? How do we create economic wealth: How do we satisfy people without destroying the system that does not satisfy them? Without ideology, in what do we find unity?

FIGURE 47   THE CHALLENGE OF HUMAN TIME: THE STAGE AND THE
ACTORS. On the top, the five figures in the center are torn between destruction
(decay and death) pulling them downhill and creation (growth and harmony)
pulling them uphill. These opposing forces, already implicit in the physical world,
become explicit in the life process, and are made conscious and become vastly
elaborated by the mind. The tension that derives from having to live with their
unresolvable conflicts lies at the root of the human experience of time.

(*Bottom left*) Albert Einstein with his sister Maja in 1884. The five-year-old
Albert holds his broad-brimmed hat and wears the kind of finery that middle-class
boys would wear to school in Bavaria. Later he created a new theory of the physical
world in which time was equated with a static, spatial dimension of an imaginary
world; this was consistent with his view that time was an illusion. As one of his
legacies, we have the atomic bomb and all its cousins.

(*Bottom right*) Winston Churchill with his mother Jeanette in 1878. The four-
year-old Winston wears the kind of Victorian finery that the son of Lord Randolph
and Lady Churchill, a noted American beauty, would wear. Later he became the
great leader of the West by insisting that time and history were the central realities
of human life. As his legacy, we possess a keen awareness of the Iron Curtain and its
many kins.

Albert Einstein and Winston Churchill, on Ingmar Bergman's stage, represent
the challenge, the power, and the burden of human time.

Upper illustration from Bergman's *The Seventh Seal,* courtesy Janus Films,
New York. The photograph of Einstein, ©Lotte Jacobi Archive, courtesy,
Department of Media Services, University of New Hampshire. Young Winston
and Lady Churchill from the Churchill Archives, Churchill College, Cambridge,
by permission of Curtis Brown, Ltd., on behalf of Winston S. Churchill.

And those are much of the same questions being asked in the
United States, Europe's maturing offspring across the seas. (*Chicago
Sun-Times,* August 27, 1985)

What single ideology, if any, could be suitable for the whole of mankind?
We have a great multiplicity of them, some very much alive, but no
particular religion or ideology is powerful enough to claim the earth as its
dominion. Hitler tried to claim Europe and Asia, and intended to join up
with the westward-moving Japanese over India, but was defeated from
the outside. His was the last ideology that could be defeated by external
forces. The Chinese had to wait until Mao died, then buried him twice.
The travels of Pope John Paul II illustrate the scope of the struggle for
man's allegiance. The smaller warlords are becoming increasingly blood-
thirsty and crazy; their actions would have long become inconsequential
were it not for the time-compact globe.

What is the social system that seems to be in need of a common
ideology? Mankind. What is mankind? The dictionary defines it as "the
human race; the totality of human beings." This is like saying that a cell is
the totality of its molecules, a man the totality of his cells: true but not
enough. Only consistent behavior through time that can give a system a
recognizable identity, and the definition I quoted misses the temporal
dimension.

With the help of our prior observations on identities and the hierarchy of
presents, a reader of this book can provide a better definition of mankind:
it is the totality of those people who were or are able to create and maintain
their common social (global) present.

It is useful to remember that an organism can have a future and a past
only with respect to its organic present, and also that a person cannot have
a future and a past unless his or her mental present is viable. From the
person, let us go to the group. What a man and woman did before they
married or got together influences their behavior, but for them as a couple,
there exists no past or future before they met.

Likewise, mankind can have neither a past nor a future before the global
present is established through the coordination of its social presents. Only
after the global present has become viable will mankind have an identity,
a future and a past. It is in our epoch that the global present and, with it,

mankind's identity are being worked out. Or will fail to be worked out.

In the case of the individual, identity is established through a process psychologists call reality testing. It begins with the small child's testing of himself against the world: "How far can I go?" The answer is: until his parents, other children, the furniture, his sleepiness, his fear of something, or the vinegar in the bottle says "Stop!" The child explores the boundaries of his world and, by so doing, establishes his identity with reference to things and other people. The same holds for collective identity. It is established and honed by other collective identities: religion versus religion, nation versus nation, "us" versus "them."

A unique difficulty arises when it comes to the identity of mankind: there are no other mankinds with respect to which our own could establish its identity. There are no cohorts that could wake us from collective fantasies and help us identify the boundaries of global possibilities. There is no "them" who would say "Stop! That's my nose!" and for the same reason, there is no "us."

In the absence of cohort control, any political, cultural, or military experimentation, if miscarried across a powerful enough part of the globe, can easily lead to socioeconomic and cultural extremes. Each extreme in its turn promotes its dialectical opposite and may force the kind of unattenuated oscillations that in complex inorganic systems are known to lead to resonance. For "resonance" in social systems read: the readiness to go into a state of chaotic change at the slightest provocation. A time-compact globe may probably be thought of as resonant.

Consider the swings of political and economic moods of contemporary America, already a time-compact society. The impression left with the observer is that public memory does not extend to the day before yesterday or public expectation to the day after tomorrow. Having been thus removed from the flow of time, any society can become the victim of its news media, concerned only with the present instant. If this kind of readiness to oscillate applies to a time-compact globe, the forecast for global behavior is one of stormy weather. Sudden political, social, or economic shocks anywhere could easily trigger sudden changes elsewhere and begin a chain reaction whose course and final outcome are unpredictable. While Americans have learned to live with these swings more or less, on the global scale rapid oscillations in political, economic, and ideological trends have no precedence.

I believe that heterogeneous as mankind is, the time-compact globe is entering the set of those systems of which there is only one each, such as the Universe, Nature, or God. The difficulties of delimiting the boundaries of effectiveness and defining the identities of these one-and-onlies are well known.

Earlier examples of time's rites of passage included biogenesis (the coming about of biotemporality) and the emergence of man (birth of nootemporality). None of the transient structures of biogenesis or those between man and ape survive: they had to evolve rapidly into stable forms, or else, as Darwin said of the hypothetical first forms of life, they would have been devoured.

If these earlier events are any guide and if we really are witnessing the coming about of a new integrative level of nature, then the transition is going to be a metastable one. Any global system-in-the-making would be metastable: It would either collapse into a tribal chaos or evolve rapidly into a social system with sufficient inner controls to maintain a global present. The symptoms of such a transition, of a turbulent act of creation, are all around us.

One of those symptoms is the interpenetration of empires, previously mentioned. The benefits of this interpenetration are great: it makes global resources in ideas, manpower, and material available to those privileged to take advantage of it. If our car would speak in all the languages of the countries where its parts were made, our garage would be a local terminal of the Tower of Babel. The globe as a single workplace and marketplace makes it possible to purchase at the local Sam Goody's a Deutsche Grammophon release of Horowitz playing Scarlatti in Moscow. Moscow is in Russia, Horowitz is a naturalized American, Scarlatti was an Italian, Deutsche Grammophon is owned by the Siemens Werke in Munich, and Sam Goody's is owned by American Can Company. Museumgoers are regularly treated to exhibits of global wealth that would have been impossible before our age.

But these noble products of human culture ride on the back of an ugly camel: the violent substructure of the global equalization of values and the reinstitutionalization of all social systems.

Thus while Americans were being kidnapped, hijacked, abused, tortured, and killed by Muslim terrorists, NASA launched a communication satellite for a consortium of nations, including those countries responsible for the hijackings and murders, and the American government was selling arms at special reduced rates to one of the terrorist states. Transnational corporations function across many borders that separate nations at war. In the global supermarket, food, feed, Asian girls, and dope flow in support of political, military, and business interests in all possible directions as do stolen scientific information, torture weapons, investments, medicine, body parts, and babies for adoption. Not only the United States but every industrialized nation makes or transports arms with little regard to their origin, destination, or intended use. In this universal wheeling and dealing, the poorer nations get poorer, the rich ones richer, the strong ones stronger, and the number of homeless increases in all lands.

A three-cornered struggle is becoming evident: there are the national governments; the transnational groups (corporations, religions, and ideological undergrounds); and the "tribal interest cells," or TICs, as I like to call them (Armenian irredentists, Irish Catholics, Japanese nationalists, terrorists of Holy Wars, the Church of Science . . . there are a million of them). Few if any of the TICs, whose behavior keeps the news media busy, would disappear if their particular grievances were remedied. Rather, their members would only find other causes to serve. When considered as a global social phenomenon, the existence of the TICs becomes frightfully intelligible. They can then be seen as the commandos of a worldwide warfare of alienation kept from exploding by the Einstein bombs and by the interpenetration of the great empires of power.

Simultaneously, the intensity of industrial use of raw material and the population explosion have brought into view certain limitations to material progress that were unnoticeable and irrelevant in earlier days. Specifically, the economy of the earth must face the irreversible march of energy from its free to its bound form. Having sufficient energy is not enough; advanced life forms—people, horses, flies—need energy with a high information content. The total information content of the biosphere has been decreasing in a process better known as the loss of renewable resources.

Those resources are of course renewable—they did after all evolve—but the tempo of their development is slower than the rate at which they are being used up. Energy alone cannot make a sheep. Nuclear power can yield unlimited amounts of energy, but there is no known way whereby the information content of the last sheep on earth could be stored and later made into wool from the chemical elements. The transition to a time-compact globe is also one to a reinstitutionalized and economically more refined but also much poorer humanity.

During the present transition period, nightmare-like fantasies have become socialized and thereby sanctioned as acceptable parts of national and international life. Civilizations have been trying to hold the violent manifestations of these fantasies in their lairs, with moderate success. But modern technology, communication, and social advances themselves helped lift the lid off the inner turmoils of people, allowing the reptile mind to act out its desires.

There are regions of the globe where this turmoil is not directly felt, but no part of the earth is isolated anymore. In every area reachable by radio, each and every large outburst elsewhere produces emotions in favor or against it, adding to the metastability of the world situation. Upon the well-charted earth, multinationals, ideologies, terrorist groups, religious movements, and military alliances are shifting around for their presumptive position in a broader order.

We have learned that the integrity of a living system or social system resides in its capacity to maintain its present. It is impossible to maintain a viable social–global present unless it is made a part of the flow of history. But of whose history? The people of the earth are at war to decide whose interpretation of the past is going to shape the future of mankind. The ethos that will eventually conquer the minds of people will be the one that succeeds in creating a framework of historical understanding for which a believable and inspiring future for mankind may be constructed.

The march today is to something more powerful than the call of any existing ideology. The new drummer has announced a crisis in needs and means without prior examples. Man the measure and measurer of time is being driven by the unresolvable conflicts of the present social systems toward a new level of incipient complexity, that of a time-compact global organization. Whether such a system can be worked out is yet to be seen. If a new integrative level does emerge, and if we can believe the theory of time as conflict, then it too will be driven by its own unresolvable conflicts. Or, the earth may relapse into a protracted state of tribal warfare, with appropriate adjustments of the sociotemporal horizons.

The poet's instruction was, "Ask my song." In this chapter a listener recorded what he has been hearing lately about time at the anthill threshold. But hearing is not a passive registration of what already exists but an act of creative participation. This listener's interpretation of what he heard is molded by whatever else he knew about the nature of time, in contexts other than that of the global family of man. If others hear the voices of contemporary men and women differently, it is for them to say.

FROM THE DIARIES OF A TIMESMITH: THE MEMORY OF A PLAN THAT BECAME THIS BOOK *

STINSFORD, DORSET, ENGLAND *This afternoon I came upon a stanza of beautiful poetry engraved on a stone in the parish churchyard. Its lines are living categories of the experience and idea of human time.*

SHALL I BE GONE LONG? *In the hazy beginning of our history, the temporal horizons of our forebears began to expand. They acquired a knowledge of time, a powerful weapon in their struggle against other species and the inanimate forces of nature, because it made possible the planning for the future, based on memories of the past.*

The evolutionary threshold to Homo sapiens *was crossed when the future came to include the fact of inevitable death, the past an awareness of the story of the tribe. Being able to think in terms of distant futures and pasts was, however, a double-edged sword, for it conferred upon the members of our species their peculiar restlessness, rooted in chronic insecurity. From then on, people very seldom, if ever, could maintain the inner peace that a satisfied animal seems to have. Along with the knowledge of time, the feelings of responsibility and guilt were born, being two sides of the same mind set.*

To be gone forever after death proved to be an unacceptable idea to the human mind. Our ancestors, therefore, began to bargain with passing, in the hope of avoiding the inevitable. That bargaining, which is still going on, created the great

---

Warning: Too heavy to be digested, therefore dangerous to be taken without reading the book first.

cultural continuities: the religions, the philosophies, the arts and letters, and the sciences.

These social institutions have one method of operation in common. They all separate out of the chaos of sense impressions those motives that seem permanent. By directing attention to them under such headings as the true, the good, and the beautiful, they lessen the insecurity born with the discovery of time.

Their collective cultural labor appeals to events, things, and feelings that are in the future or in the past, but not in the here and now, and must, therefore, be represented by symbols, such as those of the spoken and written languages. Communication through language tied the fate of the individual to that of the community, the coherence of the community to the effectiveness of its language.

FOR EVER AND A DAY. *This is an answer to the question of the first line of the poem. It invokes the skill of reckoning time by number. Everyone knows how to measure time by reading a clock, but only timesmiths know what this actually involves. For instance, no single clock or watch in itself can measure time. Each glance at a watch is a comparison of two time scales, connected by a belief that justifies the comparison. In everyday use these beliefs go unnoticed, like background music does in a supermarket. In scientific use, however, the beliefs become explicit in the form of theories expressed in mathematical relations and perceived as laws of nature.*

Each and every clock reading joins the clock watcher to the flow of time, both in its human and cosmic dimensions. Each age perceives the cosmic dimensions of time differently. There is, in fact, no time scale in nature to which one can point and say, This is the true and final rate at which time passes. There is only an ever-changing understanding of nature and a constant desire to find order among events. This state of affairs is well illustrated by the absence, even in principle, of an ultimate measure of time.

The step from clocks and watches to calendars and chronologies is a change in the methods of measuring time and in representing that measurement. The expanding time horizons of calendars and chronologies make explicit what in the use of clocks is only implicit. To wit, the role of social, political, economic, philosophical, and religious preferences in time measurement.

The largest conceivable calendrical unit is the story of the universe. Narrative cosmogonies, cosmologies, and world-endings are cultural assessments of the position of man in the world. They are the surviving ancestors of their scientific descendants.

TO WHOM THERE BELONG? *To assure their continuity in spite of their mortality, men and women must and do want to belong to places, persons, and activities. Not belonging anywhere or to anyone after death is inconceivable.*

Because of the hierarchical organization of nature, each of us belongs to the world on several levels. We are made of matter; as living beings we belong with all other living organisms; as thinking beings with all people; as social beings to social groups of many kinds. These different ties are the domains of different sciences: physics, biology, psychology, and sociology. The humanities cover all of them, as the hen covers her eggs.

The way the sciences divided the world among themselves corresponds to the way that nature is divided into stable integrative levels: those of light, of particles with non-zero restmass, of massive bodies, that of living matter, that of the works of the mind (language, art, and artifact), and that of society. The structures and processes of each of these levels are more complex than the structures and processes of the level or levels beneath it. Each organizational level has its governing principles, its peculiar logic, and its specific temporality. As living and thinking humans made of matter, our daily existence involves the temporalities of matter, life, mental work, and society.

The astronomical day, lunation, and year have been built into all living organisms, including ourselves, through organic evolution. These rhythms, together with a broad spectrum of other organic rhythms, make up the cyclic order of life. Out of the cyclic order of life arose the aging order of life with its two corollary functions: sexual reproduction and death-by-aging. The origins of life itself are believed to date to the appearance of autonomous oscillating molecular systems— miniature clockshops—possibly having arisen from crystalline ancestry.

To assure its autonomy, life must maintain an instant-to-instant control of its biochemical and biophysical simultaneities. This inner control creates the organic present, the biological "now." With respect to that "now," future and past have acquired meaning. It is thus that with life, biotemporality was born.

The limited futures and pasts of biotemporality were extended by the mind to the long-term futures and pasts of noetic time. When people speak about "time," they usually mean nootemporality. Noetic time is the human brain's way of minding the affairs of the body. It does so with the help of the symbol we know as the self. The self is the only mental image that may be attributed to both an external and an internal reality. The changing speed of experienced time, the feeling of free will, the use of human creativity and destructiveness in the name of distant goals: all these manifestations of being human may be interpreted as the efforts of the self to maintain and expand its control over behavior.

To insure the survival of the group, people create and maintain their social presents. With respect to the social present, the plans and the history of the community can acquire meaning. Thus born, social time permits the design of actions beyond the life span of any individual and the use of ideas from a collective past. To this end, social time has been used to guide the individual in his and her efforts in selecting a plan of action from among many conceivable alternatives. The guidelines are embodied in language, art, artifact, and science and in preferred ways of life.

Social time grants society degrees of freedom unavailable to its members acting alone. It also grants the group privileges of action independent of any benefit or harm to its members or even to the group at large.

ASK THE STONE TO SAY. *The task of asking nonliving matter to speak and the responsibility for interpreting its reply is that of physics. The time of the physical world is so primitive, so different from what we ordinarily mean by time, that it can only be understood through mathematics, the most abstract of all languages.*

Physics divides its concerns along the distinct temporalities of the physical world. Special relativity theory addresses the atemporal world of light. Quantum theory focuses on the prototemporal universe of particle-waves. General relativity theory speaks about the eotemporal cosmos. Thermodynamics—the art of chicken soup and entropy—straddles them all, at least as far as the discoveries of physics about time go.

As we progress along a scale of increasing complexity, from light to particle-waves to massive objects, the temporalities of the physical world become increasingly coherent. None of them, however, can define a present—that comes only with life—or, for that reason, futurity and pastness. Physical time does not "flow."

The discoveries of physics added a great deal of strangeness to what has been familiar about time. It is these novel elements, unsuspected and even unimaginable earlier, that came to elucidate the puzzle of what was to be meant by a beginning and an ending of time.

ASK MY SONG. *This is the bottom line of the stanza and also of the study of time.*

In our turbulent age, time seems to be crossing a new threshold in its rite of passage; a new temporality is emerging: that of the time-compact globe. Different and intricately linked empires are in deadly conflict, even as they approach each other in their ways of handling the affairs of state and citizens. Each hopes to write the definitive history of our species and offer it as the right foundation upon which a common, global future may be built. This is mankind's struggle for establishing its identity.

The making of this entry in the living diary of a mind took only a small part of a second. Like a bird, no sooner did that fleeting thought of an instant arrive than it was on the wing again. Also like a bird, it built itself a nest. In another small part of a second, the thoughts came to multiply and spread themselves out over the open fields of future and past. The mental present experienced in the parish churchyard of Stinsford, England, came to include anticipations concerning a book to be written. That book—so the thoughts ran—will have to be based on knowledge that, at the time of its writing, will have been gained in what then will be past.

This kind of an intricate manipulation of ideas about things and events that are not present is the privileged skill of the human mind.

From the point of view of evolutionary development, long-term futures and pasts are recently acquired categories of human reality. They make up what I have been calling time understood. Recent in their origins indeed, but with roots in the much older reality of time felt, from which they cannot be separated. Feelings that fill an instant of time may be independent, more or less, from time understood. But the way time is understood, be it through physics, biology, psychology, sociology, history, or the arts and letters, cannot be made independent of time felt. If it were, its subject would vanish.

The study of time, which is an intellectual endeavor, may therefore be compared with trying to identify the patterns on a jigsaw puzzle whose pieces grow, change, and become more complex while the puzzle is being assembled. In this case, the puzzle, its assembler, and the seeker of its design are all the same: the person himself or herself. We are charting a land that is being created by the act of its discovery.

But to keep drawing that chart seems to be our self-appointed destiny. In the pursuit of that destiny the mind searches for rest without being able to rest, and seeks predictability even in itself, without being ready or able to give up its freedom to remain unpredictable. The study of time pertains, therefore, to the very substance of the enterprise of being human.

September 10, 1986
Hickory Glen, Connecticut

APPENDIX ONE: TIME DILATION

Consider the simplest possible situation. Two identical clocks. *A* and *B*, move at a constant relative speed along the same straight line. To each object there is attached a single spatial coordinate, colinear with its motion, along which its distance from the other object may be measured. Earlier, when they were at rest with respect to each other, they were synchronized and ticked at the same rate.

Figure 48, known as a Brehme diagram, is a geometrical representation of the Lorentz transformations (R. W. Brehme, "A Geometrical Representation of Galilean and Lorentz Transformations," *American Journal of Physics* 30 (1962): 489–96). It is *not* a picture of the two systems that move with respect to each other, but a *representation* of those systems. We see two

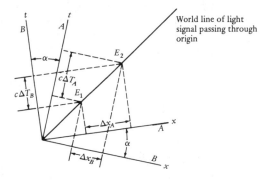

FIGURE 48

Geometrical Representation of the Lorentz Transformations, after Brehme.

coordinate systems, A and B, each with its distance and time axis. On these we record the distance and time information of events observed. To emphasize the formal symmetry of the diagram, all time readings are multiplied by c and plotted with the dimension of length. (We could have divided all length readings by c and plotted them with the dimension of time.)

It may be easily shown that the angle α relates to the relative velocity of the two systems through the equation

$$\cos\alpha = \sqrt{1 - v^2/c^2}.$$

This quantity is known as the special relativistic transformation factor.

The origin represents a position $x=0$ corresponding to the instant $t=0$ when, along their common line of motion, the two bodies are at the same place at the same time. The universe of our objects has two dimensions: length and time. An event in that universe is unambiguously identified by a point E in the plane of the paper. The time and distance coordinates of event E are the orthogonal projections of E upon the appropriate axes.

A *world line* of an object is the locus of all those points that represent the motion of the object in terms of consecutive coordinate positions of distances and time. The line that bisects the central angle of both the A and the B system is the world line of a light signal. Consider events E_1 and E_2 in the life of a photon that passed the moving bodies at the origin. Inspection will reveal that because of the symmetry of the two coordinate systems about the world line of light, the velocity of light as seen from either coordinate system will be the same. The requirement for the invariance of c is thus fulfilled.

The two diagrams of Figure 49 show the Brehme diagram put to work. The left one shows the world line of two ends of a rod stationary in system B. Those world lines are orthogonal to the x_B axis. Inspection will reveal that, as measured from system A, the rod will be foreshortened: $\Delta x_A < \Delta x_B$. The figure on the right shows that a rod stationary in system A, as measured from B, will also be foreshortened: $\Delta x_B < \Delta x_A$.

The diagram on the left of figure 50 shows the world line of a clock at rest in system B; the diagram on the right of figure 50 shows the world line of a clock at rest in system A. The intervals Δt_B and Δt_A are the proper times of clocks B and A. Inspection of the geometry of the diagrams will show that in each case, an observer in uniform relative translation will assign a greater measure to the proper time interval $E_1 - E_2$ of a clock at rest in the other system than does the clock registering that proper time. This

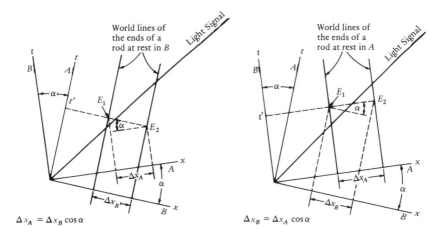

FIGURE 49

Geometrical Representations of Motional Contractions of Length

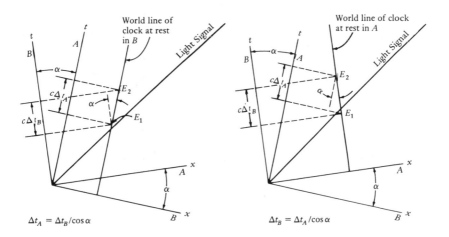

FIGURE 50

Geometrical Representations of Motional Time Dilations

phenomenon is called *time dilation.* Thus, on the left, the proper time of the clock at rest in B is Δt_B. From the system A, the measure of that same interval is $\Delta t_A > \Delta t_B$. In the diagram on the right, the relationship is reversed: $\Delta t_B > \Delta t_A$.

The Lorentz transformations in general, and their particular representations in the Brehme diagram, apply to two systems in uniform relative translation along a single straight line. If such systems once pass each other, they will never meet again. This is not the case for the "living organism in a box," described by Einstein. That organism must make a round trip. A round trip involves changes in direction, changes in direction involve acceleration, and accelerating frames are not usually contemplated within the compass of SRT. They are more naturally handled by the later, general relativity theory. But already in his 1905 paper, Einstein asserted that the time dilation effect would also apply even if the moving clock followed a circular, reentrant path.

APPENDIX TWO: THE TEMPORALITY OF THE INITIAL STATE OF THE UNIVERSE

Searching for increasingly more universal theories has been the self-appointed career of natural science. The development of a single formalism that would combine the laws that govern all the fundamental forces of nature—very much the center of concern in contemporary mainstream physical theory—is an effort in that tradition.

According to current understanding, there are four fundamental forces of nature: electromagnetic, weak nuclear, strong nuclear, and gravitational.

Electromagnetic and weak nuclear forces have already been shown to be aspects of a single force, called the *electroweak force*. The hope is that a unification of the electroweak and strong nuclear forces will also become possible in a *grand unification* and eventually that the grand unified and gravitational forces will be shown to be aspects of a single force, in a theory of *super unification*.

The prevailing trend, as I see it, points toward a total reconciliation between particle theory and cosmology, leading to an understanding of the primordial state of the universe, governed by the principles of a super-unified theory.

When the biologist Linnaeus worked out his method of the binomial classification of all living things, he believed that he was classifying the plants and animals of the day of Creation. Likewise, but corresponding to the values of a different epoch, I detect a consensus in physical cosmology, holding that once theoretical super unification has been achieved and experimentally confirmed, it will be possible to make authoritative statements about the nature of the primordial cosmos in much more detail than it is possible to do now. Such statements will surely include assertions about the nature of time in the initial singularity.

It is my belief that the eventual theorem of super unification, when applied to the conditions of the initial cosmos during the Planck era, will specify a world in which all particles shared a state of masslessness and traveled at the speed of light. For that reason, I believe that the primordial universe was atemporal, in the sense that atemporality is defined and used in this book.

GLOSSARY

For the purposes of the interdisciplinary study of time it has been found useful to assume the existence of five distinct temporalities. They form a nested hierarchy of increasing richness of content as we proceed from the level of electromagnetic radiation to that of particles with non-zero rest mass, of massive physical systems, of life, and of the mind of man. Each new temporality subsumes that or those beneath it; each permits the functioning of a qualitatively new creative freedom.

The five definitions that follow—already given and illustrated in the text—are not listed, however, from the least to the most complex temporality. They are arranged in the opposite sequence: from the most sophisticated yet most familiar (the time of the human mind) to the most primitive and strange (that of light). This is also the sequence in which the new concepts are introduced in the book.

Nootemporality, or noetic time. This is the temporal reality of the mature human mind. It is characterized by a clear distinction among future, past, and present; by unlimited horizons of futurity and pastness; and by the mental present, with its changing temporal horizons, depending on attention.

Biotemporality, or biological time. This is the temporal reality of living organisms including man, as far as his biological functions are concerned. It is characterized by a distinction among future, past, and present, but the horizons of futurity and pastness are very limited when

compared with those of noetic time. The boundaries of the organic
present are probably stable and species-specific.

Eotemporality, or the time of "the physicist's *t*." So named after Eos, the
Greek goddess of dawn, this is the simplest form of continuous time. It
is the temporal reality of the astronomical universe of massive matter.
It has also been described as the time of pure succession. It is a continuous but nondirected, nonflowing time to which our ideas of a present,
future, or past cannot be applied.

The time of the world of elementary particles bears the name *prototemporality*, for *proto-*, the first in a series, as in protoplasm. It is an undirected,
nonflowing as well as fragmented (noncontinuous) time for which
precise locations of instants have no meaning. Events in the prototemporal universe may only be located in a statistical, probabilistic
manner.

The world of electromagnetic radiation is termed *atemporal*. Atemporal
conditions do not signify nothingness but rather that the proper time of
particles that travel at the speed of light is zero. Atemporality describes
a state of energy to which none of our ordinary notions associated with
time apply.

The following two terms are also useful in the interdisciplinary study
of time.

Sociotemporality is the postulated level-specific reality of a time-compact
globe. The study of sociotemporality encompasses issues in the socialization of time and in the collective evaluation of time. Since it is
a temporality that—so it appears—is in the process of being created, it
is very difficult to sketch its hallmarks with the same authority with
which we sketched the hallmarks of the time of light, particles with
non-zero restmass, massive matter, or life.

Umwelt. Originally the circumscribed portion of the environment meaningful and effective for an animal species. In this definition, "environment" stands for our human umwelt. In the hierarchical theory of time,
umwelt signifies the level-specific realities of the different integrative
levels of nature as revealed through scientific experiment and theory.

A BIBLIOGRAPHIC SAMPLER

OR PIONEERS CAN HAVE

NO MAPS

All major languages have their literature relevant to the study of time, native to the language and its culture. Since much of that material is unavailable in English, this sampler includes only works in English, with priority given to American editions. There was no single criterion for the selection of the titles. Books were included if they were important and/or interesting, if they represented a family of publications, or because they had good bibliographies. Many of the books could have been entered under more than one heading. In such cases I opted for the earliest useful listing.

Other Books by the Author

Of Time, Passion, and Knowledge: Reflections on the Strategy of Existence. New York: Braziller, 1975. An encyclopedic overview of the study of time, with eighty pages of references.

Time as Conflict: A Scientific and Humanistic Study. Boston: Birkhäuser, 1978.

The Genesis and Evolution of Time: A Critique of Interpretation in Physics. Amherst: University of Massachusetts Press, 1982.

Books Edited

The Voices of Time: A Cooperative Survey of Man's Views of Time as Expressed by the Sciences and the Humanities. 2d ed. Amherst: University of Massachusetts Press. 1981. Twenty-seven original essays by leading scientists and humanists, with a new introduction, "Toward an Integrated Understanding of Time."

With F. C. Haber and G. H. Müller. *The Study of Time I.* New York: Springer-Verlag, 1972. Thirty-six papers from the First Conference of the International Society for the Study of Time.

With N. Lawrence. *The Study of Time II*. New York: Springer-Verlag, 1975. Thirty-two papers from the Second Conference of ISST.

With N. Lawrence and D. Park. *The Study of Time III*. Twenty-eight papers from the Third Conference of ISST.

With N. Lawrence and D. Park. *The Study of Time IV*. New York: Springer-Verlag, 1981. Nineteen papers from the Fourth Conference of ISST.

With N. Lawrence and F. C. Haber. *Time, Science, and Society in China and the West (The Study of Time V)*. Amherst: University of Massachusetts Press, 1986. Twenty-five papers from the Fifth Conference of ISST.

Time and Mind (The Study of Time VI). Amherst: University of Massachusetts Press, forthcoming. Selected papers from the Sixth Conference of ISST.

Contributions to the Multidisciplinary Study of Time

Denbigh, Kenneth G. *Three Concepts of Time*. New York, Springer-Verlag, 1981. The three concepts are those from theoretical physics, thermodynamics, and conscious awareness.

Fagg, Lawrence W. *Two Faces of Time*. Wheaton, Ill.: The Theosophical Society. 1985. A concise comparative examination of the time concepts of modern physics and those of the major religions.

Fischer, Roland, ed. "Interdisciplinary perspectives of time." *Annals of the New York Academy of Sciences* 138 (1967): 367–915.

Patrides, C. A., ed. *Aspects of Time*. Buffalo: University of Toronto Press, 1976.

United Nations Educational, Scientific and Cultural Organization. *At the Crossroads of Culture*. 3 vols. Vol. 1, *Cultures and Time*. Vol. 2, *Time and the Philosophies*. Vol. 3, *Time and the Sciences*. Paris: UNESCO, 1976–79.

Whitrow, G. J. *The Natural Philosophy of Time*. 2d ed. Oxford: Clarendon Press, 1980. A masterful introduction to, and necessary reading for the study of time.
———. *Time in History*. Oxford: Oxford University Press, 1988. A companion volume to the above, traces with authority and command of detail the mutual dependence of our ideas of history and time.

Zeman, Jiri, ed. *Time in Science and Philosophy: An International Study of Some Current Problems*. New York: Elsevier, 1971.

The Discovery of Time

Berry, W. B. N. *Growth of a Prehistoric Time Scale, Based on Organic Evolution*. San Francisco: Freeman, 1968.

Calder, Nigel. *Timescale: An Atlas of the Fourth Dimension*. New York: Viking, 1983. A well-informed and documented chronology of the world of matter, life, and man. Richly illustrated.

Marschack, Alexander. *The Roots of Civilization*. New York: McGraw-Hill, 1972. See also Marschack's relevant papers in anthropological journals.

Rosenberg, G. D., et al., eds. *Growth Rhythms and the History of the Earth's Rotation*. New York: Wiley, 1975. The history of the earth's rotation determined through measurements of biological growth rates recognized in fossils and checked against astronomical data.

Time in Religions

Brandon, S. G. F. *History, Time and Deity: A Historical and Comparative Study of the Conception of Time in Religious Thought and Practice*. Manchester, England: Manchester University Press, 1965.

———. *Man and His Destiny in the Great Religions*. Manchester, England: Manchester University Press, 1963. Two magistral works.

Eliade, Mircea. *Cosmos and History: The Myth of the Eternal Return*. New York: Harper & Row, 1959. See also Eliade's other books.

Guitton, Jean. *Man in Time*. Notre Dame, Ind.: University of Notre Dame Press, 1966.

Time in the Philosophies

Balslev, A. N. *A Study of Time in Indian Philosophy*. Wiesbaden, Germany: Harassowitz, 1983.

Capek, Milic, ed. *The Concepts of Space and Time: Their Structure and Their Development*. Boston: Reidel, 1976. An excellent sourcebook.

Gale, Richard M., ed. *The Philosophy of Time: A Collection of Essays*. Garden City, N.Y.: Anchor Books, 1967. A sourcebook.

Helm, B. P. *Time and Reality in American Philosophy*. Amherst: University of Massachusetts Press, 1985.

Mandal, K. K. *A Comparative Study of the Concepts of Space and Time in Indian Thought*. Varanasi: The Chowkhamba Sanskrit Series Office, 1968.

Mbiti, J. S. *African Religions and Philosophy*. Garden City, N.Y.: Doubleday, 1970.

Needham, Joseph. *Time: The Refreshing River*. 1943. Reprint. Atlantic Highlands, N.J.: Humanities Press, 1986. Reflections by a leading intellectual of our age on time, science, religion, organic evolution, and society.

Sherover, Charles, ed. *The Human Experience of Time: The Development of its Philosophic Meaning*. New York: New York University Press, 1975. A distinguished sourcebook. The editor's substantial summaries, read in themselves, make for a fine introduction to time in philosophy.

Smart, J. J. C., ed. *Problems of Space and Time*. New York: Macmillan, 1964. A sourcebook.

Trivers, Howard. *The Rhythm of Being: A Study of Temporality*. New York: Philosophical Library, 1985.

The Stuff that Clocks Are Made Of

Bedini, S. A. *The Scent of Time: A Study of the Use of Fire and Incense for Time Measurement in Oriental Countries.* Philadelphia: American Philosophical Society, 1963.

King, H. C. *Geared to the Stars: The Evolution of Planetariums, Orreries, and Astronomical Clocks.* Toronto: University of Toronto Press, 1978. A fascinating and definitive survey, superbly illustrated.

Needham, Joseph, et al. *Heavenly Clockwork: The Great Astronomical Clocks of Medieval China.* Cambridge: Cambridge University Press, 1960.

Nilsson, M. P. *Primitive Time-Reckoning.* 1920. Reprint. Lund, Sweden: C. W. K. Gleerup, 1960.

Price, D. J. de Solla. *Gears from the Greeks.* Philadelphia: American Philosophical Society, 1974.

Waugh, Albert E. *Sundials: Their Theory and Construction.* New York: Dover, 1973. A delightful do-it- and think-it-yourself book for the naturalist.

The Scientific Measurement of Time

Explanatory Supplement to the Astronomical Ephemeris and the American Ephemeris and Nautical Almanac. 4th ed. London: Her Majesty's Stationery Office, 1977. Prepared jointly by the Nautical Almanac Offices of the United Kingdom and the United States of America. The major reference work in scientific time reckoning.

Janich, Peter. *Protophysics of Time: Constructive Foundation and History of Time Measurement.* Boston: Reidel, 1985. A normative analysis of chronometry.

Calendars and Chronologies

Leon-Portilla, Miguel. *Time and Reality in the Thought of the Maya.* Boston: Beacon Press, 1973. An introductory and authoritative survey of the Mayan concern with chronology, time symbolism, and time as an attribute of gods.

Maurice, Klaus, and Otto Mayr, eds. *The Clockwork Universe: German Clocks and Automata, 1550–1650.* New York: Neal Watson Academic Publications. 1980. An illustrated catalog of a Smithsonian exhibit, with fourteen substantial essays of general interest.

Parise, Frank, ed. *The Book of Calendars.* New York: Facts on File, 1982.

Spier, Arthur. *The Comprehensive Hebrew Calendar: Its Structure, History, and One Hundred Years of Corresponding Dates, 5660–5760/1900–2000.* New York: Behrman House, 1952.

Zerubavel, Eviatar. *The Seven-Day Circle: The History and Meaning of the Week.* New York: The Free Press, 1985.

Ordering the Universe by Human Time

Brandon, S. G. F. *Creation Legends of the Ancient Near East.* London: Hodder and Stoughton, 1963.

Gould, S. J. *Time's Arrow, Time's Cycle.* Cambridge: Harvard University Press, 1987. An engaged and engaging work that employs the Eleatic dichotomy of change and permanence in elucidating the classic issues concerning the history of the earth.

Gunnell, John G. *Political Philosophy and Time.* Middletown, Conn.: Wesleyan University Press, 1968.

Haber, Francis C. *The Age of the World.* 2d ed. Westport, Conn.: Greenwood Press, 1978. A carefully documented study of time and historicism illustrated by the expanding time scale of the universe from the Biblical to the evolutionary–scientific.

Harrison, Edward. *The Masks of the Universe.* New York: Macmillan, 1985. An enlightened and authoritative account of the nature and structure of the universe, as seen in the changing lights of history.

Johnson, M. C. *Time and the Universe for the Scientific Conscience.* Cambridge: Cambridge University Press, 1952.

Toulmin, Stephen, and June Goodfield. *The Discovery of Time.* New York: Harper & Row, 1965.

The Cyclic Order of Life

Aschoff, Jürgen, ed. *Biological Rhythms.* New York: Plenum Press, 1981. A major reference work.

Campbell, Jeremy. *Winston Churchill's Afternoon Nap.* New York: Simon and Schuster, 1986.

Goodwin, B. C. *Temporal Organization in Cells: A Dynamic Theory of Cellular Control Processes.* New York: Academic Press, 1963.

Hastings, J. W., et al., eds. *The Molecular Basis of Circadian Rhythms.* Berlin: Dahlem Konferenzen, 1976.

Moore-Ede, M. C., et al. *The Clocks that Time Us: Physiology of the Circadian Timing System.* Cambridge: Harvard University Press, 1983.

Palmer, J. D. *Biological Clocks in Marine Organisms.* New York: Wiley, 1974.

Scheving, L. E., et al., eds. *Chronobiology.* Tokyo: Igaku Shoin, 1974.

The Aging Order of Life

Comfort, A. *Aging: The Biology of Senescence*. New York: Holt, Rinehart & Winston, 1964.

Hendricks, Jon, and M. M. Seltzer, eds. *Aging and Time*. Beverly Hills, Calif.: Sage, 1986. Volume 29, number 6, of *American Behavioral Scientist*.

Maddox, G. L., ed. *The Encyclopedia of Aging*. New York: Springer Publishing, 1987.

Strehler, B. L. *Time, Cells, and Aging*. New York: Academic Press, 1962.

Biogenesis

Bernal, J. D. *The Physical Basis of Life*. London: Routledge and Kegan Paul, 1951.

Cairns-Smith, A. G. *Genetic Takeover and the Mineral Origins of Life*. Cambridge: Cambridge University Press, 1982.

Eigen, Manfred, et al. "The Origin of Genetic Information." *Scientific American* 244 (April 1981): 88–118.

Time and Mind

Doob, Leonard W. *Patterning of Time*. New Haven: Yale University Press, 1971.

Eccles, J. C., ed. *Brain and Conscious Experience*. New York: Springer-Verlag, 1966.

Fraisse, Paul. *The Psychology of Time*. New York: Harper & Row, 1963.

Friedman, W. J. *The Developmental Psychology of Time*. New York: Academic Press, 1982.

Hartocollis, Peter. *Time and Timelessness, or the Varieties of Temporal Experience*. New York: International Universities Press, 1983. A psychoanalytic inquiry.

Melges, F. T. *Time and the Inner Future: A Temporal Approach to Psychiatric Disorders*. New York: Wiley, 1982.

Michon, J. A., and J. L. Jackson, eds. *Time, Mind, and Behavior*. Berlin: Springer-Verlag, 1986. Papers from the point of view of cognitive psychology. Read Michon's excellent summary, "The Compleat Time Experiencer."

Orme, J. E. *Time, Experience, and Behavior*. New York: Elsevier, 1969.

Piaget, Jean. *The Child's Conception of Time*. New York: Basic Books, 1970.

Schiffer, Irvine. *The Trauma of Time: A Psychoanalytic Investigation*. New York: International Universities Press, 1978.

Shalom, Albert. *The Body/Mind Conceptual Framework & the Problem of Personal Identity*. Comparative analysis of relevant theories in philosophy, psychoanalysis, and neurology. Atlantic Highlands, N.J.: Humanities Press, 1985.

Social Time

Bell, Daniel. *The Coming of Post-Industrial Society*. New York: Basic Books, 1973.

Cipolla, C. M. *Clocks and Culture, 1300–1700*. New York: Norton, 1977.

Givens, Douglas R. *An Analysis of Navajo Temporality*. Landham, Md.: University Press of America, 1977.

Glasser, Richard. *Time in French Life and Thought*. Totowa, N.J.: Rowman and Littlefield, 1972.

Grudin, Robert. *Time and the Art of Living*. San Francisco: Harper & Row, 1982.

Hall, Edward T. *The Dance of Life: The Other Dimension of Time*. Garden City, N.Y.: Doubleday, 1983.

————. *The Silent Language*. Garden City, N.Y.: Doubleday, 1973. Two rich contributions to the anthropology of time. See also Hall's many other time-related works.

Hareven, T. K. *Family Time and Industrial Time: The Relationship between the Family and Work in a New England Industrial Community*. Cambridge: Cambridge University Press, 1982.

Jacques, Elliot. *The Form of Time*. New York: Crane, Russak, 1982.

Landes, D. S. *Revolution in Time: Clocks and the Making of the Modern World*. Cambridge: Harvard University Press, 1983. A masterful review of the economic coevolution of society and the mechanical clock.

Lauer, R. H. *Temporal Man: The Meaning and Uses of Social Time*. New York: Praeger, 1981.

LeGoff, Jacques. *Time, Work, and Culture in the Middle Ages*. Chicago: University of Chicago Press, 1980.

Needham, Joseph. *Science and Civilization in China*. Vol. 49, *Intellectual and Social Factors* (tentative title). Cambridge: Cambridge University Press, forthcoming Includes a summary survey of traditional Chinese concepts of time and attitudes to change and history.

Schwartz, Barry. *Queuing and Waiting: Studies in the Social Organization of Access and Delay*. Chicago: University of Chicago Press, 1975.

Thornton, R. J. *Space, Time, and Culture among the Iraqw of Tanzania*. New York: Academic Press, 1980.

Wright, Lawrence. *Clockwork Man: The Story of Time, Its Origins, Its Uses, Its Tyranny*. New York: Horizon Press, 1969.

Time in the World of Matter

Akhundov, M. D. *Conceptions of Space and Time*. Translated from the Russian by Charles Rougle. Cambridge: MIT Press, 1986. Covers grounds all too well trodden, but unique as a guide to Marxist-Leninist thought on time, through its references to Soviet work not available in English.

Davies, P. C. W. *The Physics of Time Asymmetry*. Berkeley: University of California Press, 1974.

―――. *Space and Time in the Modern Universe*. Cambridge: Cambridge University Press, 1977.

Denbigh, K. G. *An Inventive Universe*. New York: Braziller, 1975.

Harrison, E. R. *Cosmology: The Science of the Universe*. Cambridge: Cambridge University Press, 1981. A text of precision, completeness, and clarity, written with the gifts of a superb teacher.

Landsberg, P. T., ed. *The Enigma of Time*. Bristol, England: Hilger, 1982. A sourcebook.

Marder, L. *Time and the Space-Traveller*. London: Allen & Unwin, 1971.

Park, David. *The Image of Eternity: The Roots of Time in the Physical World*. Amherst: University of Massachusetts Press, 1980.

Prigogine, Ilya. *From Being to Becoming: Time and Complexity in the Physical Sciences*. San Francisco: Freeman, 1980.

Sklar, Lawrence. *Space, Time and Spacetime*. Berkeley: University of California Press, 1974.

Szamosi, Géza. *The Twin Dimensions: Inventing Time and Space*. New York: McGraw-Hill, 1986. An original and refreshing exploration of the emergence of time and space as categories of reality, set in the background of the history of science and art.

Time in Literature and Art

Aldiss, B. W. *Trillion Year Spree: The History of Science Fiction*. New York: Atheneum, 1986. A witty, wise, and scholarly guide to this flourishing literary vision, with its ever-present interest in time.

Alkon, P. K. *Defoe and Fictional Time*. Atlanta: University of Georgia Press, 1979.

Art & Time. London: Barbican Art Gallery, 1986. Catalog of an exhibition by that title.

Blatt, S. J., and E. S. Blatt. *Continuity and Change in Art: The Development of Modes of Representation*. Hillsdale, N.J.: Erlbaum, 1984.

de Romilly, Jacqueline. *Time in Greek Tragedy*. Ithaca: Cornell University Press, 1968.

Epstein, David. *Beyond Orpheus: Studies in Musical Structure*. Cambridge: MIT Press, 1979.

Harmon, William. *Time in Ezra Pound's Work*. Chapel Hill: University of North Carolina Press, 1977.

Higdon, D. L. *Time in English Fiction*. New York: Macmillan, 1977.

Macey, S. L. *Clocks and the Cosmos: Time in Western Life and Thought.* Hamden,
Conn.: Archon Books, 1980.

———. *Patriarchs of Time: A Dualism in Saturn-Cronus, Father Time, the Watchmaker
God, and Father Christmas.* Atlanta: University of Georgia Press, 1987.

Mendilow, Adam. *Time and the Novel.* New York: Nevill, 1952.

Meyerhoff, Hans. *Time in Literature.* Berkeley: University of California Press,
1960.

Poulet, Georges. *Studies in Human Time.* Baltimore: The Johns Hopkins University Press, 1956.

Quinones, R. J. *Mapping Literary Modernism: Time and Development.* Princeton:
Princeton University Press, 1985.

———. *The Renaissance Discovery of Time.* Cambridge: Harvard University Press,
1972.

Rowell, Lewis. *Thinking about Music: An Introduction to the Philosophy of Music.*
Amherst: University of Massachusetts Press, 1983. A pioneering study of time
as music's primary dimension, in its social, ideational, and political matrix.

———, ed. "Time and Rhythm in Music." *Music Theory Spectrum* 7 (1985): 215.

Turner, Frederick. *Shakespeare and the Nature of Time: Moral and Philosophical
Themes in Some Plays and Poems of William Shakespeare.* Oxford: Clarendon Press,
1971.

The Time-Compact Globe

Carlstein, Tommy. *Time Resources, Society and Ecology.* Vol. 1, *Preindustrial
Societies.* London: Allen & Unwin, 1982.

Carlstein, Tommy, et al., eds. *Timing Space and Spacing Time.* 3 vols. Vol. 1,
Making Sense of Time. Vol. 2, *Human Activity and Time Geography.* Vol. 3, *Time
and Regional Dynamics.* London: Arnold, 1978.

Denhardt, Robert B. *In the Shadow of Organization.* Lawrence: The Regents Press of
Kansas, 1981.

Georgescu-Roegen, Nicholas. *The Entropy Law and the Economic Process.*
Cambridge: Harvard University Press, 1971.

Hendricks, Jon, and C. Davis Hendricks. *Aging in Mass Society: Myths and Realities.*
2d ed. Boston: Little, Brown & Co., 1981.

Kaufman, Herbert. *Time, Chance and Organizations: Natural Selection in a Perilous
Environment.* Chatham, N.J.: Chatham House, 1985.

Kern, Stephen. *The Culture of Time and Space, 1880–1918.* Cambridge: Harvard
University Press, 1983.

Linder, Staffan. *The Harried Leisure Class.* New York: Columbia University Press,
1970. Gives reasons for the unavoidable decline in the quality of life in the con-
sumer society due to the self-generated scarcity of time.

Lowenthal, David. *The Past Is a Foreign Country*. Cambridge: Cambridge University Press, 1985. A well-illustrated and documented treatise on the ambivalent position of the past in the time-compact societies of the West.

Lynch, Kevin. *What Time is this Place?* Cambridge: MIT Press, 1972. Explores the relationship between the quality of the personal image of time and the collective management of the environment.

Mallmann, C. A., and Oscar Nudler, eds. *Time, Quality of Life and Social Development*. San Carlos de Bariloche, Argentina: Fundacion Bariloche, 1982. Papers of a conference.

Melbin, Murray. *Night as Frontier: Colonizing the World after Dark*. New York: The Free Press, 1987.

Michelson, William, ed. *Public Policy in Temporal Perspective*. New York: Mouton, 1978.

Winston, G. C. *The Timing of Economic Activities: Firms, Households, and Markets in Time-Specific Analysis*. Cambridge: Cambridge University Press, 1982.

Young, Michael Dunlop. *The Metronomic Society: Notes on Cyclical and Linear Time*. Cambridge: Harvard University Press, 1988.

INDEX

Other Titles from Tempus Books

The World of Mathematics
*A Small Library of the Literature of Mathematics
from A'h-mosé the Scribe to Albert Einstein*

Out of print for many years, this four-volume anthology is a rich collection of 133 articles prefaced by Newman's insightful commentary, which place the essays in historical perspective and make even the most difficult concepts accessible to a wide range of readers.

James R. Newman, 2800 pages (four volumes), $55.00 softcover, Order Code WOMAB, $149.95 cloth, Order Code WOMAHB

Invisible Frontiers
The Race to Synthesize a Human Gene

INVISIBLE FRONTIERS tracks the developments of the high-stakes race to clone a human gene and engineer the mass production of the life-sustaining hormone insulin.

Stephen S. Hall, 360 pages, $8.95 softcover, Order Code INFR

Thursday's Universe
A Report from the Frontier on the Origin, Nature, and Destiny of the Universe

Cited by *The New York Times* as one of the best science books of 1987, THURSDAY'S UNIVERSE explores current ideas about the moment of creation, including the birth and death of stars, quasars, and galaxies, the composition of black holes and neutrinos, and the puzzle of the universe's missing mass.

Marcia Bartusiak, 336 pages, $12.95 softcover, Orderr Code THUN

The New Wizard War
How the Soviets Steal U.S. High Technology—And How We Give It Away

THE NEW WIZARD WAR is a riveting and timely look at the legal and illegal transfer of high technology from the U.S. to the U.S.S.R.

Robyn Shotwell Metcalfe, 288 pages, $17.95 cloth, Order Code NEWIWA

Mathematics
Queen and Servant of Science

Here—from a modern-day master of mathematics literature—is a fascinating and lively survey of the developments of pure and applied mathematics. Published in cooperation with the Mathematical Association of America.

Eric Temple Bell, 464 pages, $11.95 softcover, Order Code MAQUSE

The Tomorrow Makers
A Brave New World of Living-Brain Machines

THE TOMORROW MAKERS is a spellbinding account of visionary researchers and scientists and their modern-day quest to engineer immortality. Award-winning science writer Grant Fjermedal details the astounding work being done in robotics and artificial intelligence today. Cited by the American Library Association as one of 1987's most notable books.

Grant Fjermedal, 288 pages, $8.95 softcover, Order Code TOMA

Nobel Dreams
Power, Deceit, and the Ultimate Experiment

An intimate and colorful look at Carlo Rubbia's quest for the 1985 Nobel Prize in physics. Taubes' narrative is a very readable layperson's primer on high-energy physics, as well as a captivating modern-day adventure story.

Gary Taubes, 288 pages, $8.95 softcover, Order Code NODR

Inventors at Work
Interviews with 16 Notable American Inventors

"Invention becomes an art in these accounts of serendipitous associations and 'lateral thinking.'"
                    Publishers Weekly

A critically acclaimed collection of 16 engaging and illuminating interviews with the most notable inventors of our time.

Kenneth A. Brown, 408 pages, $9.95 softcover, Order Code INWO

Computer Lib/Dream Machines

First published in 1974, COMPUTER LIB/CREAM MACHINES became the first cult book of the computer generation, predicting the major issues of today: design of easy-to-use computer systems, image synthesis, artificial intelligence, and computer-assisted instruction. Ted Nelson's vision of a nonsequential way of storing data—hypertext—is particularly relevant today with the emergence of CD ROM.

Theodor Nelson, 336 pages, $18.95 softcover, Order Code COLIDR

Machinery of the Mind
Inside the New Science of Artificial Intelligence

Focusing on the work of giants in the artificial intelligence field—including Marvin Minsky, Roger Schank, and Edward Feigenbaum— George Johnson gives us an intimate look at the state of AI today. We see how machines are beginning to understand English, discover scientific theories, and create original works of art; how research in AI is helping us to understand the human mind. Captivating reading for anyone with an interest in science and technology.

George Johnson, 352 pages, $9.95 softcover, Order Code MAMI

Tempus Books are available at bookstores, or call 1-800-MSPRESS for ordering information or placing credit card orders.